225
Topics in Current Chemistry

Modern Mass Spectrometry

Volume Editor: Christoph A. Schalley

With contributions by
P. B. Armentrout, D. K. Bohme, M. T. Bowers,
M. Brönstrup, T. Junker, S. Petrie, D. A. Plattner,
G. Siuzdak, D. Schröder, H. Schwarz, S. A. Trauger,
F. Tureček, E. Uggerud, T. Wyttenbach

ISSN 0340-1022
ISBN 3-540-00098-4
Springer-Verlag Berlin Heidelberg New York

Library of Congress Catalog Card Number 74-644622

Cover design: KunelLopka, Heidelberg/design & production GmbH, Heidelberg

Schalley: Modern Mass Spectrometry

First Indian Reprint 2009

ISBN 978-81-8489-283-3

This edition is published by Springer (India) Private Limited, (a part of Springer Science+Business Media), Registered Office: 3rd Floor, Gandharva Mahavidyalaya, 212, Deen Dayal Upadhyaya Marg, New Delhi – 110 002, India.

Printed in India by Rakmo Press (P) Ltd., New Delhi.

Volume Editor

Dr. Christoph A. Schalley

Kekulé-Institut für Organische Chemie
und Biochemie der Universität Bonn
Gerhard-Domagk-Straße 1
53121 Bonn, Germany
E-mail: c.schalley@uni-bonn.de

Editorial Board

Topics in Current Chemistry Also Available Electronically

For all customers with a standing order for Topics in Current Chemistry we offer the electronic form via SpringerLink free of charge. Please contact your librarian who can receive a password for free access to the full articles by registration at:

http://link.springer.de/orders/index.htm

If you do not have a standing order you can nevertheless browse through the table of contents of the volumes and the abstracts of each article at:

http://www.springerlink.com/series/tcc

There you will also find information about the

- Editorial Board
- Aims and Scope
- Instructions for Authors

Preface

Mass spectrometers are used by almost all chemists and many researchers from neighboring disciplines such as physics, medicine, or biology as a powerful analytical tool. Its advantages are high sensitivity, speed, and almost no sample consumption. During the last two decades, mass spectrometry experienced a boom of new developments pushing its limits further and further at an increasing speed – just similar to the progress in NMR spectroscopy in the 1970s.

However, a mass spectrometer does not only serve as a machine for solving complicated analytical problems, it evolved meanwhile to a complete laboratory for the investigation of molecules, clusters, and other species under the environment-free conditions of the highly diluted gas phase. These special conditions existing only in high vacuum change the properties of the particles under study significantly with respect to their energetics and reaction pathways. For example, temperature is a macroscopic property of a large ensemble of particles in thermal equilibrium and is thus not defined for a single ion. This fact has severe implications for the measurement of kinetic and thermodynamic data of gas-phase species. On the other hand, the examination of gas-phase properties has the advantage that systems reduced to minimum complexity can be studied more easily without the complicated influences of solvents or counterions. In particular, the combination of isotopic labeling and mass spectrometry allows for a detailed analysis of reaction mechanisms or conformational analysis through H/D exchange experiments not only on biomolecules. A comparison of gas-phase and condensed-phase data carefully interpreted yields profound information on the effects of solvation. Consequently, gas-phase chemistry beyond mere analytics became and is still becoming more and more important in many fields of chemistry.

When assembling the contributions for this issue, the deliberate decision was made, not to fully cover the area of biochemical and biological mass spectrometry. The amount of information available in the areas of proteomics, genomics, and related fields would justify an independent volume in this series, which might appear in the future. Also, no concise treatment of recent developments in ionization methods or mass analysis has been attempted. Although all authors did an excellent job of introducing their field in a well-understandable manner, the reader is assumed to have some basic knowledge with respect to these methodological details. Rather, the present issue of "Topics in Current Chemistry" intends to give a hint at the versatility of the mass spectrometric method. Four general topics, each covered by two or even three contributions have therefore been chosen. In contrast to the usual Topics policy to publish comprehensive reviews, this issue combines shorter essays highlighting a special aspect with more complete overviews.

The first category is that of reactivity: Einar Uggerud reviews the analysis of reaction mechanisms of gaseous ions, followed by an article on the generation of elusive, highly reactive neutral intermediates by neutralization-reionization mass spectrometry authored by František Tureček. Reaction mechanisms and the fundamental questions related to ion/molecule reactions in interstellar clouds which provide similar conditions with respect to the pressure regime as compared to a mass spectrometer are the subject of Simon Petrie and Diethard Bohme.

The second topic is the examination of metal-organic species in the gas phase. In one of the essays, Detlef Schröder and Helmut Schwarz provide a brief overview on the analysis of stereochemical features through gas-phase reactions mediated by naked metal cations. Dietmar Plattner's theme of gas-phase catalysis is closely related, but quite different with respect to the methods employed.

The next section refers to mass spectrometric methodology: The investigation of structural features of gas-phase species is by no means trivial, since the primary information is that of the mass-to-charge ratio. Thomas Wyttenbach and Mike Bowers outline how gas-phase conformations can be studied with the ion mobility method. Similarly, it seems not to be straightforward to determine thermodynamic and kinetic data. Peter Armentrout demonstrates how that is nevertheless possible with great accuracy and gives insight into the guided ion beam methodology.

Finally, medicinal chemistry is an important area. Viral analysis and even the mass spectrometry of complete viruses is possible as demonstrated in the essay by Gary Siuzdak and his co-workers. This article at the same time is our tribute to biochemistry highlighting the enormous power of mass spectrometry for answering fundamental biological and biochemical questions. It is accompanied by a review article authored by Mark Brönstrup on the search for new drug candidates by high-throughput screening.

Several topics cross-link the contributions to this volume. For example, methodological aspects strongly contribute to Petrie's and Bohme's paper on the chemistry in interstellar clouds and thus it could also appear in the section on methodology. Furthermore, the question of structure plays an equally major role when discussing reaction mechanisms or when speaking of reactive neutral intermediates that are to be distinguished from their stable isomers. With the differentiation of diastereoisomeric ions, the stereochemistry, a very particular structural feature, is analyzed. Again, we encounter this subject when reading about the analysis of gas-phase conformation and in the discussion of virus structures. Structure crosses our way at very different levels of complexity and with a focus on different aspects: The detailed geometry of stereogenic centers represents an issue of the primary structure as far as atom connectivities are concerned, while analysis of the huge non-covalently bound protein shells of viruses focuses on the "secondary structure" of the mutual arrangement of the individual building blocks. Many other such links between the four sections can be found, but I would prefer to leave it to the reader to enjoy uncovering those him- or herself.

Of course, a collection like that presented here must leave open many aspects which should be included in any comprehensive overview on modern mass spectrometry. However, a single volume of Topics in Current Chemistry does not provide the space required for such a project. I nevertheless hope that the selection made here provides sufficient intriguing insights into the topic in order to fascinate not only those involved in mass spectrometric research, but also those who are interested to apply this method to their specific chemical problems.

Bonn, March 2003 Christoph A. Schalley

beyond, but I would prefer to leave it to the reader [illegible] here.

[illegible] overview [illegible] in [illegible] provide the [illegible] required [illegible] projects [illegible] the [illegible] graphs [illegible]

[illegible], March 200[illegible] [illegible]

Contents

Volume 221

Contrast Agents I

Magnetic Resonance Imaging

Volume Editor: Werner Krause

ISBN 3-540-42247-1

Contents of Volume 222

Contrast Agents II

Optical, Ultrasound, X-Ray and Radiopharmaceutical Imaging

Volume Editor: Werner Krause

ISBN 3-540-43451-8

Contents

I
Reactivity

Top Curr Chem (2003) 225: 3–36
DOI 10.1007/b10467

Physical Organic Chemistry of the Gas Phase. Reactivity Trends for Organic Cations

Einar Uggerud

Department of Chemistry, University of Oslo, P.O. Box 1033 Blindern, 0315 Oslo, Norway
E-mail: einar.uggerud@kjemi.uio.no

The existing literature in the field of physical organic chemistry of gas phase positive ions has been reviewed. The review is devided into sections covering all of the common functional groups and reaction types of organic chemistry with emphasis on substituent effects and trends in reactivity. The literature of the last ten years is covered, including comprehensive citations of older key work.

Keywords. Physical organic chemistry, Mass spectrometry, Quantum chemistry, Ion chemistry, Reaction mechanisms, Substituent effects

1 Introduction

An impressive number of gas phase chemical studies of ions have emerged during the last fifty years. Most of these studies were experimental, and a wide range of instrumentation methods, mostly mass spectrometric ones, have been used. More recently, these studies have been complemented by high level quantum chemical and other model calculations, providing firm connection between experiment and theory.

Gas phase studies reveal the intrinsic properties and reactivity of molecules free from the influence of a surrounding medium. This is valuable since it provides direct structure/property relationships, since chemists often have "naked" molecules in mind, even when they talk about solution experiments. Since most organic reactions are polar, ions– rather than neutral molecules – provide the most idealised description for some of the species involved in the corresponding solution reactions. Such basic physical organic chemistry models have been established for most of the common organic reactions, and they will be the main topic of this review. Despite the great value of these idealised reactivity models, we must never forget that they are only absolutely valid in the gas phase, and keep in mind the continuous change in going from the dilute gas phase via phases of increased pressure into the solvent phase. In this picture, solvent polarity is also a key element. In solution there are no free ions; in a polar solvent the ions are strongly solvated, while non-polar solvents prevent ion pair formation.

Physical organic chemistry is an area of research devoted to the understanding of all factors which influence the reactivity and physical properties of organic compounds. The ambitious program of this area of chemistry is to describe the behaviour of quite complicated systems quantitatively by applying simple and intuitively appealing principles and effects. In many respects, this approach has proven to be successful and instructive. However, it has become increasingly evident that there are problems with this simplistic point of view. For example, is the influence of a given substituent owing to an inductive effect, a mesomeric effect or a steric effect? Very often, the descriptors used – despite their apparent intuitive appeal – are of empirical nature and difficult to grasp at a more fundamental physico-chemical level. Gas phase studies may be very useful in unrevealing problems of this kind, and in establishing new and universally valid rules, so reliable quantitative and qualitative predictions of properties and reactivity may be given.

It has been very difficult to limit the scope of this review. After some consideration, I chose to divide it into sections in close correspondence with the tradition of text books in physical organic chemistry. I also decided to concentrate on literature from the last decade, although it has been necessary to cite classical papers of the field for the sake of completeness and to give an overview. I have to admit that I have chosen to focus on the literature of positive ion chemistry rather than that of negative ion chemistry, since excellent reviews have been published on that subject recently [1–3].

2 Solvent and Substituent Effects

The gas phase acid/base properties of molecules have been subject to equilibrium or "bracketing" measurements employing mass spectrometric techniques like ion cyclotron resonance (ICR) [4], Fourier transform ion cyclotron resonance (FT-ICR) [5, 6], Flowing afterglow (FA) and Selected ion flow tube (SIFT) [7], and high pressure mass spectrometry (HPMS) [8]. Proton transfer between neutral molecules are then investigated by measurements of reactions

$$B_1H^+ + B_2 = B_2H^+ + B_1 , \tag{1}$$

while the similar reaction of negative ions is

$$A_1H^+ + A_2^- = A_2H^+ + A_1^- . \tag{2}$$

It has been demonstrated that it is possible to determine reliable relative basicities by employing the kinetic method. By this method, a proton bonded dimer is caused to dissociate, and the difference in basicity of the constituent monomers is derived from the ratio of ionic products of the two competing dissociation channels [9]:

$$B_1H^+ + B_2 \leftarrow (B_2 \cdot B_1)H^+ \rightarrow B_2H^+ + B_1 , \tag{3}$$

This technique is especially valuable in cases where equilibrium measurement is hampered due to low volatility (peptides, zwitterions, salts etc.). Gas phase acidity (GA) and basicity (GB), and proton affinity (PA) are defined in the absolute sense through the following definitions:

$$AH \rightarrow A^- + H^+ \quad GA = \Delta G° \tag{4}$$

$$B + H^+ \rightarrow BH^+ \quad GB = -\Delta G°; PA = -\Delta H°. \tag{5}$$

Relative acidities and basicities measured by the methods described above are only liable to small errors, while conversion of these relative values to a universal and comprehensive acidity/basicity scale poses more serious problems, both with regard to evaluation of data from different laboratories and since the data have been derived using different experimental methods – and particularly since an absolute scale depends on calibration points from independent method measurements of ionic and neutral heats of formation. These problems can be overcome and one of the earliest successful attempts in this direction was made by Taft et al. [10]. More recent basicity scales are from Lias et al. [11, 12], Meot-Nier and Sieck [13] and Szulejko and McMahon [14]. The latest updated and complete list is from 1998 [15]. An evaluation of the acid/base properties of amino acids and peptides is given by Harrison [16], while some very basic compounds have been reported and reviewed by Racsynska et al. [17]. Quantum chemical methods have been of great help in evaluating acidity and basicity data and the consistency of the absolute scales, since high level *ab initio* methods predict these properties with great accuracy for small molecules [18–20].

When reliable data are at hand, it is possible to compare gas phase and solution properties. For the anions of the first-row hydrides the following order of gas phase basicities applies: $CH_3^- > NH_2^- > OH^- > F^- > Cl^- > Br^- > I^-$ – in agreement with the aqueous solution trend [21], although the gas phase data range over a considerably wider free energy range than the aqueous data. The reason is that solvation energies are substantial for both the neutral and its ionised counterpart, and that solvation energies tend to be larger for the charged species. It has been claimed that the big gap in solution pK_a values between HF and HCl should be due to the unique ability of F^- in forming species such as HF_2^- and H_3O^+[22]. This claim seems unjustified since a plot of new gas phase vs aqueous acidities of the hydrogen halides does not display this feature (Fig. 1).

The gas phase basicity order of the neutral molecules $NH_3 > H_2O > HF$ is the same as that found in aqueous solution. Also in this case the solvation energies are substantial, both for acid and base. For this reason it is difficult to know whether the solution behaviour reflects the intrinsic electronic properties of the molecule, or if the sum of solvent/solute interactions partially cancel out in a pattern which resembles the acidity and basicity orders.

Early gas phase data represented an important contribution to the understanding of solvent effects on acidity/basicity. In aqueous solution the basicity order was known to be $(CH_3)_3N < (CH_3)_2NH > CH_3NH_2$, and it was recognised that this is not the anticipated inductive order. The advent of Munson's gas phase order, $(CH_3)_3N > (CH_3)_2NH > CH_3NH_2$, settled the case by pointing out that only differences in solvation energies could explain the irregular solution behaviour [24].

On the basis of gas phase and solution data from their own and several other laboratories Aue, Webb and Bowers published a paper in 1976 in which they were able to assess solvent effects on the basicities of amines in quantitative terms by applying the Born electrostatic model of solvation [25]. By separating the enthalpic and entropic contributions, they noted that solvation attenuates gas

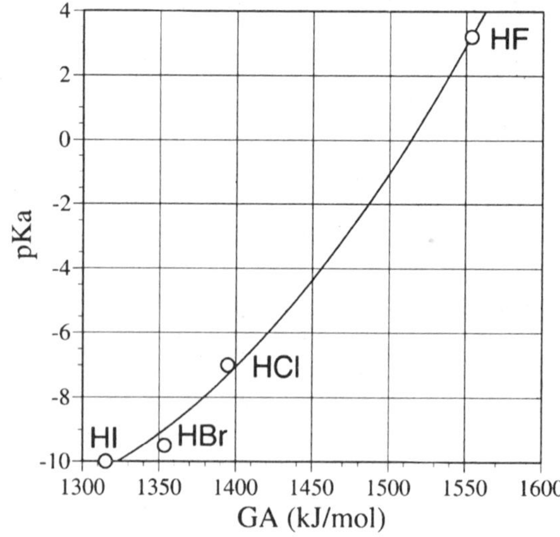

Fig. 1. Aqueous acidity (*pKa*) as a function of gas phase acidity (*GA*) for the hydrogen halides. The gas phase data have been taken from Ref. [23]

phase behaviour. While the gas phase proton affinities vary considerably upon substitution, the variation in solution is smaller by a factor of approximately six. However, the enthalpies of solvation are proportional to the gas phase proton affinities within the primary, secondary and tertiary amine series. Small, but significant entropy terms alter these orders. Similar patterns have also emerged for the acidities of alcohols, and for other compound classes.

Despite the fact that the solvent is an important factor in determining the properties of molecules in solution, it should be realised that moving to the gas phase does not solve all the conceptual problems of physical organic chemistry. We will consider this for the rest of this chapter.

Table 1 gives the gas phase acid and base properties of alcohols and amines, and it is evident that there are clear trends in the data with respect to the size of the alkyl group. The larger group is, the more basic and acidic is the corresponding amine and alcohol. Taft has given an in-depth discussion of substituent effects on acidity/basicity of these compounds in the gas phase [26]. Using a simple electrostatic model he divided substituent effects into three parts: field or inductive effect (F), polarisability effect (P), and resonance effect (R). It was demonstrated that in the case of the alcohols it was possible to separate the F and P contributions, and thereby correlate both the acidities and the basicities. The inductive F contribution is stabilising for RO^- relative to ROH, while it destabilises ROH_2^+. The P contribution stabilises both through polarisation.

Despite the success in parameterizing acid/base and many other properties for a range of different compounds, it is obvious that the simple electrostatic model used by Taft and extended by others [27, 28] have fundamental weaknesses – both with regards to the domain of validity and at a more fundamental level. The model is more intuitive than physical, in the sense that the inductive effect, polarisation effect, resonance effect, mesomeric effect and steric effect have no proper quantum mechanical definition, and can therefore not be derived directly from the system's wave function [29, 30].

One good example is provided by the carboxylic acids. It is generally believed, as in Taft's work, that the relative high acidity of these compounds originates from the truism that carboxylate ions are "resonance stabilised". Siggel et al. suggested on the basis of O1 s electron spectroscopy and *ab initio* calculations that the high relative instability of the acid rather than the stability of the anion is the key factor [31, 32]. On the other hand, more recent theoretical analysis has re-

Table 1. Gas phase basicities (*GB*) and acidities (*GA*) of alkyl alcohols and amines. The data are from Ref. [23]

Substituent R	GB ROH (kJ mol^{-1})	GB RNH_2 (kJ mol^{-1})	GA ROH (kJ mol^{-1})	GA RNH_2 (kJ mol^{-1})
H	660	819	1607	1661
CH_3	725	865	1569	1656
C_2H_5	746	878	1555	1639
i-C_3H_7	763	889	1542	1631
tert-C_4H_9	772	900	1540	–

vealed that this is probably not the case, and that the properties of the carboxylate anion are most important [33]. In any case, it turns out to be impossible to account for the contribution of resonance stabilisation to the anion stabilisation, since this effect is not uniquely defined in terms of quantum mechanics.

In water the basicity of 2,6-dialkylpyridines decreases upon increasing the size of the alkyl group, a fact which has been explained by the steric effect. The disturbing observation is that the gas phase trend is opposite. By analysing ab initio basicity data of 2,6-dialkylpyridines using the concept of isodesmic reactions, Roithova and Exner showed that the substituent effect is greatest in the protonated forms, and in the uncharged bases it is three times smaller [34]. In both cases, it is stabilizing in terms of polarization, and most importantly there is no sign of steric effects. The use of the term steric effect was also criticized recently in a study on the preferred conformation of ethane[35] and in a study of S_N2 reactions [36].

Recently, a paper was published in which the acidities of aliphatic carboxylic acids were analysed in terms of a quantum mechanical model. The inductive effect of Taft was shown to be composed of one electrostatic and one polarization term [37]. We will discuss examples of similar approaches based on the hardness and electronegativity as strictly defined concepts of density functional theory in the chapter on amides and peptides, and also discuss the problem of ill-defined terms a bit further in the chapter on carbocations, coming next.

3 Carbocations

Although perfectly free carbocations (in a strict sense) may only exist in the dilute gas phase, weakly bonded ions of this kind have been inferred from solution and solid state data [38–40]. It is one hundred years since the existence of carbocations in solution was recognised [41–47]. A startling change from colourless to yellow was observed upon adding aluminium trichloride to chlorotriphenylmethane. It was soon realised that the colour of this and analogous substances is due to the formation of salt-like compounds including triarylmethyl cationic structures. It took sixty years until the first X-ray crystallographic structure was presented in the literature [48]. In the meantime, compelling evidence from NMR, IR and ESCA spectroscopy had established the existence of alkyl cations, like the *tert*-butyl cation in superacidic solution [49]. Ten years ago, the crystal structure of (*tert*-$C_4H_9^+$)($Sb_2F_{11}^-$) was determined [50]. The idea of carbocations as intermediates during substitution and elimination reactions was put forward in the 1930s [51, 52]. Before looking further into this, we will consider the properties of carbocations in some more detail.

It has been understood for a long time that more substituted carbenium ions are more stable than their less substituted counterparts. To put this in more quantitative terms, and in order to avoid the complications of the solvent, it has proved worthwhile to examine the energetics of the following reactions [53]

$$B + R^+ \rightarrow RB^+. \quad (6)$$

In this equation, B is an electron pair donor (Lewis base) while R^+ is an alkyl cation (R = H, CH_3, CH_3CH_2, $(CH_3)_2CH$ and $(CH_3)_3C$). The corresponding enthalpies of reaction were obtained from accurate experimental data from the literature by employing the set B = H_2O, NH_3, I^-, Br^-, Cl^-, F^-, OH^-, NH_2^-. For a given R and B the negative of the enthalpy of equation 6 defines the alkyl cation affinity of B, $ACA(B) = -\Delta H°$ (note that in the case of R = H the alkyl cation affinity reduces to the proton affinity). It has been known for some time that there is a linear relationship between the methyl cation affinity and the proton affinity, but it has also turned out that the same is valid also for the other alkyl cation affinities. Plots of the various *ACA*s versus the corresponding *PA*s give straight lines for all R groups. This is particularly significant since a wide range of Lewis bases with both neutral and negatively charged species are included in the data set. The slopes of these lines were taken as a measure of the stability of each R (alternatively the instability of RB^+), and the following set of data was obtained reflecting the stability of R^+ : $a = 1.000$ (hydrogen), 0.938 (methyl), 0.915 (ethyl), 0.895 (*i*-propyl), and 0.883 (*tert*-butyl). In this respect a low a value means that R^+ is a poor Lewis acid. Whether the intrinsic stabilisation of the increasing number of methyl groups attached to a carbocationic centre is due to a hyperconjugative interaction or an inductive effect has been heftily debated in the literature [54], but from the point of view of this author the discussion is exhaustive and non-productive. It should be mentioned that these a values correlate well with the polarizability volumes of the corresponding alkanes in the form of an exponential relationship [53].

While anions are bonded strongly to an alkyl cation (e.g. in CH_3F), neutral molecules are less strongly bonded. This quite obvious point can be inferred from the data above, and has also been investigated theoretically [55]. This forms the basis for activation by proton transfer.

$$RX + AH^+ \rightarrow RXH^+ + A \tag{7}$$

It was shown around 25 years ago that alkyl cations may exist as distinct moieties in intermediates during unimolecular decomposition of organic ions [56, 57]. This was an important discovery since alkyl ions complexed in this way may serve as better models for intermediates in elimination and substitution reactions in solution than free alkyl ions. Complex formation may be enforced upon protonation.

$$RXH^+ \rightarrow R^+ \cdots XH \tag{8}$$

Protonated alcohols are central in this respect. One early example is protonated *n*-propanol [58]. En route to unimolecular decomposition of this ion (made by chemical ionisation) a rearrangement to protonated *iso*-propanol takes place, as evidenced by the translational energy release associated with the ultimate water loss. This rearrangement appears to take place within the confinements of a complex of water and the propyl cation. Protonated ethanol is the best studied case, and may serve as another good example [59–73]. There are two low energy isomers; the covalent $CH_3CH_2OH_2^+$ and the hydrogen bonded $C_2H_4 \cdots H^- OH_2^+$(a symmetric structure where the proton of H_3O^+ points towards the mid-point of the C=C bond). The former structure is 67 kJ mol^{-1} lower in energy (high level (G2) ab initio value). Since the barrier for isomerisation is lower than that of dissociation into either $CH_3CH_2^+$ or H_3O^+, and since the proton affinity of water is 17 kJ mol^{-1} higher than that of ethylene, both ionic dissociation products are formed from energetic ions of these structures. For partially deuteriated systems multiple H/D exchange between the carbon and oxygen atoms indicate that reversible isomerisation is a more facile process than dissociation.

For the protonated forms of higher alcohols the tendency for H/D exchange diminishes with the size [63]. This has been explained to be a consequence of the larger proton affinity difference between water and the alkene in these systems. The higher proton affinity of more substituted and larger alkenes (PA(isobutene) >PA(propene) >PA(ethylene)) leads to formation of the alkyl cation rather than H_3O^+ upon dissociation, and the alkene/H_3O^+ product becomes less and less favourable as the size increases. This means that in the case of the protonated forms of higher alcohols than ethanol the hydrogen bonded isomer $R^{+}\cdots OH_2$ is dominating form of the complex. For example, *tert*-butanol may isomerize swiftly from the covalent *tert*-$C_4H_9OH_2^+$ to the hydrogen bonded *tert*-$C_4H_9^{+}\cdots OH_2$, which is less than 30 kJ mol^{-1} higher in potential energy [63, 74–76]. The most remarkable point is perhaps the unusual mechanism of the isomerization; simple rotation of two methyl groups, each by 60°, induces this dramatic change in bonding character. This is illustrated by the two structures displayed in Fig. 2. Probably, there is a third isomer in-between – the result of rotating only one of the methyls.

The same general situation has been shown to exist for the protonated forms of alkyl halides, amines, arenes and similar compounds [63, 74, 77–80].

It is noteworthy that even methyl compounds may give rise to ion/neutral complexes, despite the fact that there is no corresponding alkene [81]. It was demonstrated that the hydrogen bonded species $CH_3^{+}\cdots OH_2$ and $CH_3^{+}\cdots NH_3$ correspond to stationary points on the respective potential energy surfaces. In the lat-

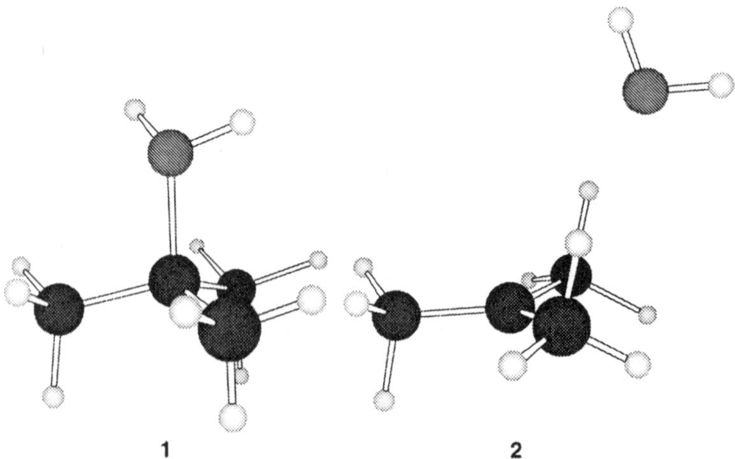

Fig. 2. Isomerization of protonated tertiary butanol (1) to a hydrogen bonded complex between the tertiary butyl cation and water (2) is provided by the successive rotations of two methyl groups. Results of MP2/6–31G* calculations

ter case, there is also a minimum for $CH_2{\cdots}H^-NH_3^+$, and it is relevant that experiment shows that in addition to loss of NH_3 and H_2, a small fraction of protonated methyl amine molecules also loose CH_2. Singlet methylene is slightly less basic than ammonia in the gas phase. Only a questionable shallow minimum was found for $CH_3^+{\cdots}F$. Similar complexes have been found for methanol and dimethyl ether [70]. Also in these cases experimental results are supported by quantum chemistry.

One important difference between aqueous solution and alkyl cations complexed by a single water molecule is that bulk water is far more basic than an isolated molecule [82]. For this reason alkene formation is thermodynamically more favourable *in aquae*, providing a more favourable all-over route for acid-catalysed elimination from protonated alkyl compounds. Gas-phase studies of water clustered protonated alcohols provides the link between the isolated alkyl cation and the water solution [83–85].

The examples discussed above not only show that complexed alkyl cations may be intermediates during ionic reactions in the gas phase, but also provide a dynamical machinery for rearrangements between substructures – either by atomic transfer between the two entities, or by reorientation leading to isomerisation or racemisation [86].

Solvolysis reactions are of particular concern nowadays. It is often difficult to envisage the detailed dynamics of a solution reaction on the basis of kinetic and mechanistic measurements, simply because the system under investigation is too complicated. This has led to ambiguity and indifference with regard to how kinetic data should be interpreted mechanistically, and the traditional S_N1 mechanism has been challenged by a number of recent studies [87–92]. The gas phase situation provides a far simpler picture, involving only a few molecular entities. For ionic gas phase reactions mimicking solvolysis, involving chiral substrates and putative carbocationic intermediates, normal mass spectrometric methods are of limited value, and have to be supplemented with techniques for analysis of the enantiomeric composition of the neutral end products [93–95]. The diagram in Fig. 3 illustrates a simplified view of acid-catalysed solvolysis of optically active phenyl ethanol derivatives, demonstrating how the timing of the different events involved determines the chirality of the end product. By using radiolysis for gas phase ion production and performing detailed analysis of the reaction kinetics based on the observed enantiomeric product composition, problems of this type have been addressed recently. For example, incomplete racemisation has been explained by a mechanism for hindered inversion involving an intimate ion-dipole pair [96]. In another study of similar compounds it was elegantly demonstrated how the degree of racemisation depends both on the temperature and substitution of the aryl group [97]. The term "troposelectivity" was introduced to illustrate how the substrate itself may direct the reagent towards one particular chiral or prochiral face and thereby determine the outcome of the reaction. The lack of troposelectivity could therefore mean that the lifetime of the substrate/reagent complex allows for reorientation and complete racemisation.

The above mentioned gas phase and solution studies are examples of nucleophilic substitution reactions, and will be the topic in the next section.

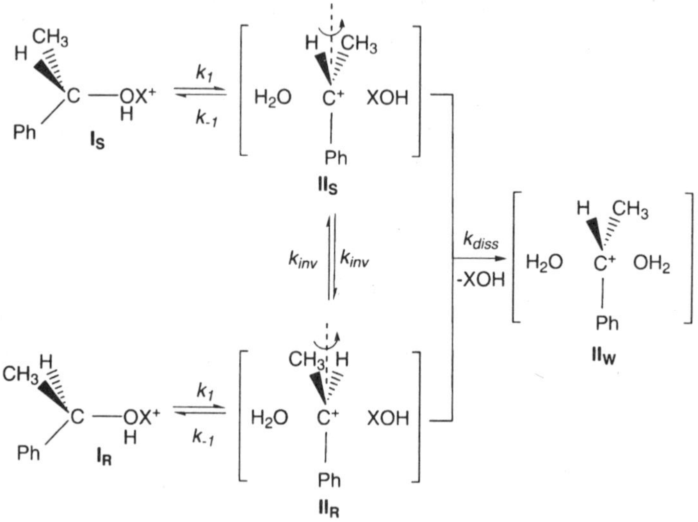

Fig. 3. A possible mechanistic scenario for acidic solvolysis (Reproduced from Ref. [96] with permission from the authors, copyright American Chemical Society)

4
Nucleophilic Aliphatic Substitution

The gas phase counterpart of the classical S_N2 reaction, which we call S_NB (all ionic nucleophilic substitutions in the dilute gas phase are by necessity of second kinetic order; B denotes "back-side"), proceeds via a mechanism where the incoming nucleophile attacks the face opposite that of the leaving group (Fig. 4, upper part). The transition state corresponds to a bipyramidal structure with partial bonds to the nucleophile (entering group) and the nucleofuge (leaving group). As a result of this, the reaction proceeds with inversion of the configuration around the central carbon atom (Walden inversion).

There exist a large number of gas phase studies of anionic substitution reactions ($Y^- + RX \rightarrow X^- + RY$), but unfortunately only a few have been conducted for the cationic counterpart. The whole field of gas phase nucleophilic substitution was reviewed recently [82].

Generally, cationic substrates, RX^+ are less strongly bonded than neutral, and – as mentioned in the previous chapter – the bond energies are related through the *a*-parameters of the R groups, also for the more strongly bonded neutral substrates RX. Despite the regular patterns in $R\text{-}X^+$ energies, it appears to be no simple physical relationship valid for S_NB barrier heights. The general idea that the barrier height is connected to the "looseness" of the TS in some way [98], cannot be justified. A quantum chemical study of a wide range of identity S_NB reactions (X = Y, R = CH_3), comprising a variety of nucleophiles ranging from He to F^-, concludes that there seems to be no universal principle which covers all cases [99]. Interestingly, for the poorest nucleophiles X = Y = He or Ne, the two

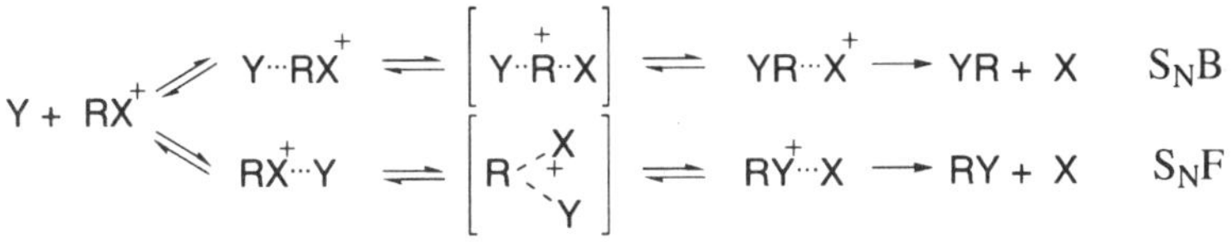

$$Y + RX^{+} \rightarrow Y\cdots RX^{+} \rightleftharpoons \left[Y\cdots \overset{+}{R}\cdots X\right] \rightleftharpoons YR\cdots X^{+} \rightarrow YR + X \qquad S_NB$$

$$Y + RX^{+} \rightarrow RX^{+}\cdots Y \rightleftharpoons \left[R\overset{X}{\underset{Y}{<}}{}^{+}\right] \rightleftharpoons RY^{+}\cdots X \rightarrow RY + X \qquad S_NF$$

Fig. 4. The reaction steps during nucleophilic substitution in the dilute gas phase. The upper route corresponds to backside displacement (the traditional S_N2 mechanism) and the lower route is the frontside displacement mechanism. The latter is possible in weakly bonded RX^+

potential energy minima Y···R–X and YR···X degenerate into one single symmetrical minimum with no transition structure separating them. Even in these extreme cases the properties of the "sandwiched" methyl cation of the central X···R···X intermediate cannot be described without taking the significant interaction with the X atoms into full account.

A previous theoretical study of many of the same nucleophiles/nucleofuges had the same general conclusion, but that author pointed out that there are regularities within sub-groups of similar nucleophiles [100]. Linear rate-energy relationship in the form of a correlation between the critical energy and the proton affinity of X was found within the following three groups: 1) NH_3, H_2O, HF; 2) NH_2^-, OH^-, F^-; and 3) F^-, Cl^-, Br^-, I^-.

Consideration of the effect of the alkyl group in reactions between water and protonated alcohols ($H_2O + ROH_2^+$) has revealed several surprising facts [36]. It was found that the relative rates of these identity substitution reactions are $(CH_3)_3C > (CH_3)_2CH > CH_3 > CH_3CH_2$, which is in good agreement with the theoretical calculations. This is different from the situation in solution where the trend $CH_3 > CH_3CH_2 > (CH_3)_2CH > (CH_3)_3C$ traditionally is explained by the notion that increased methyl substitution at the α-carbon reduces the rate constant for S_NB reactions due to increased "steric hindrance". A similar irregular pattern was found for $HF + RFH^+$,[101] but in the case of $NH_3 + RNH_3^+$ the expected decrease in reactivity with increased substitution was found [102]. Furthermore theoretical calculations of $H_2O + ROH_2^+$ in water clusters demonstrate that the medium is a key factor, since "normal" behaviour was found in this case [103]. The stark contrast between the gas and aqueous phases seems to be related to the effective basicity of the nucleophile. Bulk water is far more basic than a single water molecule, and the basicity order between different nucleophiles in the gas phase is $NH_3 > H_2O > HF$, as mentioned in the second chapter. The interplay between all groups interacting with the carbon centre (including nucleophile and the fixed alkyl substituents) is quite complex, and strong bases interact differently than weak. The details of the electron reorganisation during an identity S_NB reaction are at present poorly understood.

For thermoneutral identity reactions, there is no thermochemical driving force. In the case of non-identity nucleophilic substitution reactions – when the nucleophile and nucleofuge are different – reaction exothermicity may be taken quantitatively into account. This can be quite elegantly considered by applying the simple Marcus equation [104–109]. For cationic reactions, where interactions with the neutral nucleophile and nucleofuge are quite weak, it turns out that the barrier disappears for sufficiently exothermic reactions, even when the corresponding identity reactions each posses an intrinsic barrier [110].

A second significant feature of the study of the reactions between water and the protonated alcohols was the finding that frontside substitution, S_NF (Fig. 4, lower part) becomes more and more feasible the higher the degree of alkyl substitution there is on the carbon center [36]. In the case of the reaction H_2O + $(CH_3)_3COH_2^+$ the difference in barrier heights of S_NB and S_NF is small, the latter TS being only 10 kJ mol^{-1} higher in potential energy. Similar trends are found for the analogue reactions HF + RFH^+ [101] and NH_3 + RNH_3^+ [102]. The division between S_NB and S_NF is stereochemically essential since the former gives inversion of configuration, while the latter gives retention.

It is a fact that many solution phase reactions give enantiomeric mixtures other than the expected 100:0 (S_N2) or 50:50 (S_N1). This has traditionally been explained by a cross-over between S_N2 and S_N1 [111–114], but there is now a lot of experimental evidence which seems to be explained better on the basis of a competition between the S_NB and S_NF routes [88, 89, 91, 115–118].

Support for the S_NB/S_NF paradigm has now been obtained from quantum chemical studies of the gas phase analogues of some cationic cyclic substrates of this kind, including the norbornyl ring system [119, 120]. In some cases it was even found that S_NF is of lowest potential energy, in agreement with the corresponding experimental data of the corresponding solution reactions [119].

We conclude this section by noting that, at least in the gas phase with probable implications to solution, the general dynamic picture of Fig. 4 seems to be generally valid. To what degree this scheme is compatible to the traditional S_N2 and S_N1 picture is not only a question of semantics. According to Ingold the number 2 in the term S_N2 "... designates the molecularity of the reaction and not the kinetic order", and ".... molecularity... is a salient feature of the mechanism, meaning the number of molecules undergoing covalency change... in the rate determining step" [52].

5 Carbonyl Additions and Related Reactions

Carbonyl addition reactions include hydration, reduction and oxidation, the aldol reaction, formation of hemiacetals and acetals (ketals), cyanohydrins, imines (Schiff bases), and enamines [54]. In all these reactions, some activation of the carbonyl bond is required, despite the polar nature of the C=O bond. A general feature in hydration and acetal formation in solution is that the reactions have a minimum rate for intermediate values of the pH, and that they are subject to general acid and general base catalysis [121–123]. There has been some discussion on how this should be interpreted mechanistically, but quantum chemical calculations have demonstrated the bifunctional catalytic activity of a chain of water molecules (also including other molecules) in formaldehyde hydration [124–128]. In this picture the idealised situation of the gas phase addition of a single water molecule to protonated formaldehyde (first step of Fig. 5) represents the extreme low pH behaviour.

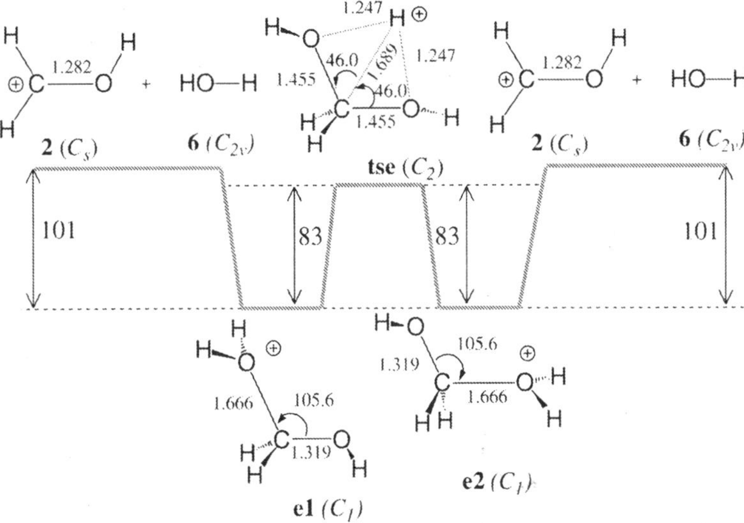

Fig. 5. Theoretical potential energy profile for the identity reaction between protonated formaldehyde and water. (Reproduced from Ref. [129] with permission, copyright Royal Society of Chemistry)

It is natural to extend this to the more general addition followed by elimination reaction ($A_N + D_N$), of which one prototype is depicted in Fig. 5. Intermediate level ab initio calculations (MP2/6–31G(d,p)) predict that the key transition structure of this identity addition/elimination reaction, corresponding to a 1,3-intramolecular proton transfer is slightly below the energy of the reactants/products [129]. Since experiments with $^{18}OH_2$ show no oxygen exchange [76, 130], the exact barrier must be higher than that calculated by MP2/6–31G(d,p). By comparison with experiment and more accurate quantum chemical calculations it has recently been established that this type of addition reactions are underestimated by ca. 30 kJ mol^{-1} with MP2/6–31G(d,p). In any instance, it is clear that the proton provides highly effective C=O bond activation.

In contrast to protonated formaldehyde itself, the proton bound dimer between unlabelled formaldehyde and ^{18}O labelled water reacts with a second molecule of $^{18}OH_2$ by slow exchange of the carbonyl ^{16}O [131].

Quantum chemical calculations have shown that the reaction depicted in Scheme 1 (addition followed by elimination) has a lower barrier than that in Fig. 5 [129, 132]. Since ammonia is more basic than formaldehyde, a swift proton transfer is kinetically favoured. For this reason, the prototype Schiff base forming reaction of Scheme 1 was overlooked for a long time, but a few years ago it was demonstrated that 1% of the gas phase collisions give protonated methylene imine, while 99% give protonated ammonia [132].

$$NH_3 + CH_2\overset{+}{O}H \longrightarrow H_3\overset{+}{N}CH_2OH \xrightarrow{1,3\ H^+} H_2NCH_2\overset{+}{O}H_2 \longrightarrow CH_2\overset{+}{N}H_2 + H_2O$$

Scheme 1

In a series of theoretical calculations of addition to protonated formaldehyde it was also revealed that the intrinsic properties of the attacking nucleophile (X) are reflected in the bond energy of the $X{-}CH_2{-}OH^+$ intermediate ($X = NH_3 > H_2O > HF > H_2$), and in the barrier for the subsequent 1,3-intramolecular proton transfer leading to water elimination [129]. In comparing different substituents X and Y, it was also found that for the same X, but different CH_2Y^+ ions the reactivity order is $Y = NH_2 < OH < F < H$. In reactions with formamide, it has been found that $Y = OH < Cl < F$ [133].

In order to protect the proton, and thereby suppress the kinetically favoured proton transfer route, it has been found out that gas-phase addition followed by elimination can be enhanced by reacting the proton bound dimer of the carbonyl compound rather than the protonated monomer [134]. In cases where the carbonyl compound has a higher proton affinity than the nucleophile, proton transfer is of course no problem. Alternatively, if the nucleophile already is protonated, as in the reactions between NH_4^+ and various carbonyl compounds, proton catalysed addition/elimination is possible as demonstrated experimentally by observation of immonium ion formation [135–137]. Likewise, the hydrazonium ion has been found to react with formaldehyde and a wide range of other aldehydes and ketones [138].

Reaction between protonated acetone and methanol gives slow reaction at low pressure in the ICR, while it is unobserved at the higher pressure of the flowing afterglow, in agreement with accompanying ab initio data[139]:

$$(CH_3)_2COH^+ + CH_3OH \rightarrow (CH_3)_2COCH_3^+ + H_2O\,, \tag{9}$$

while the reaction

$$(CH_3)_2CHOH^+ + CH_3OH \rightarrow (CH_3)_2COCH_3^+ + H_2O\,, \tag{10}$$

is quite fast as evidenced by a number of experimental techniques [140–142]. Experimental data from one laboratory for both reactions appear not to exist, but the experimentally determined barrier supported by a quantum chemical model calculation [142] indicates a lower barrier for the latter reaction. The reaction between protonated formaldehyde and water only gives swift proton transfer. The observation [143] of a fast reaction for

$$CH_2O + CH_3OH_2^+ \rightarrow CH_2OCH_3^+ + H_2O\,, \tag{11}$$

clearly indicates that the hypothetical reaction between protonated formaldehyde and methanol will have an even lower barrier, since methanol is more basic than formaldehyde.

Dissociation of protonated dimers of acetaldehyde/ketone (homogenous or heterogenous clusters) leads to water loss [144, 145], which is also the result of the corresponding ion molecule reactions [130]. Rearrangement of at least one of the reaction partners to the (protonated) enol seems likely. No comparable rates or barriers can be inferred from these data.

Another attractive alternative to protonation is to activate the C=O bond by alkylation. The simplest activated molecule of this kind is the methoxymethyl cation, $CH_2OCH_3^+$ [146–149], which has been the subject of numerous gas phase reactivity studies with a wide range of nucleophiles (H_2O, NH_3, H_2S, alcohols, amines, carboxylic acids, alkenes, aldehydes, ketones, benzene and other aromatics, C_{60}, peptides, amino acids, nucleotides etc.) [150–177]. The methoxymethyl cation is a versatile reagent also for analytical purposes and typically reacts and gives products according to a reaction sequence where addition is followed by elimination ($A_N + D_N$). The competing proton transfer is of course not possible in this case, but instead and in analogy, alkyl transfer to the attacking nucleophile is observed, consistent with an S_N2 mechanism. In this sense the methoxymethyl ion (and in general all alkoxymethyl ions) are ambident electrophiles by having two centres disposed for nucleophilic attack. Since the S_N2 reaction has higher enthalpic and entropic requirements than proton transfer, the addition/elimination products seems to dominate as long as the nucleophile has a transferable hydrogen bonded to the attacking atom (in order to accomplish the critical 1,3 proton shift).

Also with $CH_2OCH_3^+$ it is found that ammonia reacts faster than water, both in addition/elimination and substitution [173]. These observations were supported by quantum chemical model calculations. The methoxymetyl cation does not react with water, as implied in Eq. 11, but it reacts slowly with the lower alcohols [130, 171]. The ion reacts with dimethyl ether, acetone and acetaldehyde (only giving substitution products) [139, 155, 173, 174]. In gas phase reactions between amines and $CH_2OCH_3^+$ the following irregular rate order for addition/elimination has been found: $CH_3NH_2 > NH_3 > (CH_3)_2NH$ [165]. However, another study has shown a regular pattern for the reactions with primary amines [178]. The rates of the addition/elimination, the substitution, and a competing α-hydride abstraction reaction all increase with size: $(CH_3)_2CHNH_2 > CH_3CH_2NH_2 > CH_3NH_2 > NH_3$. In all respects the latter experimental and theoretical reactivity data are consistent with the regular pattern in Lewis basicity, as reflected in the a values. In reactions between the methoxymethyl cation and various aldehydes and ketones, the total rate increases with the size (and proton affinity) of the aldehyde or ketone, indicating that nucleophilicity and basicity are closely related in these gas phase reactions [174]. This type of basicity/nucleophilic relationship has also been suggested by others [169].

Also in reactions with amines we see that increasing alkyl substitution at the carbonyl carbon results in reduced reactivity, both with respect to addition/elimination and substitution, a finding which is perfectly consistent with increasing electron donating ability with the size of the alkyl group.

It is possible to alter the alkyl group giving rise to a series of methoxyalkyl cations, CH_2OR^+ [132, 179]. In this case reactivity decreases upon reaction with ammonia, as a result of the decrease in electrophilicity of both the carbonyl carbon and the alpha carbon of the alkyl group.

In all these cases it has been found that for a given alkyl group the relative potential energies of all stationary points along the reaction co-ordinate correlate linearly with the alkyl group's a value, clearly pointing to electronic control.

From electron impact ionisation mass spectra of alcohols and amines, it is well known that onium ions – oxonium ions: $(CR^1R^2OR^3)^+$ and immonium ions: $(CR^1R^2NR^3R^4)^+$, respectively – play important roles in determining fragmentation and thereby the detailed appearance of the spectra. Onium ions are usually the results of α-cleavages of the molecular ions, and they may fragment further to give smaller onium ions by alkene elimination [57]. The importance of hydrogen bonded ion-molecule complexes during the unimolecular decomposition of onium ions was pointed out in 1978 [56]. Interestingly, the same year, a paper appeared indicating that a similar process must be operative during the reverse reaction, namely addition of alkenes to small onium ions like $CH_2OCH_3^+$ [153]. Alkene metathesis products and product ions typical of oxonium ion decomposition were observed. Quantum chemical calculations have been valuable in substantiating the intermediacy of H-bonded alkene complexes of the type $(CR^1R^2OH^+)\cdots$(alkene) in the decomposition of $(CR^1R^2OR^3)^+$ and other onium ions [180–186]. On the other hand, when the alkene which is split off an onium ion originates from the carbon and not the oxygen of the C=O bond it seems that H bonded intermediates are not involved [183, 187, 188]. One example showing several of these features is the low pressure reaction between protonated formaldehyde and ethylene:

$$CH_2OH^+ + CH_2CH_2 \rightarrow C_3H_5^+ + H_2O\,. \tag{12}$$

No systematic data have been found on how reactivity varies with substitution in reactions between alkenes and onium ions.

The reactivities of activated aldehydes and ketones are exactly as we may deduce from our knowledge about the inherent electronic properties of the molecules involved. It is well known that the proton affinity increases with size and branching of the alkyl groups attached to the carbonyl carbon [15]. This may be directly attributed to the electron donating ability of the alkyl groups, and it is interesting that the proton affinities are linear in the *a* values [189], as well as in the O1s electron ionisation energies, at least to the extent such data are available [190]. The latter is of great significance, and the same seems to apply for the corresponding C1s electron ionisation energies. It is found that these core electron binding energies (determined for both atoms of the carbonyl group) decrease upon substitution. This is noteworthy, since it is often reasoned that the reduced electrophilicity upon increased alkyl substitution is due to steric hindrance, and the idea is that the alkyl groups physically hinder the nucleophile to approach. However, the 1s ionisation energies, and recent ab initio calculations [191] point in another direction, namely, it is the electronic effect which is dominant. To the degree it is meaningful at all to use the term "steric hindrance", it seems to be completely entangled with the electron donating property of the substituent.

6
Carboxylic Acid Derivatives and Esterification

At a first glance it may appear eccentric to show interest in the *basicities* of carboxylic *acids*. However, this is not as curious as it seems, since protonated carboxylic acids and their derivatives play the role as important intermediates during acid-catalysed reactions of these compounds. As a matter of fact, Scheele showed as early as in 1782 that addition of sulphuric acid to a mixture of acetic acid and ethanol accelerated esterification [192]. In this respect it is a key point that carboxylic acids, esters and anhydrides possess two oxygens which are potential sites of protonation; they are ambident bases. Of these two, the carbonyl oxygen is the more basic, as implied by the very good correlation between proton affinities and the corresponding core electron binding energies [193–197]. It is gratifying to notice that the advent of highly accurate schemes for quantum chemical calculations during the last decade or so, has substantiated this empirical relationship, by showing that the carbonyl oxygen indeed is more basic than the hydroxyl/ester oxygen by roughly 80 kJ mol^{-1} [198–201].

There appear to be few studies related to substituent effects on the properties of protonated carboxylic acid and related molecules. It has been found that the proton affinities of the saturated methyl esters increase with increasing chain length (n) according to the equation (in units of kJ mol^{-1})

$$PA = (40.0 \pm 2.5)^{*} \log(n) + (784.7 \pm 3.9) \text{ [202].}$$

Proton affinity data in the literature can be analysed on a physicochemical basis by applying the aforementioned a constants. Figure 6 shows plots of PA vs a for different classes of carboxylic compounds (RCOOH). The first panel shows that the plot for the three lowest carboxylic acids is perfectly linear, again indicating that the influence of the alkyl group is well reproduced by the a constants. The second panel shows the methyl esters ($RCOOCH_3$), for which there exist reliable PA values for all five esters for which an a value is defined. The third panel shows how the ester substituent R influences the PA values of formic esters (HCOOR) and acetic esters (CH_3COOR), and that also these homologue series correlate properly with the a constants.

The gas phase basicities of methyl benzoates have been studied systematically using ICR mass spectrometry. For electron withdrawing ring substituents there are very good correlations between the GBs of methyl benzoate and the GBs of α-methylstyrenes and acetophenones, respectively [203, 204]. On the other hand, there are noticeable deviations for electron donating substituents bearing free electron pairs. This was explained by a polarisability effect. To this reviewer, however, it seems that the possibility should have been considered that protonation in these cases may take place on the ring substituent rather than on the carbonyl oxygen. Irrespective of these complications, it is interesting and relevant that when a two-parameter Yukawa-Tsuno analysis is used, the PAs turn out to be linear in a modified set of σ constants. This was interpreted as being the result of different classes of compounds having different resonance energy demands upon protonation.

The proton affinities of alicyclic carboxylic acids are identical within 5 kJ mol^{-1}, with cyclohexane >cyclopropane >cyclopentane >cyclobutane [205]. This quite surprising situation seems not to be fully explained yet.

Although chemical ionisation spectra have been recorded for a large number of carboxylic acids and esters, there are few detailed studies of the unimolecular reactivity of protonated molecules of these classes of compounds. Early investigators [206] reported the general reaction:

$$[RCOOH]H^+ \rightarrow RCO^+ + H_2O\,, \qquad (13)$$

which at higher pressures was shown to be reversible [207]. The ambident character of carboxylic compounds is demonstrated by the observation that loss of

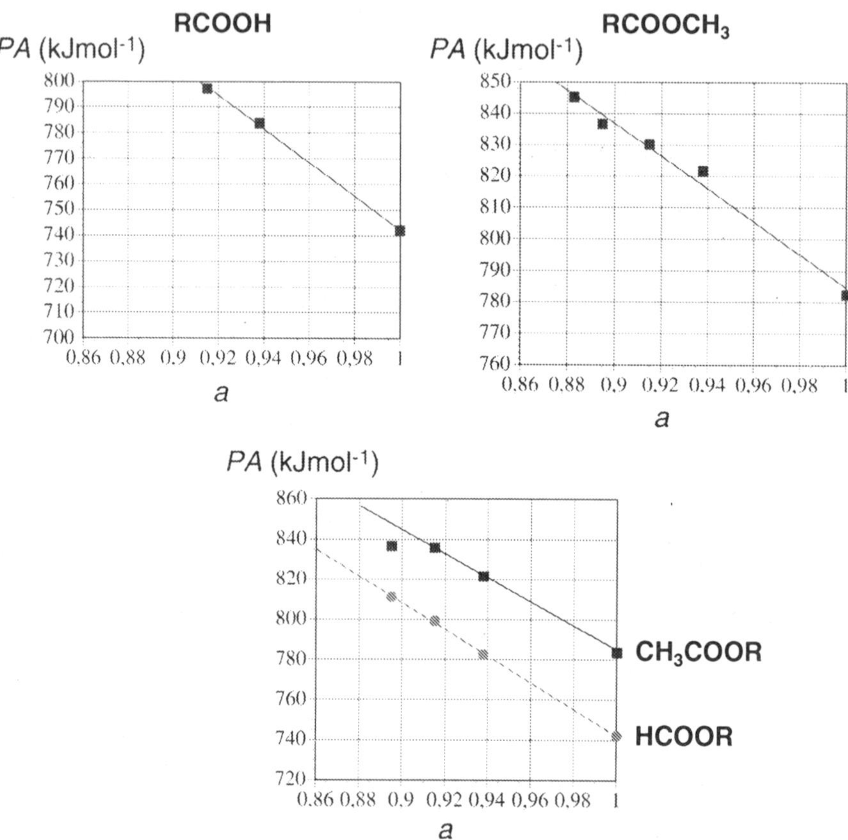

Fig. 6. Plot of proton affinities (*PA*) of carboxylic acids and esters versus the *a* value of the substituent R. These values are 1.000 (hydrogen), 0.938 (methyl), 0.915 (ethyl), 0.895 (*i*-propyl), and 0.883(*tert*-butyl), as explained in the chapter on carbocations. Proton affinity data have been taken from Ref. [15]

water is accompanied by a bimodal translational energy release distribution [208]. The low energy component was suggested as being due to direct dissociation from the hydroxy protonated form ($RCOOH_2^+$), while the high energy component results from the barrier resulting from passage from $RC(OH)OH^+$ to $RCOOH_2^+$; a necessary prerequisite for water loss from the carbonyl protonated form. Although the energy difference between the two isomeric forms seems to increase slightly with the size of the R group, it may be an oversimplification to state that the translational energy release of the high energy component follows directly from this. Firstly, the method for peak shape analysis leading from the experimental raw data to the translational energy release distribution could introduce artefacts. Secondly, the detailed reaction dynamics are generally unknown, so it is difficult to determine the fraction of the available energy that enters the reaction co-ordinate, thereby ending up as relative translational of the two fragments formed.

It should also be noted that in the case of formic acid, two products are formed, the formyl cation plus water (as discussed above), plus and hydronium ion plus carbon monoxide [206]. The latter product seems to be the result of intracomplex proton transfer within the dissociating $RCO\cdots OH_2^+$. Likewise, the higher homologue, protonated methylformiate gives $CH_3OH_2^+ + CO$ as the only product [209, 210].

Esterification in the gas phase was studied by the use of ICR in the 1970s [130, 211]. These studies revealed an unexpected richness in chemical reactivity. There are two different ways to promote the gas phase analogue of acid catalysed esterification; depending on whether the carboxylic acid or the alcohol is protonated initially:

$$R'C(OH)_2^+ + ROH \rightarrow R'C(OH)(OR)^+ + H_2O\,, \tag{14}$$

$$R'COOH + ROH_2^+ \rightarrow R'C(OH)(OR)^+ + H_2O\,. \tag{15}$$

Even though these reactions appear similar, there are principal differences, since the relative proton affinities of the acid and the alcohol are essential. For example, if we assume that condensation only can take place if the proton can be transferred to the carboxylic acid before the actual reaction takes place, it is difficult to envisage a reaction between a protonated alcohol and a carboxylic acid if the alcohol has higher proton affinity. On the other hand, it is a great advantage if the proton is situated at the carboxylic acid from the beginning, since proton transfer from the alcohol to the acid often is too exothermic to allow for a long-lived intermediate necessary for esterification, and the reaction product is consequently protonated carboxylic acid. This is exactly what has been observed experimentally. In reactions between protonated acetic acid it has been shown that reactivity follows the order primary >secondary >tertiary alcohol. In this case ^{18}O-labelling unequivocally shows fragmentation of the O-acyl bond, in favour of the traditional $A_N + D_N$ type of mechanism rather than of the O-alkyl bond which would have indicated an S_N2 type of mechanism. Apparently, the opposite reactivity order is observed for protonated formic acid. However, in this case the low proton affinity of formic acid gives rise to predominant proton transfer being most dominant for tertiary alcohols, so the observed reactivity trend may not reflect relative barriers to esterification. Protonated acetic esters only react with sufficiently nucleophilic (and therefore of sufficient proton affinity) alcohols and the trend is tertiary >secondary >primary. There is, however, one important difference between the reactions of protonated acetic acid and protonated acetic acid esters with alcohols in that ^{18}O-labelling shows that, in the latter case, the O-alkyl bond is broken, pointing towards S_N2. A third mechanism, acyl transfer is an alternative which seems to be operative in other systems [212]. Despite the reaction's apparent simplicity, at present it seems difficult to rationalise and fully understand the mechanistic scenario of gas phase esterification and similar reactions.

7
Amides, Peptides and Proteins

Mass spectrometry is one of the major techniques in the interdisciplinary field of proteomics. It provides a rapid, sensitive and reliable means of protein identification and structural determination, allowing for development in this newly baptised but yet classical field of biochemistry and biomedicine. The use of electrospray ionisation in conjunction with a tandem mass spectrometer (MS/MS) provides essential amino acid sequence information from the *m/z* values of the so-called *b* and *y* ions formed from cleavage of the amide bond of a protonated peptide. This reaction requires proton catalysis, and the mechanism is of interest in the present context, since it is closely related to the processes occurring in other protonated carboxylic acid derivatives.

The simplest model of an amide bond is found in formamide, and several features of protonated formamide are highly relevant to the cleavage of protonated peptides into *b* and *y* ions. Amides are bidentate bases, and it has been demonstrated from correlations between core electron energies and proton affinities [213] and from quantum chemical calculations [214] that the carbonyl oxygen is more basic than the amide nitrogen. As demonstrated by FT-ICR, metastable ion dissociation, and RRKM and quantum chemical model calculations [214], the unimolecular dissociation of a protonated formamide molecule depends on which site the proton is attached to:

$$HC(OH)NH_2^+ \rightarrow HCO^+ + NH_3\,, \tag{16}$$

$$HCONH_3^+ \rightarrow CO + NH_4^+\,, \tag{17}$$

since the upper reaction (a *b* type dissociation), takes place from relatively energetic ions of the more stable structure, while the least stable N-protonated isomer gives rise to the thermochemically favoured products of the lower reaction (a *y* type dissociation).

The proton affinities of all the 20 most common amino acids ($NH_2CHRCOOH$) have been determined in several laboratories, using equilibrium, bracketing or kinetic methods [15, 16, 215–226]. In addition, there exist quantum chemical calculations at an intermediate level predicting relative *PA*s with good confidence [227–233], although the sizes of most of the amino acids do yet not allow for benchmark calculations. Except for lysine, histidine and arginine, for which the side chain terminus is the most basic site, the most basic site is the nitrogen of the amino group directly bonded to the alpha carbon [234]. The data reveal that for amino acids with aliphatic side chains, proton affinity increases with size and branching. By using the concepts of hardness and electronegativity, as they are rigorously defined in the framework of density functional theory, it is possible to consider the amino acid residues of peptides as functional groups (-NHCHRCO-) [230]. It was shown that the proton affinity of an amino acid correlates well with these two quantities in the form of a linear two-parameter model, having the group's hardness as the dominating parameter. It is also notable that the hardness is closely related to the polarizability, which is a measurable property [235–237], indicating a link between the theoretical fundament and the language of the practising chemist.

A critical reader would perhaps note that the proton affinities of the individual amino acids are not directly relevant to peptides, since the basic amino groups of amino acid are incorporated in amide bonds, for which the nitrogen basicity is far lower. For whole peptides or proteins, even more basic sites are found at the termini of side chains, as in the cases of lysine, histidine and arginine. However, as we will see below, the basicity of the free amino group created during peptide bond dissociation is crucial in determining which of the two dissociation products that will end up with the transferable proton, and thereby the charge.

The fragmentation of protonated amino acids was the subject of several early investigations [238–240]. The most pronounced reaction is loss of the elements CO_2H_2. Detailed quantum chemical calculations of the potential energy hypersurface of protonated glycine demonstrated that the sequence of events preceding this dissociation are [231, 241, 242]:

$$^{+}H_2NCH_2COOH \rightarrow H_2NCH_2COOH^{+} \rightarrow H_2O\cdots H\text{-}NHCH_2^{+}\cdots CO\,. \quad (18)$$

There are several points to be made. Firstly, the hydroxy protonated isomer formed in the first step is of transient existence, since its potential energy well is only 6 kJ mol^{-1} deep. The isomer resulting from the second step is only bonded by 14 kJ mol^{-1}, giving loss of CO. According to the calculation, subsequent loss of water will require an extra 92 kJ mol^{-1}. Since the combined energy of the dissociation products $CO + H_2O + H_2NCH_2^{+}$ is only marginally above the energy of the transition state of the rate determining step (the first isomerisation), the combined loss of carbon monoxide and water is observed. However, a metastable ion decomposition study has also revealed noticeable CO loss without the loss of water [243]. A completely analogous mechanism of sequential CO + NH_3 loss was unveiled a couple of years before in protonated glycinamide [244].

It appears surprising that glycine and most other amino acids do not loose water upon protonation, since this would correspond to simple acylic cleavage catalysed by a proton – a process observed for amides and carboxylic acid derivatives in general, as discussed above. The reason for this lies in the fragile nature of the product ion $H_2NCHRCO^{+}$, which according to the calculations has the structure of a complex, $H_2NCHR^{+}\cdots CO$. The driving force is the ultimate formation of the highly stabilised immonium ion, $^{+}CH_2NHR$ [245]. This is also reflected in the quite substantial translational energy releases associated with the process [246]. In striking contrast to this, it was noticed already in 1976 that water is lost from protonated amino acids having side chains with nucleophilic functional groups capable of attacking the protonated amide group intramolecularily, probably in an acylium transfer mechanism [240]. More recently, this idea has been elaborated on and extended to the fragmentation mechanism of protonated peptide responsible for the important *b* and *y* fragments used for amino acid sequence determination [245, 247, 248]. The mechanism, which has been supported by theoretical calculations [249–251], is illustrated in Fig. 7. After amide bond dissociation, an intermediate ion-molecule complex is formed. If this complex breaks directly, a *b* ion is the result. If a proton is transferred from

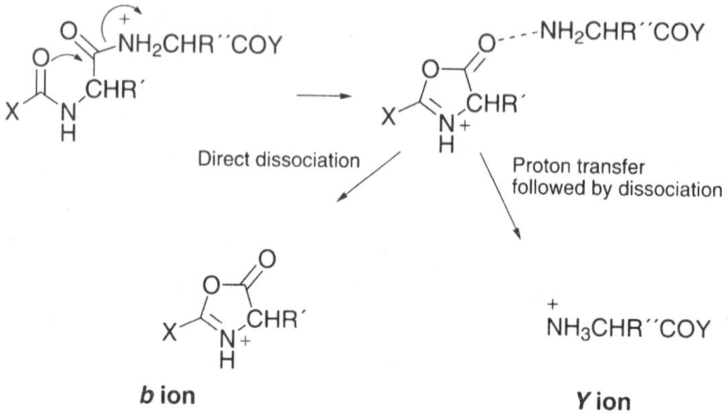

NH2CHR′′COY CHR′ X N H → O---NH2CHR′′COY CHR′ X N+ H

Direct dissociation

Proton transfer followed by dissociation

CHR′ X N+ H

NH3CHR′′COY

b ion

Y ion

Fig. 7. The mechanism for formation of *b* and *y* ions during fragmentation of protonated peptides and protein. The mass spectral pattern of *b* and *y* ions are used to determine the amino acid sequence

the *b* ion part of the complex to the neutral part before the complex dissociates, the complementary *y* ion is formed. As indicated in Fig. 7 the proton is then transferred to the newly formed amino terminus. The ratio between *b* and *y* ions according to this mechanism should therefore depend on the basicity of the newly formed amino group of the neutral part of the complex. This appears to be the case, since tandem mass spectrometry experiments of dipeptides and tripeptides where one of the amino acid residues is variable show linear relationships between log (*b*/*y*) and the *PA* of the variable amino acid [252, 253].

8 Electrophilic Aromatic Substitution

8.1 General

The simplest and most general mechanism for electrophilic aromatic substitution in solution is the so-called arenium ion mechanism, depicted in Scheme 2 [54, 254].

It corresponds to addition followed by elimination, and is symbolised by $A_E + D_E$. The departing X^+ is often a proton, while Z is a general substituent. The key step in this scheme is the formation of an intermediate arenium ion (Wheland intermediate, σ complex), and the relative stability of this species is crucial to the outcome of the reaction. Isolable arenium ions are known, and the benzenonium ion itself; $C_6H_7^+$ has been inferred from NMR of strongly acidic solutions [255],

Scheme 2 + Y^+ → → + X^+

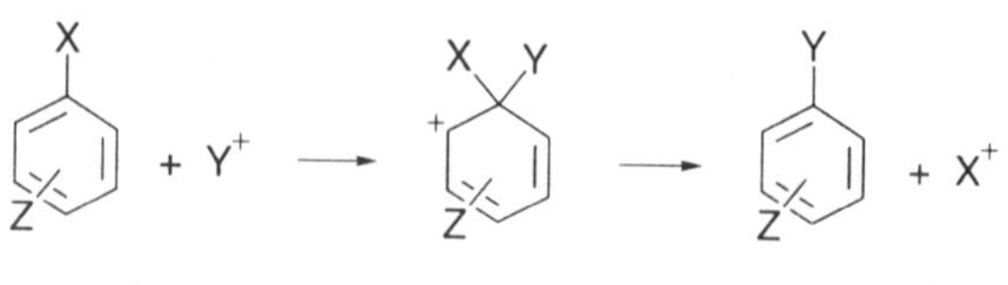

and on the basis of X-ray crystallography [256]. The *ortho*, *meta*, *para* and *ipso* positions of a monosubstituted benzene are all susceptible to attack from Y^+, and the properties of the substituent Z determine the reaction rate and the product isomer distribution. It is well known that aromatic molecules form charge-transfer complexes with many electrophilic molecules, so-called π complexes. To which extent such π complexes may act as precursors to arenium ion formation has been debated frequently in the literature, since Dewar's proposition in 1946 [257]. This will be discussed below.

There are several in-depth reviews on gas phase electrophilic aromatic substitution. The literature up to 1990 was thoroughly reviewed by Kuck [258], up to 1996 by Fornarini [259], up to 1998 by Fornarini [260], and up to 2000 by Gronert [3].

8.2
The σ complex and protonation of aromatic molecules

The prototype arenium ion, the benzonium ion **i**, corresponds to an energy minimum structure having C_{2v} symmetry (Scheme 3), as demonstrated by *ab initio* calculations at high level [261, 262].

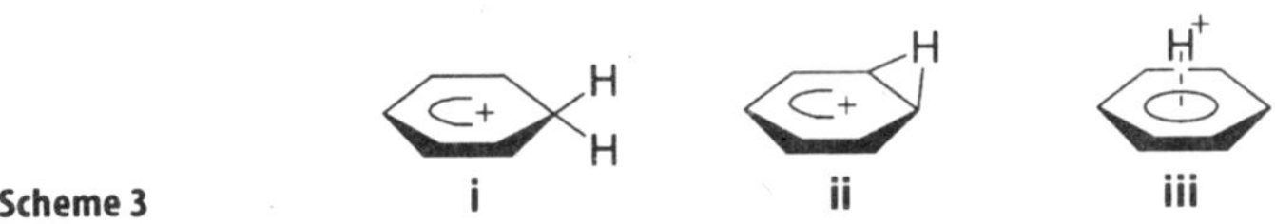

Scheme 3

The isomeric structures **ii** and **iii** do not correspond to minimum structures of the $C_6H_7^+$ potential energy surface, but are critical points of first and second rank, respectively. The transition structure, represented by **ii**, corresponds to a 1,2-proton shift, 35 kJ mol^{-1} higher in energy than **1**, and it provides a mechanism for the fast proton/deuterium scrambling observed in the gas phase and in acidic solution. Structure **iii** would correspond to a π complex between a proton and benzene. On the basis of the quantum chemical calculations it is clear that this is not a stable structure, and it is 199 kJ mol^{-1} above **1**.

A monosubstituted benzene, Ph-Y, has at least five different sites available for protonation, including the substituent. Out of the four ring positions, the *ipso* position is of particular importance since the *ipso*-protonated form of a monosubstituted benzene is identical to the σ complex obtained upon electrophilic attack of Y^+ on benzene. On the basis of a plot of gas phase proton affinities versus the corresponding Brown and Okamoto σ^+ substituent constants [263] for a series of monosubstituted benzenes, it was suggested that it is possible to distinguish between benzenes with a thermodynamic preference for protonation at the substituent (Y = CHO, CN and NO_2) and those with a preference for ring protonation (NH_2, OCH_3, OH, CH_3, C_2H_5, Cl, F) [264]. A neutralisation-reionisation study of protonated aromatics showed the same tendency [265]. Interestingly, we observe that the tendency for protonation at the substituent is observed for substituents which are known from solution chemistry to be *meta* directing and deactivating through electron withdrawal.

Quantum chemical studies have given detailed insight by providing site specific basicities. From calculations with the MP2 and HF wave functions for Y = CHO, NO_2 and NO it has been revealed that all ring positions in these compounds are less basic than benzene, but for all three, the *ortho* and *para* positions are slightly more basic than the *meta* position [266], in contrast to the known *meta* directing ability of these substituents in solution. On the other hand, the gas phase basicity of the *ipso* position is lower, while that of the substituents is higher; the latter is in accordance with the experimental evidence cited above. In solution Y = F, Cl are deactivating, but *ortho* and *para* directing. This diverges from the gas phase basicity order; the experimental gas phase proton affinities of fluorobenzene and chlorobenzene are 6 kJ mol^{-1} and 2 kJ mol^{-1} higher than that of benzene, respectively. Quantum chemical calculations do unequivocally show that the *PA* order is $p > o > m > i >$F(Cl) [267–269]. The quantum chemically calculated local proton affinities in phenol and aniline (Y = OH, NH_2) show that $p > o > m > i$, with all positions, except *i* is more basic than benzene [270, 271]. In the case of aniline, N is the most basic site. This is in good agreement with the available experimental evidence [272–274]. In the case of toluene (Y = CH_3) the solution behaviour (activating, *ortho/para* directing) is reflected in the gas phase with $p > o > m$, and *PA* above benzene [269].

In text books, substituent effects on reactivity in electrophilic aromatic substitution are explained in terms of valence bond theory by examination of possible resonance forms of the intermediate σ complex, without any consideration of the neutral reactant. As evident from the discussion above this type of analysis can be misleading. The properties of both the reactant and the intermediate are of course of importance, and the medium must be taken into account. In fact, closer analysis of the site-specific proton affinity data cited above and available core electron binding energies [275] reveals good correlation, while ^{13}C-NMR shifts show far poorer correlation. This is in line with the considerations of several authors [276–278], and gives a hint that the neutral reactant's properties are reflected to the intermediate. It is highly relevant in this respect that core electron binding energies of the *para* carbons correlate well with the aforementioned σ^+ substituent constants [275].

8.3
The initial stage and the role of the π complex

While the methyl and silyl cations behave like the proton in only forming σ complexes with benzene [279], there is considerable experimental and quantum chemical evidence that alkali metal ions, SF_3^+ and NO^+ only form π complexes [280–284]. It has been demonstrated by X-ray crystallography that there is a continuous transition in going between these extremes (Fig. 8) [285]. It seems that the more electrophilic the cation is, the closer is the complex to the σ form, while the more electron deficient the ring is, the more it seems to tend towards the limiting π complex character. The latter has also been inferred from the appearance of the collisional mass spectra of a variety of aromatic complexes [286]. It is quite surprising that so far we have not discovered any documented examples for which an ion may form both a σ complex *and* a π complex. This

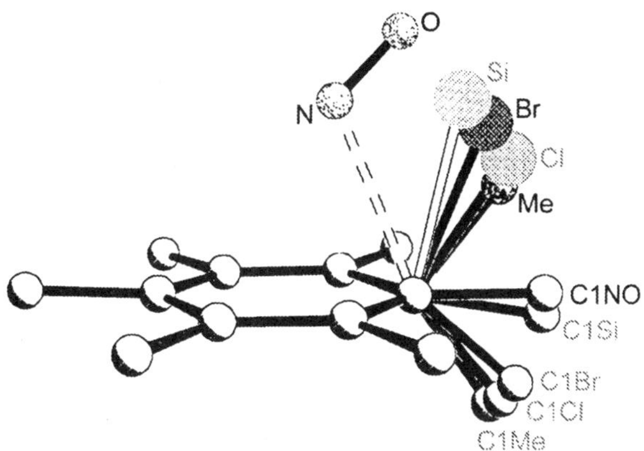

Fig. 8. Superpositioned pictures of X-ray crystallographically determined structures showing that poorer electrophiles have an enhanced tendency for π rather than σ complexation (Reproduced from Ref. [285] with permission from the authors, copyright American Chemical Society)

puts some doubt about the role of the π complex as a necessary precursor to the σ complex, as indicated in the usual text book mechanism.

There are, however, examples indicating that in ion molecule reactions between a protonated species (AH^+) and benzene (B), two isomeric forms of the intermediate complex may exist; (AH+)(B) and (A)(BH+) [74, 286]. In the cases of water [287] and propene [74], quantum chemical calculations clearly indicate that the former corresponds to a π complex where A-H acts as a hydrogen bond donor towards the centre of the benzene ring, while the latter is a hydrogen bonded complex between the benzenium ion and A. In neither case has a barrier been located, but is probably rather low in both cases. The role of the π complex has still not been clarified, since direct downhill routes from the reactants to the σ complex exist. It has been pointed out that π complex formation between a proelectrophile and the substrate may be important in solution and in biological systems for molecular recognition purposes. In such cases the proelectrophile is activated to form the actual electrophile subsequent to π complexation, thereupon giving rise to the σ complex. This has been shown by quantum chemistry to provide a reasonable scenario for the reaction between HF and benzene, in which BF_3 is ultimately required to promote ion formation of the HF/benzene π complex [288].

8.4 Alternative routes to the σ complex and the final stage

Gas phase arenium ions can be generated in many different ways as depicted in Fig. 9. The energy content of the arenium ion in the diluted gas phase will depend strongly on the exothermicity of the reaction used to form the ion, which in turn determines the fate of the arenium ion. Highly exothermic formation processes may even give rise to unimolecular dissociation. On the other hand, if

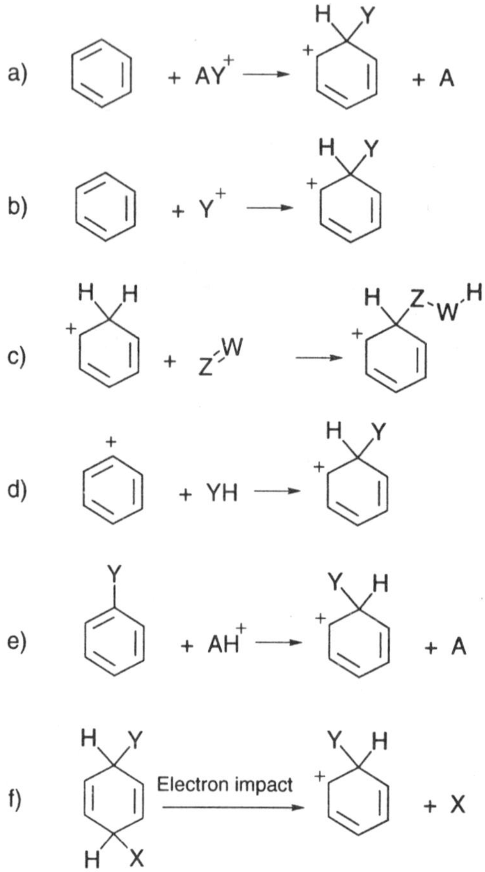

Fig. 9. Some alternative routes to arenium ions. a) Ref [289], b) Ref [290], c) Ref [291], d) Refs [292, 293], e) Ref [294, 295], f) [296]

a base is present and the arenium ion sufficiently long-lived, proton abstraction may occur, thereby completing an electrophilic aromatic substitution reaction. In this respect, a radiolytic technique is quite unique in bridging the gap between the gas and solvent phases [259]. The high pressure of an inert gas (typically atmospheric pressure) allows near thermal conditions, and careful control of the conditions opens up the possibility for kinetic studies of the intermediate arenium ions. The final neutral products are collected and analysed using ordinary organic spectroscopic techniques. It has been demonstrated that monosubstituted benzenes react with protonated methyl nitrate [297], in good agreement with the solution reactivity trend, but in poorer accord with the gas phase basicity trend reported above. I am not in a position to be able to give a fully satisfactory explanation, except that NO_2^+ may have slightly different affinity to the various sites of monosubstituted benzenes than H^+. It has also been possible to study the kinetics of for example hydrogen/deuterium exchange and other rearrangement processes [298]. In one case it has even been demonstrated that deprotonation is rate determining, expressing a kinetic isotope effect [299].

9
Hydride abstraction and Hydrogenation/Dehydrogenation

One is not used to thinking of H_2-additions or eliminations as polar reactions, but this seems to be the case. The following reactions:

$$CH_3NH_3^+ \rightarrow CH_2NH_2^+ + H_2\,, \tag{19}$$

$$CH_3OH_2^+ \rightarrow CH_2OH^+ + H_2\,, \tag{20}$$

$$CH_3FH^+ \rightarrow CH_2F^+ + H_2\,, \tag{21}$$

$$CH_3SH_2^+ \rightarrow CH_2SH^+ + H_2\,, \tag{22}$$

$$CH_3ClH^+ \rightarrow CH_2Cl^+ + H_2\,, \tag{23}$$

have been studied experimentally [300–304], and all but the last have also been investigated by quantum chemical calculations [302, 304–306]. The accurate data for the series X = NH_2, OH, F (Eqs. 19–21) has shown how the electronic factors govern the barrier heights. The barriers for Eqs. 19–21 are linear in the proton affinities of CH_3NH_2, CH_3OH and CH_3F. This has been interpreted in favour of a polar mechanism where a proton moves intramolecularily from the hetero atom (N, O, F) towards one of the hydrogens of the methyl group which is abstracted in the form of a hydride [307]. The corresponding intermolecular proton to hydride transfer is observed in many ion molecule reactions [308]. The reactant ions can be considered analogues of CH_5^+. The latter is a hypervalent adduct between H_2 and CH_3^+, while the former are adducts between HX and CH_3^+. The product ions are all of the type CH_2X^+. In the case X = H, there is no barrier for the reverse reaction which is addition of H_2 to CH_2X^+, while increasing *p* electron donation from X to C increases the barrier height. It has been demonstrated that the barriers for reactions of this kind are significantly diminished upon introduction of a separate catalyst molecule, e.g. water. This molecule acts as a mediator for the proton transfer, avoiding the unfavourable tight transition state. It has been shown that this type of non-metal activation of the hydrogen molecule is found in living systems [309] and in zeolites [310].

Acknowledgements. I wish to thank everybody who has contributed to my understanding of the phenomena described in this review. In addition I would like to acknowledge the NFR (The Norwegian Research Council) for support of our own studies in this field. I would like to thank Dr. Jon K. Lærdahl for valuable discussions and for giving comments on the manuscript.

10 References

1. DePuy CH (2000) Int J of Mass Spectrom 200:79–96
2. DePuy CH (2002) J of Org.Chem 67:2393–2401
3. Gronert S (2001) Chem Rev 101:329–360
4. Lehman TA, Bursey MM (1976) Ion Cyclotron Resonance Mass Spectrometry. Wiley-Interscience, New York
5. Comisarow MB, Marshall AG (1996) J Mass Spectrom 31:581–585
6. Marshall AG, Hendrickson CL, Jackson GS (1998) Mass Spectrom. Rev 17:1–35
7. Adams NG, Smith D (1988) In: Farrar JM, Saunders WH (eds) Techniques for the Study of Ion-Molecule Reactions (Techniques of Chemistry XX). Wiley-Interscience, p 165–220
8. Briggs JP, Yamdagni R, Kebarle P (1972) J Am Chem Soc 94:5128–5130
9. Cooks RG, Patrick JS, Kotiaho T, McLuckey SA (1994) Mass Spectrom Rev 13:287–340
10. Wolf JF, Staley RH, Koppel I, Taagepera M, McIver RT, Jr., Beauchamp JL, Taft RW (1977) J Am Chem Soc 99:5417–29
11. Lias SG, Liebman JF, Levin RD (1984) J Phys Chem Ref Data 13:695
12. Lias SG, Bartmess JE, Liebman JF, Holmes JH, Levin RD, Mallard WG (1988) J Phys Chem Ref Data 17:1
13. Meot-Ner M, Sieck LW (1991) J Am Chem Soc 113:4448–4460
14. Szulejko JE, McMahon TB (1993) J Am Chem Soc 115:7839–7848
15. Hunter EP, Lias SG (1998) J Phys Chem Ref Data 27:413–656
16. Harrison AG (1997) Mass Spectrom Rev 16:201–217
17. Raczynska ED, Decouzon M, Gal J-F, Maria P-C, Gelbard G, Vielfaure-Joly F (2001) J Phys Org Chem 14:25–34
18. Smith BJ, Radom L (1993) J Am Chem Soc 115:4885–4888
19. Smith BJ, Radom L (1994) Chem Phys Lett 231:345–351
20. Smith BJ, Radom L (1995) J Phys Chem 99:6468–6471
21. Bartmess JE, McIver JRT (1979) In: Bowers MT (ed) Gas Phase Ion Chemistry, vol 2. Academic Press, New York
22. Cotton FA, Wilkinson G, Murillo CB, Bochmann M (1999) Advanced Inorganic Chemistry. John Wiley, New York
23. Lias SG, Rosenstock HM, Deard K, Steiner BW, Herron JT, Holmes JH, Levin RD, Liebman JF, Kafafi SA, Bartmess JE, Hunter EF, Linstrom PJ, Mallard WG (2002) NIST Chemistry Webbok (http://webbook.nist.goc/chemistry)
24. Munson MSB (1965) J Am Chem Soc 87:2332–6
25. Aue DH, Webb HM, Bowers MT (1974) J Am Chem Soc 98:318–29
26. Taft RW (1983) Prog Phys Org Chem 14:247–350
27. Yukawa Y, Tsunao Y (1959) Bull Chem Soc Japan 32:965–971,971–981
28. Charton M (1999) Adv Mol Struct Res 5:25–88
29. Exner O (1999) J Phys Org Chem 12:265–274
30. Galkin V (1999) J Phys Org Chem 12:283–288
31. Siggel MR, Thomas TD (1986) J Am Chem Soc 108:4360–3
32. Siggel MRF, Thomas TD, Saethre LJ (1988) J Am Chem Soc 110:91–6
33. Exner O, Carsky P (2001) J Am Chem Soc 123:9564–9570
34. Roithova J, Exner O (2001) J of Phys Org Chem 14:752–758
35. Pophristic V, Goodman L (2001) Nature 411:565–8
36. Bache-Andreassen L, Uggerud E (1999) Chem Eur J 5:1917–1930
37. Ponec R, Girones X, Carbó-Dorca R (2002) J Chem Inf Comput Sci 42:564–570
38. Olah G (1995) Angew Chem, Int Ed Engl 34:1393
39. Olah GA (2001) J Org Chem 66:5943–5957
40. Vogel P (1985) Carbocation Chemistry. Elsevier, Amsterdam
41. Norris JF, Sanders WW (1901) Am Chem J 25:54–62
42. Norris JF (1901) Am Chem J 25:117–122
43. Kehrmann F, Wentzel F (1901) Chem Ber 34:3815–3819

44. Baeyer A, Villiger V (1902) Chem Ber 35:1189
45. Baeyer A, Villiger V (1902) Chem Ber 35:3013
46. Gomberg M (1902) Chem Ber 35:2397–2408
47. Walden P (1902) Chem Ber 35:2018
48. Gomez de Mesquita AH, MacGillary CH, Eriks K (1965) Acta Crystallogr 18:437–443
49. Olah GA, Kuhn SJ, Tolgyesi WS, Baker EB (1962) J Am Chem Soc 84:2733
50. Hollenstein S, Laube T (1993) J Am Chem Soc 115:7240–7245
51. Hughes ED, Ingold CK (1933) J Chem Soc 526
52. Ingold CK (1953) Structure and Mechanism in Organic Chemistry. Cornell University Press, Ithaca, NY
53. Uggerud E (2000) Eur Mass Spectrom 6:131–134
54. Smith MB, March J (2001) March's advanced organic chemistry: reactions, mechanisms and structure. John Wiley, New York
55. Esseffar M, El Mouhtadi M, Lopez V, Yanez M (1992) Theochem 87:393–408
56. Bowen RD, Williams DH, Hvistendahl G, Kalman JR (1978) Org Mass Spectrom 13:721–728
57. Bowen RD (1993) Org Mass Spectrom 28:1577–1595
58. Schwarz H, Stahl D (1980) Int J Mass Spectrom Ion Phys 36:285–9
59. Bohme DK, Mackay GI (1981) J Am Chem Soc 103:2173–2175
60. Dawson PH (1983) Int J Mass Spectrom. Ion Phys 50:287–97
61. Jarrold MF, Bowers MT, DeFrees DJ, McLean AD, Herbst E (1986) Astrophys J 303:392–400
62. Jarrold MF, Kirchner NJ, Liu S, Bowers MT (1986) J Phys Chem 90:78–83
63. Harrison AG (1987) Org Mass Spectrom 22:637–41
64. Meot-Ner M, Sieck LW (1989) Int J Mass Spectrom. Ion Processes 92:123–33
65. Smith SC, McEwan MJ, Giles K, Smith D, Adams NG (1990) Int J Mass Spectrom Ion Proc 96:77–96
66. Bouchoux G, Hoppilliard Y (1990) J Am Chem Soc 112:9110–15
67. Wesdemiotis C, Fura A, McLafferty FW (1991) J Am Soc Mass Spectrom 2:459–63
68. Swanton DJ, Marsden DCJ, Radom L (1991) Org Mass Spectrom 26:227–234
69. Mason RS, Parry A (1991) Int J Mass Spectrom. Ion Proc 108:241–53
70. Audier HE, Koyanagi GK, McMahon TB, Tholmann D (1996) J Phys Chem 100:8220–3
71. Matthews KK, Adams NG, Fisher ND (1997) J Phys Chem A 101:2841–2847
72. Fairley DA, Scott GBI, Freeman CG, Maclagan RGAR, McEwan MJ (1997) J Phys Chem A 101:2848–2851
73. Mason RS, Naylor JC (1998) J Phys Chem A 102:10090–10098
74. Berthomieu D, Brenner V, Ohanessian G, Denhez JP, Millie P, Audier HE (1995) J Phys Chem 99:712–20
75. Bell AJ, Giles K, Moody S, Underwood NJ, Watts P (1997) Int J Mass Spectrom Ion Proc 165:169–178
76. Uggerud E (unpublished data)
77. Audier HE, Monteiro C, Berthomieu D, Tortajada J (1991) Int J Mass Spectrom Ion Proc 104:145
78. Bouchoux G, Nguyen MT, Longevialle P (1992) J Am Chem Soc 114:10000–10005
79. Audier HE, Morton TH (1993) Org Mass Spectrom 28:1218–1224
80. Tu Y-P, Holmes JL (2001) J Chem Soc, Perkins 2:1378–1382
81. Uggerud E (1994) J Am Chem Soc 116:6873
82. Laerdahl JK, Uggerud E (2002) Int J Mass Spectrom 214:277–314
83. Hiraoka K, Kebarle P (1977) J Am Chem Soc 99:360–6
84. Hiraoka K, Takimoto H, Kebarle P (1986) J Am Chem Soc 108:5683–89
85. Karpas Z, Eiceman GA, Ewing RG, Harden CS (1994) Int J Mass Spectrom Ion Proc 133:47–58
86. Morton TH (1992) Org Mass Spectrom 27:353–368
87. Toteva MM, Richard JP (1996) J Am Chem Soc 118:11434–11445
88. Dale J (1998) J Chem Edu 75:1482–1485
89. Meng Q, Thibblin A (1999) J Chem Soc, Perkin Trans 2:1397–1404

90. Müller P, Rossier J-C, Abboud J-LM (2000) J Phys Org Chem 13:569–573
91. Müller P, Rossier J-C (2000) J Chem Soc, Perkin Trans 2:2232–2237
92. Gajewski JJ (2001) J Am Chem Soc 123:10877–10883
93. Morton TH (1982) Tetrahedron 38:3213
94. Speranza M (1992) Mass Spectrom Rev 11:73–118
95. Morton TH (2001) In: Adams NG, Babcock LM (eds) Adv Gas Phase Ion Chem, vol 4. JAI Press, Greenwich CT, p 213–256
96. Filippi A, Gasparrini F, Speranza M (2001) J Am Chem Soc 123:2251–2254
97. Filippi A, Speranza M (2001) J Am Chem Soc 123:6077–6082
98. Wolfe S, Kim C-K (1991) J Am Chem Soc 113:8056–8061
99. Ruggiero GD, Williams IH (2002) J Chem Soc, Perkin Trans 2:591–597
100. Uggerud E (1999) J Chem Soc, Perkin Trans 2:1459–1463
101. Civcir PU, Bache-Andreassen L, Laerdahl JK, Faegri Jr K, Uggerud E (manuscript in preparation)
102. Laerdahl JK, Uggerud E (unpublished results)
103. Uggerud E (1999) Int J Mass Spectrom 182/183:13–22
104. Marcus RA (1968) J Phys Chem 72:891–899
105. Murdoch JR (1972) J Am Chem Soc 94:4410
106. Wolfe S, Mitchell DJ, Schlegel HB (1981) J Am Chem Soc 103:7694–7696
107. Dodd JA, Brauman JI (1984) J Am Chem Soc 106:5356–5357
108. Wladkowski BD, Brauman JI (1993) J Phys Chem 97:13158–13164
109. Glukhovtsev MN, Pross A, Radom L (1996) J Am Chem Soc 118:6273–6284
110. Uggerud E (1999) J Chem Soc, Perkin Trans 2:1465–1467
111. Bentley TW, Bowen CT (1978) J Chem Soc, Perkin Trans 2:557
112. Bentley TW, Bowen CT, Parker W, Watt CIF (1979) J Am Chem Soc 101:2486
113. Bentley TW, Bowen CT, Morten DH, Schleyer PvR (1981) J Am Chem Soc 103:5466
114. Bentley TW, Llewellyn G (1990) Prog Phys Org Chem 17:121–158
115. Dauner H, Lenoir D, Ugi I (1979) Z. Naturforsch 34b:1745–1749
116. Stohrer W-D, Schmieder KR (1976) Chem Ber 109:285
117. Schöllkopf U, Fellenberger K, Patsch M, Schleyer PvR, Su T, van Dine GW (1967) Tetrahedron Lett 37:3639–3644
118. Schreiner PR, Schleyer PvR, Hill RK (1994) J Org Chem 59:1849–1854
119. Uggerud E (2001) J Org Chem 66:7084–7089
120. Sauers RR (2002) J Org Chem 67:1221–1226
121. Jencks WP (1972) Chem Rev 72:705–718
122. Lowry TH, Richardson KS (1981) Mechanism and Theory in Organic Chemistry Harper & Row, New York
123. Grunwald E (1985) J Am Chem Soc 107:4715–4720
124. Minyaev RM, Starikov AG, Lepin EA (1998) Russ Chem Bull 47:2078–2086
125. Wolfe S, Kim C-K, Yang K, Weinberg N, Shi Z (1995) J Am Chem Soc 117:4240–60
126. Ventura ON, Coitino EL, Lledos A, Bartran J (1992) J Comput Chem 13:1037–46
127. Ventura ON, Coitino EL, Irving K, Iglesias A, Lledos A (1990) Theochem 69:427–40
128. Williams IH (1988) Bull Soc Chim Fr: 192–8
129. Uggerud E (1996) J Chem Soc, Perkin Trans 2:1915
130. Pau JK, Kim JK, Caserio MC (1978) J Am Chem Soc 100:3831–7
131. Van der Rest G, Morgues P, Fossey J, Audier HE (1997) Int J Mass Spectrom Ion Proc 160:107–115
132. Bache-Andreassen L, Uggerud E (2000) Int J Mass Spectrom 195/196:171–184
133. O'Hair RAJ, Gronert S (2000) Int J Mass Spectrom 195/196:303–317
134. Pykäläinen M, Vainiotalo A, Pakkanen T, Vainiotalo P (1996) J Mass Spectrom 31:716–726
135. Maquestiau A, Flammang R, Nielsen L (1980) Org Mass Spectrom 15:376–379
136. Tabet JC, Fraisse D (1981) Org Mass Spectrom 16:45–47
137. Burinsky DJ, Campana JE, Cooks GR (1984) Int J Mass Spectrom Ion Proc 62:303–315
138. Morrison GC, Howard CJ (2001) Int J Mass Spectrom 210/211:503–509
139. Sheldon JC, Currie GJ, Bowie JH (1986) Aust J Chem 39:839

140. Dolnikowski GC, Heath TG, Throck Watson J, Shrivens J, Rolando C (1990) J Am Soc Mass Spectrom 1:481–488
141. Koenig S, Kofel P, Reinhard H, Saegesser M, Schlunegger UP (1993) Org Mass Spectrom 28:1101–5
142. Fridgen TD, Keller JD, McMahon TB (2001) J Phys Chem A 105:3816–3824
143. Karpas Z, Meot-Ner M (1989) J Phys Chem 93:1859
144. Aviyente V, Varnali T (1991) J Am Soc Mass Spectrom 2:113–119
145. Aviyente V, Varnali T (1993) J Mol Struct 299:191–195
146. Meerwein H, Bodenbenner K, Borner P, Kunert F, Wunderlich K (1960) Ann 632:38–54
147. Olah GA, Bollinger JM (1967) J Am Chem Soc 89:2993–6
148. Harrison AG, Ivko A, Van Raalte D (1966) Can J Chem 44:1625–32
149. Shannon TW, McLafferty FW (1966) J Am Chem Soc 88:5021–2
150. Dunbar RC, Shen J, Melby E, Olah GA (1973) J Am Chem Soc 95:7200–2
151. Staley RH, Corderman RR, Foster MS, Beauchamp JL (1974) J Am Chem Soc 96:1260–1
152. Matsumoto A, Okada S, Taniguchi S, Hayakawa T (1975) Bull Chem Soc Jpn 48:3387–3388
153. Van Doorn R, Nibbering NMM (1978) Org Mass Spectrom 13:527–34
154. Pau JK, Kim JK, Caserio MC (1978) J Am Chem Soc 100:3838–3846
155. Nibbering NMM (1979) Philos. Trans R Soc London, Ser A 293:103–15
156. Kim JK, Bonicamp J, Caserio MC (1981) J Org Chem 46:4236–42
157. Caserio MC, Kim JK (1982) J Org Chem 47:2940–2944
158. Kinter MT, Bursey MM (1986) J Am Chem Soc 108:1797–1801
159. Longevialle P (1987) Rapid Commun Mass Spectrom 1:122–4
160. Okada S, Abe Y, Taniguchi S, Yamabe S (1987) J Am Chem Soc 109:295–300
161. Karpas Z, Meot-Ner M (1989) J Phys Chem 93
162. Brodbelt J, Liou C-C, Donovan T (1991) Anal Chem 63:1205–7
163. Eichmann ES, Brodbelt JS (1993) Org Mass Spectrom 28:737–44
164. Audier HE, Bouchoux G, McMahon TB, Milliet A, Vulpius T (1994) Org Mass Spectrom 29:176–85
165. Wilson PF, McEwan MJ, Meot-Ner M (1994) Int J Mass Spectrom Ion Proc 132:149–152
166. Audier HE, McMahon TB (1994) J Am Chem Soc 116:8294–9
167. Tang M, Isbell J, Hodges B, Brodbelt J (1995) J Mass Spectrom 30:977–84
168. O'Hair RAJ, Freitas MA, Gronert S, Schmidt JAR, Williams TD (1995) J Org Chem 60:1990–8
169. Alvarez E, Brodbelt JS (1996) J Mass Spectrom 31:901–907
170. Freitas MA, O'Hair RAJ, Dua S, Bowie JH (1997) J Chem Soc, Chem Commun: 1409–1410
171. Audier HE, McMahon TB (1997) J Mass Spectrom 32:201–208
172. Freitas MA, O'Hair RAJ, Williams TD (1997) J Org Chem 62:6112–6120
173. Freitas MA, O'Hair RAJ (1998) Int J Mass Spectrom Ion Processes 175:107–122
174. Van der Rest G, Bouchoux G, Audier HE, McMahon TB (1998) Eur Mass Spectrom 4:339–347
175. Mosi AA, Skelton RH, Eigendorf GK (1999) J Mass Spectrom 34:1274–1278
176. Ma L, Liu Z, Wang W, Guo X, Liu S (1999) J Phys Chem A 103:8634–8639
177. Ramos LE, Cardoso AM, Correia AJF, Nibbering NMM (2000) Rapid Commun Mass Spectrom 14:408–416
178. Bache-Andreassen L, Uggerud E (2002) In preparation
179. Bache-Andreassen L, Uggerud E (2001) Int J Mass Spectrom 210/211:459–468
180. Nobes RH, Rodwell WR, Bouma WJ, Radom L (1981) J Am Chem Soc 103:1913–1922
181. Nobes RH, Radom L (1983) Chem Phys Lett 99:107–111
182. Nobes RH, Radom L (1984) Org Mass Spectrom 19:385
183. Tortajada J, Audier HE (1991) Org Mass Spectrom 26:913–914
184. Bouchoux G, Penaud-Berruyer F, Audier HE, Mourgues P, Tortajada J (1997) J Mass Spectrom 32:188–200
185. Chalk AJ, Radom L (1998) J Am Chem Soc 120:8430–8437
186. Chalk AJ, Mayer PM, Radom K (2000) Int J Mass Spectrom 194:181–196
187. Veith HJ, Gross JH (1991) Org Mass Spectrom 26:1097–1108

188. Sølling TI, Hammerum S (2001) J Chem Soc, Perkin 2:2324–2328
189. Bache-Andreassen L, Uggerud E (2002) In preparation
190. Jolly WL, Bomben KD, Eyerman CJ (1984) Atomic Data and Nuclear Data Tables 31:433–94
191. Homan H, Herreros M, Notario R, Abboud JLM, Esseffar M, Mo O, Yanez M, Foces-Foces C, Ramos-Gallardo A, Martinez-Ripoll M, Vegas A, Molina MT, Casanovas J, Jimenez P, Roux MV, Turrion C (1997) J Org Chem 62:8503–8512
192. Scheele CW (1782) Neue Abhandl. der Königl. Schwed. Akademie der Wissenschaften 3:32
193. Martin RL, Shirley DA (1974) J Am Chem Soc 96:5299–5304
194. Davis DW, Rabelais JW (1974) J Am Chem Soc 96:5305–5310
195. Benoit FM, Harrison AG (1977) J Am Chem Soc 99:3980–4
196. Smith SR, Thomas TD (1978) J Am Chem Soc 100:5459–66
197. McQuaide BH, Banna MS (1988) Can J Chem 66:1919–22
198. Von Nagy-Felsobuki EI, Kimura K (1990) J Phys Chem 94:8041–4
199. Bordeje MC, Mo O, Yanez M, Herreros M, Abboud JLM (1993) J Am Chem Soc 115:7389–96
200. Bagno A, Scorrano G (1996) J Phys Chem 100:1536–44
201. Bouchoux G, Djazi F, Houriet R, Rolli E (1988) J Org Chem 53:3498–501
202. Evans J, Nicol G, Munson B (2000) J Am Soc Mass Spectrom 11:789–796
203. Mishima M, Fujio M, Tsuno Y (1986) Tetrahedron Lett 27:951–4
204. Mishima M, Mustanir, Fujio M, Tsuno Y (1996) Bull Chem Soc Jpn 69:2009–2018
205. Nourse BD, Cooks RG (1991) Int J Mass Spectrom Ion Proc 106:249–72
206. Mackay GI, Hopkinson AC, Bohme DK (1978) J Am Chem Soc 100:7460–7464
207. Davidson WR, Kau YK, Kebarle P (1977) Can J Chem 56:1016
208. Middlemiss NE, Harrison AG (1979) Can J Chem 57:2827–33
209. Hopkinson AC, Mackay GI, Bohme DK (1979) Can J Chem 57:2996–3004
210. van Baar B, Halim H, Terouw JK, Schwarz H (1986) J Chem Soc, Chem Commun: 728–730
211. Tiedemann PW, Riverós JM (1974) J Am Chem Soc 96:185–189
212. Kim JK, Caserio MC (1981) J Am Chem Soc 103:2124–7
213. Catalan J, Mó O, Pérez, Yánez M (1982) J Chem Soc, Perkin Trans 2:1409
214. Lin HY, Ridge DP, Vulpius T, Uggerud E (1994) J Am Chem Soc 116:2996
215. Meot-Ner M, Hunter EP, Field F (1979) J Am Chem Soc 101:686
216. Locke MJ, Hunter RL, McIver RT (1979) J Am Chem Soc 101:272
217. Bojesen G (1987) J Am Chem Soc 109:5557–8
218. Gorman GS, Speir JP, Turner CA, Amster IJ (1992) J Am Chem Soc 114:3986–8
219. Li X, Harrison AG (1993) Org Mass Spectrom 28:366–71
220. Zhang K, Zimmerman DM, Chung-Phillips A, Cassady CJ (1993) J Am Chem Soc 115: 10812–22
221. Wu J, Lebrilla CB (1993) J Am Chem Soc 115:3270
222. Wu Z (1994) Proton affinities of amino acids and peptides determined using the kinetic method. Thesis, Univ. Maryland, Baltimore County
223. Bojesen G, Breindahl T (1994) J Chem Soc, Perkin Trans 2:1029–37
224. Wu J, Lebrilla CB (1995) J Am Soc Mass Spectrom 6:91
225. Carr SR, Cassady CJ (1996) J Am Soc Mass Spectrom 7:1203
226. Afonso C, Modeste F, Breton P, Fournier F, Tabet JC (2000) Eur J of Mass Spectrom 6:443–449
227. Jensen F (1992) J Am Chem Soc 114:9533
228. Bouchonnet S, Hoppilliard Y (1992) Org Mass Spectrom 27:71–6
229. Yu D, Rauk A, Armstrong DA (1995) J Am Chem Soc 117:1789–96
230. Baeten A, De Proft F, Geerlings P (1996) Int J Quantum Chem 60:931–940
231. Uggerud E (1997) Theoret Chem Acc 97:313–316
232. Topol IA, Burt SK, Toscano M, Russo N (1998) Theochem 430:41–49
233. Maksic ZB, Kovacevic B (1999) Chem Phys Lett 307:497–504

234. Campbell S, Beauchamp JL, Rempe M, Lichtenberger DL (1992) Int J Mass Spectrom Ion Proc 117:83–99
235. Ghanty TK, Ghosh SK (1993) J Phys Chem 97:4951–3
236. Itskowitz P, Berkowitz ML (1998) J Chem Phys 109:10142–10147
237. De Luca G, Sicilia E, Russo N, Mineva T (2002) J Am Chem Soc 124:1494–1499
238. Milne GWA, Axenrod T, Fales HM (1970) J Am Chem Soc 92:5170
239. LeClerq PA, Desiderio DM (1973) Org Mass Spectrom 7:515
240. Tsang CW, Harrison AG (1976) J Am Chem Soc 98:1301–8.
241. Rogalewicz F, Hoppilliard Y (2000) Int J Mass Spectrom 199:235–252
242. Balta B, Basma M, Aviyente V, Zhu C, Lifshitz C (2000) Int J Mass Spectrom 201:69–85
243. Beranová S, Cai J, Wesdemiotis C (1995) J Am Chem Soc 117:9492
244. Kinser RD, Ridge DP, Hvistendahl G, Rasmussen B, Uggerud E (1996) Chem Eur J 2: 1143–1149
245. Yalcin T, Khouw C, Csizmadia IG, Peterson MR, Harrison AG (1995) J Am Soc Mass Spectrom 6:1165–74
246. Ambihapathy K, Yalcin T, Leung H-W, Harrison AG (1997) J Mass Spectrom 32:209–215
247. Yalcin T, Csizmadia IG, Peterson MR, Harrison AG (1996) J Am Soc Mass Spectrom 7: 233–42
248. Nold MJ, Wesdemiotis C, Yalcin T, Harrison AG (1997) Int J Mass Spectrom Ion Proc 164:137–153
249. Paizs B, Lendvay G, Vwkey K, Suhai S (1999) Rapid Commun Mass Spectrom 13:525–533
250. Paizs B, Suhai S (2001) Rapid Commun Mass Spectrom 15:651–663
251. Paizs B, Suhai S (2002) Rapid Commun Mass Spectrom 16:375–389
252. Morgan DG, Bursey MM (1995) J Mass Spectrom 30:290–5
253. Morgan DG, Bursey MM (1994) Org Mass Spectrom 29:354–9
254. Taylor R (1990) Electrophilic aromatic substitution. John Wiley, Chichester
255. Olah GA, Staral JS, Asenico G, Liang G, Forsyth DA, Mateescu GD (1978) J Am Chem Soc 100:6299
256. Stasko D, Reed CA (2002) J Am Chem Soc 124:1148–9
257. Dewar MJS (1946) J Chem Soc 406:777
258. Kuck D (1990) Mass Spectrom Rev 9:583–630
259. Fornarini S (1997) Mass Spectrom Rev 15:365–389
260. Fornarini S, Crestoni ME (1998) Acc Chem Res 31:827–834
261. Hehre WJ, Pople JA (1972) J Am Chem Soc 94:6901
262. Glukhovtsev MN, Pross A, Nicolaides A, Radom L (1995) J Chem Soc, Chem Commun: 2347–2348
263. Brown HC, Okamoto Y (1958) J Am Chem Soc 80:4979
264. Lau YK, Kebarle P (1976) J Am Chem Soc 98:7452–3
265. McMahon AW, Chadikun F, Harrison AG, March RE (1989) Int J Mass Spectrom Ion Proc 87:275–285
266. Eckert-Maksic M, Hodoscek M, Kovacek D, Maksic ZB, Primorac M (1997) J Molec Struct (Theochem) 417:131–143
267. Hrusak J, Schröder D, Weiske T, Schwarz H (1993) J Am Chem Soc 115:2015–2020
268. Mason RD, Parry AJ, Milton DMM (1994) J Chem Soc, Faraday Trans 90:2015–2020
269. Wiberg KB, Rablen PR (1998) J Org Chem 63:3722–3730
270. Raos G, Gerratt J, Karadakov PB, Cooper DL, Raimondi M (1995) J Chem Soc, Faraday Trans 91:4011–4030
271. Russo N, Toscano M, Grand A, Mineva T (2000) J Phys Chem A 104:4017–4021
272. Nold MJ, Wesemiotis C (1996) J Mass Spectrom 31:1169–1172
273. Karpas Z, Berant Z, Stimac RM (1990) Struct Chem 1:201–4
274. Harrison AG, Tu Y-A (2000) Int J Mass Spectrom 195/196:33–43
275. Takahata Y, Chong DP (2000) Bull Chem Soc Jpn 73:2453–2460
276. Catalan J, Yanez M (1979) J Chem Soc, Perkin Trans 2:741–746
277. Brown RS, Tse A (1980) Can J Chem 58:694
278. Saethre LJ, Thomas TD (1991) J Phys Org Chem 4:629

279. Miklis PC, Ditchfield R, Spencer TA (1998) J Am Chem Soc 120:10482–10489
280. Yoshida M, Tsuzuki S, Goto M, Nakanishi F (2001) J Chem Soc, Dalton Trans: 1498–1505
281. Sparrapan R, Mendes MA, Eberlin MN (1999) Int J Mass Spectrom 182/183:369–80
282. Cacace F, Ricci A (1996) Chem Phys Lett 253:184–188
283. Reents WD Jr, Freiser BS (1980) J Am Chem Soc 102:271–6
284. Skokov S, Wheeler RA (1999) J Phys Chem A 103:4261–9
285. Hubig SM, Kochi JK (2000) J Org Chem 65:6807–6818
286. Holman RW, Eary T, Whittle E, Gross ML (1998) J Chem Soc, Perkin Trans 2:2187–2194
287. Kryacko ES, Nguyen MT (2001) J Phys Chem A 105:153–5
288. Heidrich D (1999) Phys Chem Chem Phys 1:2209–2211
289. Aschi M, Attina M, D'Arcangelo G (1998) Chem Phys Lett 283:307–312
290. Kotiaho T, Shay BJ, Cooks RG, Eberlin MN (1993) J Am Chem Soc 115:1004–14
291. Aschi M, Attina M, Cacace F (1998) Chem Eur J 4:1535–1541
292. Filippi A, Grandinetti F, Occhiucci G, Speranza M (2000) Int J Mass Spectrom, 195/196: 21–31
293. Iganatyev IS, Sundius T (2001) J Phys Chem A 105:4535–4540
294. Schröder D, Oref I, Hrusak J, Weiske T, Nikitin EE, Zummack W, Schwarz H (1999) J Phys Chem A 103:4609–4620
295. Lorquet JC, Lorquet AJ (2001) J Phys Chem A 105:3719–3724
296. Kuck D, Bäther W, Grützmacher HF (1985) J Chem Soc, Perkin 2:689
297. Attina M, Cacace F (1986) J Am Chem Soc 108:318–19
298. Chiavarino B, Crestoni ME, Di Rienzo B, Fornarini S (1998) J Am Chem Soc 120:10856–10862
299. Crestoni ME, Fornarini S (1994) J Am Chem Soc 116:5873–9
300. Bowers MT, Chesnavich WJ, Huntress WT (1973) Int. J. Mass Spectrom. Ion Phys 12:357–382
301. Huntress WT, Bowers MT (1973) Int. J. Mass Spectrom. Ion Phys 12:1–18
302. Day RJ, Krause DA, Jorgensen WL, Cooks RG (1979) Int. J. Mass Spectrom. Ion Proc. 30:83–92
303. Gilbert JR, van Koppen PAM, Huntress WT, Bowers MT (1981) Chem Phys Lett 82: 455–457
304. Øiestad EL, Øiestad ÅML, Skaane H, Ruud K, Helgaker T, Uggerud E, Vulpius T (1995) Eur Mass Spectrom 1:121–129
305. Nobes RH, Radom L (1982) Org Mass Spectrom 17:340–344
306. Nobes RH, Radom L (1983) Chem Phys 74:163–169
307. Uggerud E (1999) Mass Spectrom Rev 18:285–308
308. Harrison AG (1992) Chemical Ionization Mass Spectrometry. CRC, Boca Raton
309. Scott AP, Golding BT, Radom L (1998) New J Chem 1171–1173
310. Senger S, Radom L (2000) J Am Chem Soc 122:2613–2620

Top Curr Chem (2003) 225: 37–75
DOI 10.1007/b10469

Mass Spectrometric Approaches to Interstellar Chemistry

Simon Petrie[1] · Diethard K. Bohme[2]

[1] Department of Chemistry, the Faculties, Australian National University, Canberra ACT 0200, Australia. *E-mail: spetrie@rsc.anu.edu.au*

[2] Department of Chemistry, Centre for Research in Mass Spectrometry, Centre for Research in Earth and Space Science, York University, Toronto, ON, Canada M3J 1P3
E-mail: dkbohme@yorku.ca

In this review we focus both on major developments in new mass spectrometric techniques and on novel chemical applications of existing mass spectrometric techniques that have been reported since 1990. Emphasis is given to the application of these techniques to the study of bimolecular ion/molecule reactions, radiative association, and dissociative recombination of positive ions. Particular attention is given to the emerging field of interstellar metal-ion chemistry and recent studies of fullerene-ion chemistry and the influence of charge state on this, and related, chemistry. Mass spectrometric studies of the photochemistry of interstellar ions are briefly considered as is interstellar negative-ion chemistry. We conclude with a brief description of the use of mass spectrometry to examine interstellar material that has made the long journey to our solar system.

Keywords. Mass spectrometry, Interstellar chemistry, Ion/molecule reactions

List of Abbreviations

AISA	Advanced integrated flowing afterglow
CRESU	Cinétique de réaction en écoulement supersonique uniforme
DR	Dissociative recombination
FA	Flowing Afterglow
FALP	Flowing afterglow/Langmuir probe
FT	Flow-tube
ICP	Inductively coupled plasma
ICR	Ion cyclotron resonance
IS	Interstellar
JPL	Jet Propulsion Laboratory
LIF	Laser-induced fluorescence
MS	Mass spectrometry
PAH	Polycyclic aromatic hydrocarbon
RA	Radiative association
REMPI	Resonant excitation multiphoton ionisation
SIFT	Selected-Ion Flow Tube
TOF	Time-of-flight
VUV	Vacuum ultraviolet

1 Introduction

The proposal in the early 1970s that homogeneous ion-molecule reactions play a role in the formation of the small molecules detected by radio-astronomers in interstellar (IS) clouds [1, 2], as well as the direct observation of small ions such as CH^+, HCO^+, DCO^+, N_2H^+ and $HOCO^+$ in these environments [3], spawned much interest in the measurement of chemical and physical processes involving these and related ions. The challenge for gas-phase ion chemists was to identify and characterize the IS processes in which these ions are featured, either as reactants or products. The involvement of these ions in the formation of the diverse range of ambient neutral molecules was also scrutinized. Mass spectrometry, of course, became the experimental method of choice. Many different mass spectrometry (MS) techniques have been applied since the mid 1970s in studies of the large variety of proposed IS processes. Table 1 summarizes the types of processes that involve ions and that have some relevance to the chemistry of astrophysical environments. A smaller subset of reaction types, comprising bimolecular ion/molecule reaction, radiative association and dissociative recombination of positive ions, has received the greater proportion of experimental investigation to date. It is believed that these three broad reaction classes are

Table 1. Interstellar chemistry reaction classes involving ions

Class	General form of reaction	Example	MS techniques used
Cosmic ray ionization	$cr+AB \rightarrow A^++B+e+cr$	$cr+H_2 \rightarrow H^++H+e+cr$	
UV photoionization	$AB+h\nu \rightarrow AB^++e$	$C_{24}H_{12}+h\nu \rightarrow C_{24}H_{12}{}^++e$	
Associative ionization	$A+B \rightarrow AB^++e$	$CH+O \rightarrow HCO^++e$	
Radiative association	$A^++B \rightarrow AB^++h\nu$	$CH_3{}^++HCN \rightarrow CH_3NCH^++h\nu$	ICR, ion trap
Bimolecular ion/ molecule chemistry	$A^++B \rightarrow C^++D$	$H_3{}^++CO \rightarrow HCO^++H_2$	SIFT, ICR
Dissociative recombination	$AB^++e \rightarrow A+B$	$HCO^++e \rightarrow H+CO$	FALP, ion storage ring
Radiative attachment	$AB+e \rightarrow AB^-+h\nu$	$C_5N+e \rightarrow CCCCCN^-+h\nu$	
Dissociative attachment	$AB+e \rightarrow AB^-+h\nu$	$HNCCC+e \rightarrow CCCN^-+H$	
Associative detachment	$A^-+B \rightarrow AB+e$	$C_6H^-+H \rightarrow C_6H_2+e$	SIFT

responsible for most of the ion-induced chemical processing within dense IS clouds, circumstellar envelopes, and other cold molecular environments. Thousands of MS investigations in the 1970s and 1980s yielded a data base for ion/molecule reaction kinetics of substantial proportion, certainly sufficient for the crafting of advanced chemical models to explain the gas-phase ionic synthesis of IS ions and many of the IS molecules detected by radio-astronomers [4–7]. Table 2 serves to demonstrate the range of conditions under which complex molecule formation occurs in space. Almost all of the known IS molecules have been detected in one or more of the sources given in Table 2, and these are also the objects for which the most detailed chemical models have been developed.

The 1990s saw further developments and refinements in MS techniques. They were directed to allow studies of ionic processes in more chemical detail and under conditions more closely resembling the conditions in astrophysical environments. Also, existing MS techniques were applied to the study of novel chemistry so as to advance the scope of our understanding of fundamental aspects of interstellar processes. In this review, we focus both on developments in new MS methods and on novel chemical applications of existing MS approaches. By surveying these fields, we hope to convey a sense of the many different ways in which mass spectrometry has met the challenge of mimicking chemical processes which are truly 'out of this world'. However, not everything that is born in space remains there: we conclude with a brief exploration of the use of MS to examine interstellar material that has made the long journey to our own solar system.

Table 2. Characteristics of important molecular astrophysical sources

Parameter	Object			
	TMC-1	IRC +10216	OMC-1	Sgr B2
Classification	Cold dense cloud	Outflowing C-rich circumstellar envelope	Warm molecular cloud/star-forming region	Giant molecular cloud
Constellation	Taurus	Leo	Orion	Sagittarius
Distance/pc	140	150	460	10,000
Temperature/K	10–20	1000 → 10	50–300	100
n/cm^{-3}	10^4	$10^7 \rightarrow 10^3$	10^5–10^7+	10^6–10^{10}
Representative trace constituents	HNC, c-C_3H_2, polyacetylenic compounds	Cyanopolyynes, polycarbon sulfides, metal cyanides, silicon compounds	Methylated organics	Alcohols, ethers, esters, carboxylic acids, amino acids (?)

2 Positive Ion/Neutral Interactions

2.1 Longstanding Experimental Techniques

The laboratory study of IS ion chemistry has its origins in measurements of bimolecular ion/molecule reaction kinetics. Experimental conditions that are found in most of the mass spectrometers employed in the measurement of bimolecular ion/molecule reactions are far from those found in the cold low-pressure environments of space. Nevertheless, because of the pressure independence of bimolecular reactions and the absence of significant activation energies in most bimolecular ion/molecule reactions, MS measurements performed here on earth do have relevance for the chemistry in space. The substantial database available in the early 1990s on the kinetics of bimolecular ion/molecule reactions important in IS chemistry [8–10] was obtained almost entirely using ion cyclotron resonance (ICR) and flow-tube (FT) mass spectrometry techniques. Both techniques are well established and continue to be used extensively for ion/molecule reaction measurements generally.

In the ICR technique [11, 12], ions are constrained in cyclotron motion (see Fig. 1) by the application of magnetic and electric fields to electrodes which form the boundaries of a small low-pressure chamber serving as both ion source and reaction region. The ions' motion is mass-dependent, and calibration of the electric field strength ensures that only ions of a particular mass are effectively trapped within the ion chamber. Transverse drift of ions, towards the chamber's end cap, is another aspect of the ion motion which can be controlled as required. ICR chambers of widely different geometry (see Fig. 2) have been used for studying a variety of ion properties. Characteristic operating conditions are a reaction

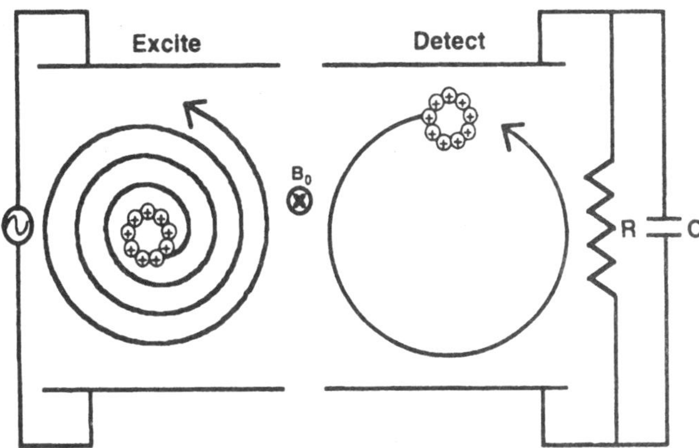

Fig. 1. An illustration of the use of a rotating magnetic field to induce coherent (and therefore detectable) ion cyclotron orbital motion in an ion packet within an ICR mass spectrometer [12]

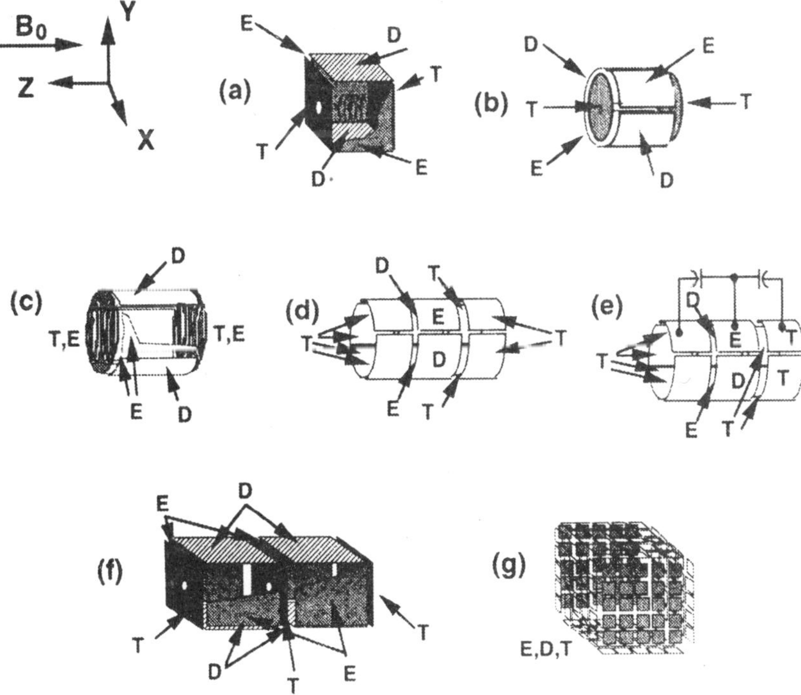

Fig. 2a–g. A diversity of ICR reaction chamber configurations, with excitation, detection, and trapping (end cap) electrodes indicated as E, D and T. Configurations shown are: **a** cubic; **b** cylindrical; **c** the 'infinity' trap with segmented end caps; **d** open-ended; **e** open-ended with capacitative rf coupling between sections; **f** dual; **g** 'matrix-shimmed'. From [12]

chamber pressure of 10^{-7} to 10^{-3} Torr and a temperature of 80–450 K (although 'ion temperature' is typically viewed as being significantly higher than the temperature of the chamber itself). Ion residence times (of the order of several tens of milliseconds, or higher) are usually long enough to ensure that molecular ions are not internally excited. The pressure range encompassed is often sufficient to distinguish between bimolecular and termolecular reaction processes. At higher pressures, an inert 'bath gas' is generally employed to dilute the reactant neutral. ICR measurements of relevance to IS chemistry were first reported in the 1970s [13–15] and, as we shall see, the instrument still has a valued role to play as interstellar ion/molecule kinetics progresses into the twenty-first century.

The most useful FT mass spectrometer for interstellar chemistry has been the Selected-Ion Flow Tube (SIFT) tandem mass spectrometer [16]. A schematic of the basic instrument is shown in Fig. 3. The reaction region in this instrument is a flow tube that separates two quadrupole mass filters. Ions are normally produced in an electron-impact ion source, mass selected with a quadrupole mass filter and injected into the upstream end of the flow tube against a pressure gradient through a Venturi-type interface. Ions are carried down the tube in a bath gas, usually He at about 0.4 Torr, in which they thermalize. Neutral reagents are added downstream, although ions can be chemically manipulated first by adding selected neutrals upstream. Reactant and product ions are sampled still further downstream with the second quadrupole mass filter and counted. Kinetic measurements are usually performed by following the reactant and product ions as a function of the flow of reactant gas. This allows the determination of both reaction rate coefficients and product distributions for ion/molecule reactions. Variations of the basic SIFT instrument include the use of other ion sources and the use of a triple quadrupole downstream for collision-induced dissociation experiments. The flow tube itself can be heated or cooled with the use of appropriate jackets. Variations in temperature from 80 to 1800 K have now

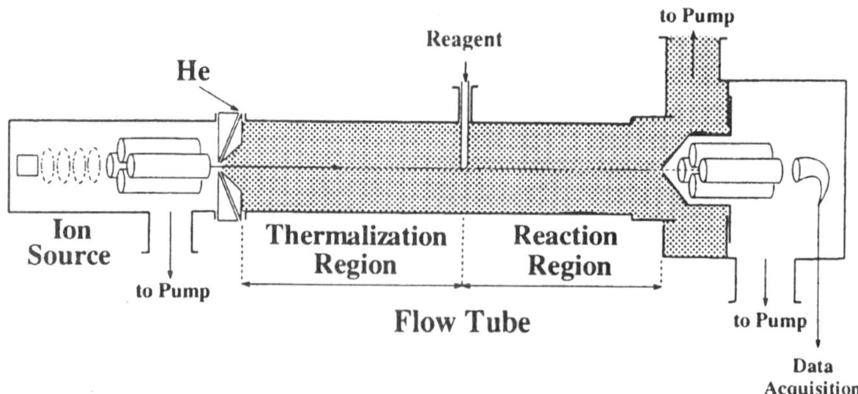

Fig. 3. Schematic diagram of the Selected-Ion Flow Tube (SIFT) tandem mass spectrometer. Ions from the ion source are selected with a quadrupole mass filter before injection into the flow tube. A second quadrupole mass filter is used downstream to separate reactant and product ions in the sampled reaction mixture. The He pressure in the flow tube is ca. 0.4 Torr

been achieved with this instrument. The birth, evolution and achievements in the twentieth century of flow-tube mass spectrometry employed in the study of positive ion chemistry recently has been reviewed by Bohme [17].

2.2 Bimolecular Ion/Molecule Reactions

A number of research groups have used SIFT instruments for measurements directed toward IS chemistry. The Birmingham group of Adams and Smith, the inventors of the SIFT technique [16], was particularly active in this regard and a major focus of their SIFT measurements was the systematic study of reactions of hydrogenated ions, e.g. CH_n^+, $C_2H_n^+$, NH_n^+, H_nS^+, H_nCO^+ etc., with numerous molecular species [18]. Further contributions by this group include detailed studies of isotope exchange in ion-neutral reactions, studies for which the SIFT is eminently suited, since the ion source gas and the reactant gas are not mixed. From these studies and detailed kinetic models of interstellar ionic reactions, it is now understood that the observed enhancement of the rare isotopes (e.g. D, ^{13}C) in some IS molecules is due to the process of isotope fractionation in ion-neutral reactions [19].

The emphasis on interstellar/circumstellar ion chemistry in Bohme's laboratory at York University has been on the synthesis of hydrocarbon-chain and organonitrogen molecules [20], on ionic origins of carbenes [21] and on the formation of organosilicon molecules [22]. In the studies of silicon chemistry a state-selected beam of ground-state $Si^+(^2P)$ ions, produced in a conventional electron-impact source containing both a parent silane molecule and deuterium, was injected into the flow tube of a SIFT instrument. The deuterium in the ion source removes the excited ^4P state of Si^+ by deuterium-atom transfer so that only the ground state emerges from the source. $Si^+(^2P)$ was observed to initiate a variety of ion-molecule reactions resulting in Si-H, Si-C, Si-N, Si-O and Si-S bond formation. Indications are that sequential chemistry initiated by $Si^+(^2P)$ with oxygen and carbon containing molecules such as H_2O, CH_3OH, HCOOH, C_2H_2, HC_3N and C_2H_4 ultimately may lead to silicate or silicon carbide particles, respectively, in the gas phase. Further experiments along these lines seem warranted given the high abundance of silicon in space. SIFT experiments have also shown that gaseous benzene and naphthalene can form adduct ions with $Si^+(^2P)$ in the presence of He gas by collisional stabilization and so presumably by radiative stabilization at the low interstellar gas densities. These adduct ions are non-reactive with H_2 and CO but can react with H_2O, NH_3, C_2H_2 and C_4H_2 to promote the synthesis of novel silicon-containing molecules such as HSiOH and the cyclic adducts of atomic silicon with acetylene and diacetylene [22]. It has been suggested that Si^+ and other atomic cations may also be trapped by polycyclic aromatic hydrocarbon (PAH) molecules embedded in hydrogenated amorphous carbon (HAC) grains that may be present in the interstellar medium, by grains of pure graphite (the end members of the PAH series), and even by fullerene molecules such as C_{60} [23].

McEwan and co-workers have continued to use the Canterbury SIFT (also known as 'the Southern Hemisphere SIFT') to characterize product-ion structure by assessing the subsequent reactivity of product ions with other reagents added further downstream [24–27]. This 'monitor gas' approach, necessitated by the lack of structural information conveyed by MS detection of products, is a technique which has also seen widespread use among other SIFT laboratories [28]. Recent research at Canterbury [24] has shown that association of $C_2H_3^+$+CO, previously proposed as the route to IS propynal HCCCHO [29], yields protonated propadienone (H_2CCHCO) rather than protonated propynal: dissociative recombination of the resulting $C_3H_3O^+$ ion should favour formation of propadienone (H_2CCCO) rather than propynal, and propadienone is not seen in IS clouds. Petrie has suggested that processes other than ion/molecule chemistry are necessary to account for the observed IS propynal [30]. McEwan and associates have also investigated the structure of the $C_2H_5O^+$ ion produced in the reaction of H_3O^++C_2H_2 [25], a possible route to IS vinyl alcohol: their theoretical studies suggest that the reaction should give O-protonated vinyl alcohol $H_2CCHOH_2^+$ rather than the lower-energy isomer H_3CCHOH^+ (O-protonated acetaldehyde). The product ion reactivity seen in the Canterbury SIFT [25] shows that two different $C_2H_5O^+$ isomers are produced, most likely $H_2CCHOH_2^+$ and a $C_2H_2{\cdot}H_3O^+$ complex, providing support for the proposal of vinyl alcohol as an eventual product. One complicating factor which can impinge on the utility of the 'monitor gas' approach is that the monitor gas can sometimes catalyse *isomerization* of the product ion under investigation [28, 31]: obviously, the mass spectrometer is blind to this reaction pathway, and careful investigation (involving, perhaps, the use of a further monitor gas) is required to uncover such processes.

Shortcomings in the monitor gas approach (it may, for example, be difficult or impractical to find a reagent with the right combination of properties) can often be overcome if physical, rather than chemical, means of ion-structure determination are employed. Bond connectivity measurements for gas-phase ions produced in ion/molecule reactions are generally performed using collision-induced dissociation (CID), although photodissociation is also employed [32]. When performed under single-collision conditions, CID measurements also provide information about the energetics of the bond being broken. This has been achieved with a SIFT triple quadrupole mass spectrometer built in the laboratory of Bob Squires [33, 34]. In the SIFT experiments performed by Bohme's group to investigate IS ion/molecule reactions, multi-collision induced dissociation is used to explore bond connectivities in product ions simply by raising the potential of the sampling nose-cone without introducing mass discrimination [35]. Figure 4 shows multi-collision CID results that illustrate the cyclization of three acetylene molecules after having been attached sequentially to Fe^+ into benzene [35].

2.3 Ion/Atom Reactions

An important facet of IS cloud composition is that laboratory-unstable species (e.g. H, C, N, O, CN, HNC, c-C_3H_2) are often highly abundant, and the reaction chemistry of such species is a challenging field of study. This is particularly true

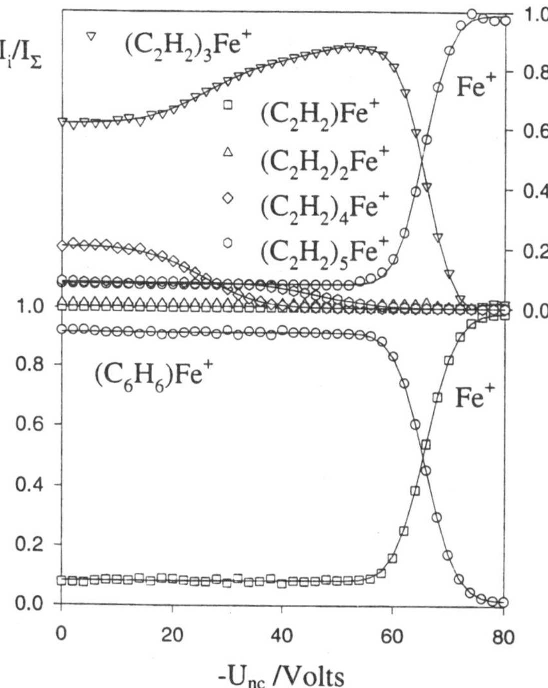

Fig. 4. Multi-collision CID results for $(C_2H_2)_3Fe^+$ and $(C_6H_6)Fe^+$ in the laboratory energy scale with He as the collision gas. $(C_2H_2)_3Fe^+$ is produced by sequential addition of C_2H_2 to Fe^+ and $(C_6H_6)Fe^+$ is produced by direct addition of C_6H_6 to Fe^+. The similarity in the threshold for dissociation of $(C_2H_2)_3Fe^+$ and $(C_6H_6)Fe^+$ suggests that Fe^+ has mediated the cyclotrimerization of acetylene to benzene [35]

with regard to ion/neutral chemistry. Reactions of IS ions with the atoms H [36–38], N [39–42] and O [43–45] have been one main focus of the Canterbury SIFT in recent years, and McEwan and co-workers hope ultimately to be able to extend this effort to the ion chemistry of neutral atomic carbon. While production of C, in a SIFT-usable reactant inlet, remains elusive, the methods for generating H, N and O are rather more well-established [46–48]. A representative reaction inlet used for the study of N-atom reactions is shown in Fig. 5. Nevertheless, the still small list of ion/atom reactions to have been studied over the past 37 years [48] is testament to the intrinsic difficulties in studying such reactions. While the production of free atoms, generally by action of a microwave discharge on an appropriate gas, may in itself be comparatively straightforward, the atoms so produced will tend to recombine or to react with the walls of the reaction vessel. Against the background of this unwanted chemistry the remaining unreacted atoms must somehow be assessed and quantified so that the kinetics of any ion/neutral reactions may finally be established. These difficulties, and others, are exemplified in the measurements of the H_3^++N reaction to which McEwan and co-workers have now devoted much attention [39–42]. In these studies (for which N is generated by microwave discharge of N_2, typically

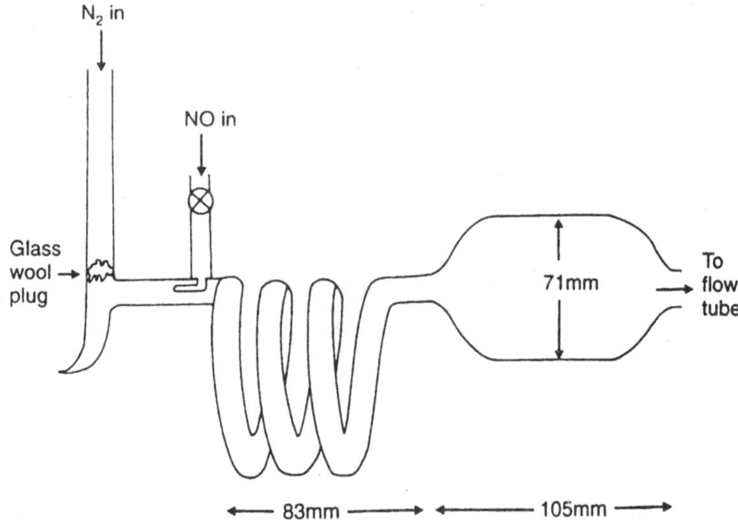

Fig. 5. Schematic diagram of an N-atom reactant inlet as used by Scott et al. [41]. Nitrogen atoms are produced by the action of a microwave discharge applied upstream of the glass wool plug. The design of this particular inlet also allows absolute determination of the N-atom concentration by titration of N with NO (yielding $O+N_2$); the titration endpoint is marked by detection of NO^+, from reaction of excess NO with O_2^+ as a reactant ion

yielding only 1% N in N_2), the existence of rapid proton transfer from H_3^+ to N_2 is a severe obstacle to the detection of any direct reaction between $H_3{}^+$ and N, and in practice the sole practical indication of any H_3^+/N reaction is in the observation of product channels not attributable to the H_3^+/N_2 reaction. In an initial study in which H_2, rather than He, was used as the SIFT buffer gas, and in which H_3^+ was produced in the flow tube by injection of Kr^+ into this hydrogen buffer gas, a significant product signal due to NH_4^+ was interpreted as arising from the occurrence of $H_3^+ + N \rightarrow NH_2^+ + H$ with a rate coefficient of $k=(4.5\pm1.8)\times10^{-10}$ cm^3 molecule^{-1} s^{-1} (NH_2^+ itself was not directly observable, due to further hydrogenation of this ion by reactions with the hydrogen buffer gas) [39, 40]. Subsequent augmentation of the Canterbury SIFT to include a flowing-afterglow (FA) ion source (as shown in Fig. 6) has now permitted low-energy injection of H_3^+ from the FA ion source into He buffer gas, and with this experimental configuration it has now been established that the H_3^+/N reaction is at least an order of magnitude slower than the previous study had suggested [42]. Such a result has important implications for the IS synthesis of ammonia, and the sensitivity of the inferred reaction rate to the experimental configuration serves to illustrate that the topic of ion/atom reactivity remains a vital area for continued research.

The SIFT group of Bierbaum at the University of Colorado has also recently been active in the field of ion/atom chemistry, as a facet of their exploration of the IS chemistry of PAH cations. Rate coefficient measurements for reactions of the naphthalene radical cation, $C_{10}H_8^+$, the closed shell naphthylium cation,

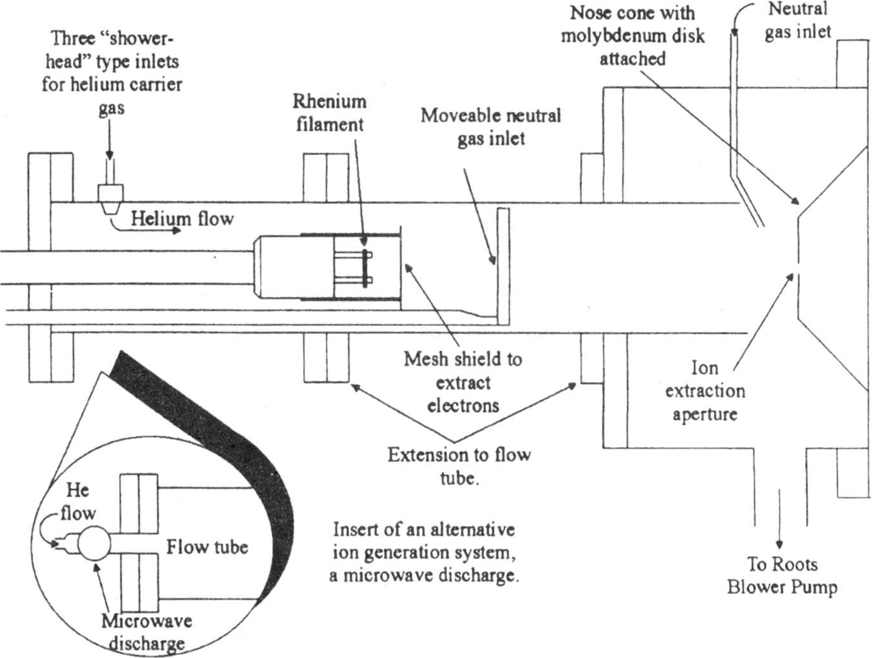

Fig. 6. A flowing afterglow ion source used by Milligan et al. [42] in the study of the H_3^++N reaction. Ions are generated by electron impact, or by the action of a microwave discharge

$C_{10}H_7^+$, and the naphthyne radical cation, $C_{10}H_6^+$, with H, O and N have been reported, as well as reactions of these ions with H_2, CO, H_2O and NH_3 [49]. Differences in reactivity were observed that could be correlated with the multiplicities of the reactant $C_{10}H_x^+$ cation. The radical $C_{10}H_8^+$ cation was found to react with H, O and N atoms but not with H_2, CO, H_2O and NH_3. In contrast, the closed-shell $C_{10}H_7^+$ cation was found to be essentially unreactive with atoms but was seen to associate via nucleophilic addition with most of the molecules studied. The atom and molecule reactions of the naphthyne radical cation $C_{10}H_6^+$ were found to be similar to those observed for the $C_{10}H_8^+$ radical cation, except for the reaction with ammonia which proceeded at a moderate rate.

2.4 Low-Temperature Reaction Measurements

Temperatures can be very low in molecular astrophysical environments, as low as 10 K in dense IS clouds. Many of the SIFT reactions were investigated by the Birmingham group down to temperatures of 80 K with the use of suitable cooling and heating jackets. In the early 1980s Rowe and coworkers developed the CRESU technique to allow measurements of ion-molecule reaction rates down to 7 K [50]. CRESU is the French acronym for "Cinétique de réaction en écoulement supersonique uniforme" which means kinetics of reaction in uniform su-

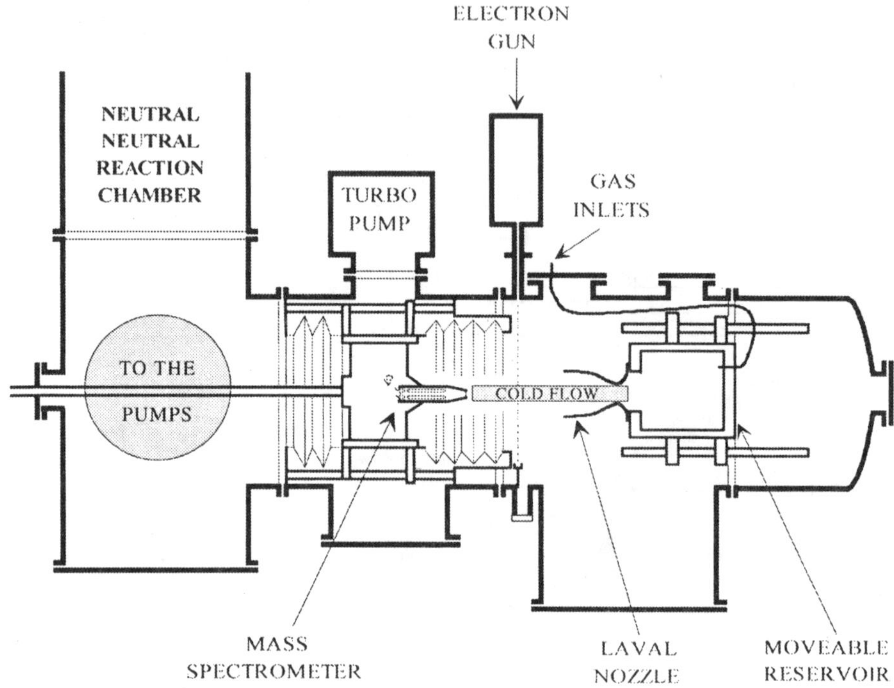

Fig. 7. Schematic of the CRESU apparatus devoted to the measurement of ion/molecule reactions at low temperatures [50]

personic flow. The ion-molecule reaction is made to occur in a uniform supersonic flow in the absence of wall effects, thus avoiding the heterogeneous condensation inherent in flow reactors that involve cryogenic cooling. The supersonic expansion in the CRESU apparatus is achieved with axisymmetric, converging-diverging Laval nozzles. The expansion generates a uniform flow for some 100–500 μs during which pressure, density and temperature (local thermodynamic equilibrium) remain constant. The lowest temperature achieved in a CRESU is 7 K, but this required pre-cooling of the gases to liquid N_2 temperatures. Without pre-cooling, temperatures as low as 15 K are achievable. Typical operating conditions of gas flow limit the measurements to rate coefficients larger than 10^{-12} cm^3 $molecule^{-1}$ s^{-1}. A sketch of the apparatus is shown in Fig. 7. Reactant ions are created using an electron beam (generally operating at 12 kV, 10 μA). Reactant and product ions are mass selected with a quadrupole mass filter and detected with a channeltron electron multiplier. For rate coefficient measurements reactant ions R^+ are monitored as a function of the density of the added reagent [N] at a fixed distance x between the sampling MS orifice and the electron beam. As the jet is uniform, the decrease in R^+ ion density resulting from reaction with the reagent N at a rate coefficient k is given by

$$[R^+]=[R_0^+]\exp(-k[N]x/v)$$

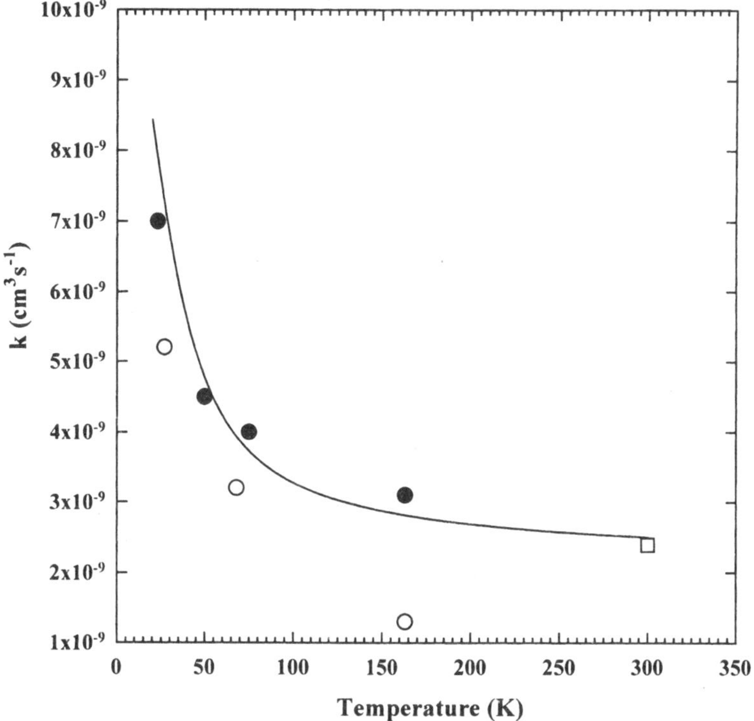

Fig. 8. Variation of the rate coefficient with temperature as measured with the CRESU apparatus for the reaction of N^+ with ammonia [50]. The *open circles* represent early CRESU (at Meudon) results [52] while the *solid circles* are newer CRESU (at Rennes) results [50]. The *open square* is a room-temperature result obtained by Adams et al. [53] with a SIFT apparatus. The *solid line* is a theoretical prediction by Troë using the statistical adiabatic channel model [54]

where v is the constant jet velocity derived from Pitot measurements. Diffusion is negligible. The experiments are generally carried out as a function of [N] at two different positions in the jet to correct for shock wave effects in front of the 50 μm diameter sampling orifice.

Rate coefficients for both bimolecular and termolecular ion/molecule reactions have been measured with the CRESU technique in the temperature range 10–80 K. These have been summarized by Smith and Rowe [51] and include reactions of a doubly-charged ion, Ar^{2+} and the negative ion Cl^-. Important IS reactions of He^+, C^+, N^+, O^+, Ar^+, N_2^+, O_2^+, H_3^+ and CH_3^+ with various molecules have been investigated. Emphasis has been on those reactions whose rates show large temperature dependencies. Figure 8 provides kinetic results for the reaction of N^+ with ammonia which illustrates a rise in rate with decreasing temperature. CRESU measurements have confirmed experimentally the theoretically expected rise in the rate of ion/polar molecule reactions with decreasing temperature and so have had a significant impact on the development of chemical models of IS clouds.

Rowe and co-workers are developing a so-called diffusion technique to extend the temperature and pressure range. The technique will use the conversion of the initial kinetic energy (per unit volume) of the jet into a pressure increase downstream of the mass spectrometer, when the flow is brought from a supersonic to a subsonic regime through suitably shaped tubing. Also, it has been shown that the use of pulsed Laval nozzles reduces the appreciable amounts of gas that are consumed in the continuous flow CRESU apparatus [55].

The CRESU approach is capable of attaining the very low temperatures, down to ~10 K, which are pertinent to IS cloud chemistry. The free jet flow reactor device [56, 57] constructed by Mark Smith, in Tucson, can go lower still, below 1 K – in fact, this device has an *upper* effective temperature limit of ~20 K. The free jet flow reactor, depicted in Fig. 9, has been used for the study of both ion/molecule and radical/neutral reactions [58], and, when employed for the ion chemistry of polyatomic cations such as $C_2H_2^+$ [59], features laser-induced Resonant Excitation Multiphoton Ionization (REMPI) to ensure that ions are produced exclusively in the ground vibrational state. The nascent ions, in the surrounding reactant neutral mixture, are able to undergo ion/molecule reactions in the expanding jet; reactant and product ions are analysed by ejection from the rarefied jet into a time-of-flight (TOF) mass spectrometer. The uncertainties in rate coefficients and product distributions obtained using this technique are broadly comparable to those delivered by the other flow techniques such as SIFT and CRESU. Advantages of using supersonic expansion, as in CRESU or the free jet flow reactor, over cryogenic cooling (e.g. a liquid-nitrogen-cooled flow tube) are that generally lower effective temperatures can be obtained, using an experimental setup in which relatively nonvolatile reagents such as NH_3 [59] can nevertheless be used. One disadvantage inherent in the free jet flow reactor is that the neutral molecules have a nonthermal rotational distribution (where 'ther-

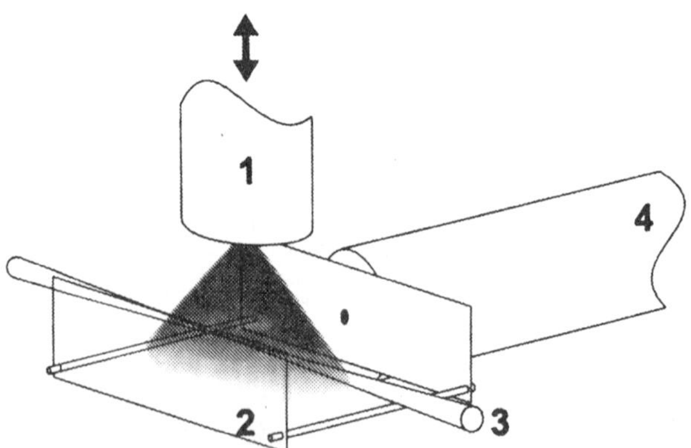

Fig. 9. Schematic diagram of the free jet flow reactor used by Mark Smith and co-workers for very low temperature reaction kinetic measurements [58]. The jet originates from a pulsed beam valve **1**, and ions are produced by REMPI using a focussed pulsed laser. The reaction zone is bounded by a repeller plate **2** and an endplate **3**: ions are propelled, by a pulsed voltage on the repeller, towards a sampling aperture in the endplate which leads to a TOF-MS **4**

mal' is taken to mean the effective temperature described by the spread in collision energies within the expanding jet) [59]. This makes problematic the assignment of a precise temperature to measurements involving molecular reactants. Nonetheless, the reactor has provided some important insights into astrochemically-relevant reactions, having shown, for example, that the product distribution seen in the reaction $C_2H_2^+$+C_2H_2 is effectively unchanged between 300 K and 10 K [59]. Smith and Atkinson have more recently constructed a pulsed uniform Laval flow reactor [55, 60] which to date has been employed only for the study of radical-neutral reactions, although mass spectrometry is used for product detection in one configuration of this instrument [58].

2.5
Radiative Association Reactions

The very low temperatures and very small molecule number densities typical of molecular IS environments favour the occurrence of radiative association (RA) processes, and such reactions are thought to be important in promoting the development of molecular complexity in dense clouds. However, the laboratory replication of these pathways is not straightforward: RA reaction kinetics are typically much more dependent on temperature than is the case for reactions producing bimolecular products. This temperature dependence can be rationalized as arising from the competition between collision complex dissociation and the radiative dissipation of thermal and bond energy, with dissociation becoming increasingly probable as the available thermal energy increases. Furthermore, association reactions are generally highly pressure-dependent, due to competition between (bimolecular) RA and collisional (termolecular) association. Even with helium as a buffer gas, efficient termolecular association will very often mask the contribution from true radiative association at a given operating pressure. Consequently, experimental approaches to the study of RA reactions have involved two strategies: use of techniques which involve intrinsically very low operating pressure, such as ion traps, in which the low probability of termolecular collisions ensures that three body kinetics can effectively be discounted, or the use of instruments in which a wide range of operating pressures can be accommodated, so that the relative contributions of radiative and termolecular association can be unravelled. The latter approach has been widely pursued, and with considerable success, using the ion cyclotron resonance (ICR) technique. ICR is a venerable but powerful method which in recent years has often been coupled with Fourier transform techniques [12, 61–63]. One of the most important features of the ICR is the wide range of operating pressure, from $\sim 10^{-7}$ to $\sim 10^{-3}$ Torr: this four-orders-of-magnitude range is often sufficient to allow the precise mechanism for the occurrence of a particular association process to be established. Dunbar and associates have performed several studies using ICR to explore the efficiency of RA for various series of homologous reactions [64–67]: while the reactions investigated are not in themselves highly pertinent to IS chemistry models, these detailed investigations of RA kinetics are valuable as 'test sets' against which theoretical models of interstellar RA reactions can be assessed [68, 69].

In studies spanning Pasadena (JPL) and Canterbury, collaborations between the McEwan and Anicich groups have focussed on RA reactions of small nitriles [70–73], principally with the important IS ion CH_3^+. The Canterbury ion-molecule laboratory is currently unique in boasting both ICR and SIFT instruments [73], effectively giving an even larger range in operating pressure than that obtainable using ICR alone: however, comparison of ICR and SIFT results is complicated by the fact that the SIFT method undoubtedly features much better thermalization of the reactant ions than does ICR, and 'high pressure' mass spectrometrists often comment that the chief drawback of ICR is the uncertainty regarding the effective ion temperature. Despite this limiting feature, the ICR technique remains attractive for the study of several classes of reaction, and other recent results of interest include the studies by Filippi et al. [74] on reactions of PF^+ with small IS molecules, and of Schwarz and co-workers [75] on the association reaction of Al^+ with benzene and its deuterated isotopomers. The latter study nicely illustrates the influence of D/H substitution on the RA kinetics with a 300 K rate coefficient for C_6D_6, almost four times as large as that seen for C_6H_6.

While ICR is probably the most widely-used methodology in the context of ion/molecule radiative association, ion traps also have a long history of application in this area. The ion traps employed by Barlow et al. [76] and by Gerlich and co-workers [77–80] have a perceived advantage over ICR that they permit occurrence of reactions not only at low pressure but also at very low temperature, nominally as low as ~10 K, and in this context they probably come closest, of any of the many ion/molecule techniques devised to date, to genuine interstellar cloud conditions. However, while such measurements are very valuable the range of significant interstellar RA processes studied to date is limited to the reactions of CH_3^+ [76, 77], $C_2H_2^+$ [79] and C_3H^+ [80] with H_2: in the latter case [80], production of $C_3H_2^+$+H competes with RA, and bimolecular product channels also feature in the reaction of N^++D_2 which has also been studied in a low-temperature ion trap [81].

2.6 Metal-Ion Chemistry

The chemistry of metals in molecular astrophysical environments is an emerging field. Little more than a decade ago, no metal-containing species had been identified in cold interstellar or circumstellar space [82]; currently, NaCl, MgNC, AlF, MgCN, AlCl, KCl, Na(CN) and AlNC have all been detected in the outflowing envelope of the carbon-rich, mass-losing star ICR+10216 [32, 83–86], with the first three of these species also seen in other sources [87]. Tentative detection of FeO (seen towards the Galactic centre) has also been reported [88]. Ion/molecule reactions are seen as the most plausible route to MgNC and MgCN, and are also thought important in AlNC formation [89]. While the precise reactions considered most effective in producing the observed metal cyanides [90] – i.e. association of metal ions with the cyanopolyynes HC_5N and HC_7N – have not yet been studied experimentally, a closely analogous reaction, involving Mg^+ and HC_3N, has been subjected to SIFT study by Bohme and co-workers [91]. The SIFT technique does not yield direct information on the occurrence of radiative association in this case, since the association product seen is likely to result principally from termolecular association, but the absence of competing bimolecular product channels is at least encouraging for the proposed metal-cyanide formation mechanism [89, 90]. Another York SIFT study [92] has thrown light on the reaction chemistry of Fe^+ with small unsaturated hydrocarbons; again, information on relevant IS radiative association rates could not be extracted directly, but the occurrence of efficient collisional stabilization and the absence of competing product channels suggest that RA rate coefficients for these reactions, at cold IS temperatures, should be significant. Subsequent neutralization products of the chemistry initiated by such Fe^+/hydrocarbon reactions could include FeC_2H_n and FeC_4H_n (n=1, 2) species. Observable association (again, under lab conditions, presumably termolecular) between Mg^+ and NH_3 [93] provides an apparent route to $MgNH_2$ in circumstellar environments such as IRC+10216, although subsequent theoretical determinations of the relevant RA rate coefficient cast doubt on the detectability of this species [94].

The ICR and ion trap techniques are, as noted above, well-suited for the study of RA reactions, and their contributions in the field of IS metal ion chemistry have to date concerned reactions with PAH molecules. Boissel and co-workers have used a Penning ion trap coupled with Fourier transform MS to characterize the addition of Fe^+ to naphthalene; in related experiments, Fe^+ was seen to undergo charge transfer, rather than addition, to the larger homologue anthracene ($C_{14}H_{10}$) [95–98]. The Dunbar group has employed ICR to study the reactions of first-row transition-metal ions (and some main-group ions) with benzene, naphthalene, anthracene and coronene [99–102]. The efficient occurrence of association with coronene, with almost all of the 25 different metal ions surveyed [102], is notable given the low-pressure, heated-ion environment characteristic of ICR instruments. In fact, secondary addition to form $M^+(C_{24}H_{12})_2$ is also generally seen with transition metal ions when $C_{24}H_{12}$ is coronene, but is not seen in some instances when $C_{24}H_{12}$ is the isomer tribenzocyclyne, which has a much larger central cavity: it appears that some metal ions are capable of in serting directly into this cavity [102].

The York SIFT has also been employed to study the reactions of metal ions (notably Fe^+) with benzene, with the bowl-shaped PAH corannulene ($C_{20}H_{10}$), and with C_{60} [103]; the latter two ligands are unusual among terrestrially-known polyaromatic compounds in possessing surface curvature, which may nevertheless be a feature of interstellar PAHs, and with corannulene it is also interesting to explore whether ligation occurs preferentially at the convex or the concave face of the 'bowl' [104]. The general indication from all of the M^+/PAH association studies is that metal ion adsorption onto the surfaces of large PAHs will be an important IS loss process for gas-phase metals [105–108] as discussed earlier for the Si^+ cation. In a sense, the recent experimental studies on this topic represent the first steps towards quantification of a generalized process, of depletion of metals by adsorption onto IS dust grains, which has been qualitatively recognised for a considerable time.

As can be appreciated from the examples of sequential addition of two coronene molecules to Cr^+, Mn^+ and other transition-metal ions [99], placing a metal ion on a surface does not automatically negate all of its reactivity. One of the most interesting aspects to emerge from the chemistry initiated by M^++PAH reactions is in the subsequent reactivity of the adsorbed metal ion. Bohme has coined the term 'gas-phase surface chemistry' to describe such processes [109, 110], and there are now several different reaction classes of this type which have been demonstrated. The adsorption-induced modification in the metal ion's reactivity is often manifested in a diminished propensity to undergo further association reactions – for example, studies using the York SIFT [103] have found that while Fe^+ is capable of sequentially adding up to five small inorganic ligands, $Fe(C_{60})^+$ adds a maximum of four and $Fe(C_{20}H_{10})^+$ only three. There are other examples where the metal ion uses the PAH substrate as 'feedstock' to produce a novel product, for example [96, 111]:

$$Fe(C_6H_6)^+ + O_2 \rightarrow Fe(C_6H_4O)^+ + H_2O$$

$$Fe(C_{10}H_8)^+ + O_2 \rightarrow Fe(C_{10}H_6O)^+ + H_2O\,;$$

Very recent SIFT experiments by Caraiman and Bohme [111] that explored the chemistry of Fe^+ coordinated to benzene and the extended aromatic coronene with a number of interstellar gases provided two additional, although minor, reaction channels for the reactions of $Fe(C_6H_6)^+$ with O_2. Both $Fe(C_6H_6)^+$ and $Fe(coronene)^+$ reacted to produce FeO_2 and, when trapped by benzene, Fe^+ was found also to initiate the catalytic oxidation of benzene, possibly to catechol (1,2-dihydroxybenzene), by regenerating Fe^+ [111]. There are even instances in which the metal ion (or the M^+/PAH adduct) acts as a 'template' for formation of a new aromatic ring [108] or for the fusing of two smaller PAHs [97]. These are all processes which may, in many instances, have their analogies in the surface chemistry occurring on IS dust grains, and there is clearly much scope for further interesting research here.

A new technique that promises to provide extensive kinetic data for metal-ion and organometallic-ion chemistry is the Inductively Coupled Plasma/Selected-Ion Flow Tube (ICP/SIFT) tandem mass spectrometer that has been constructed at York University and reported early in 2000 [112]. ICP-MS was originally developed for elemental and isotopic analysis, but in this application the ICP is intended to provide a universal source for atomic ions for injection into a flow tube and the study of their chemistry at room temperature. The ICP ion source and its interface with a flow tube are shown schematically in Fig. 10. Ions are produced in an rf argon plasma operating at atmospheric pressure and passed through differentially pumped sampling and skimmer cones into a quadrupole pre-filter and resolving filter combination and are then injected through a Venturi-type inlet into the flow tube containing helium at 0.35 Torr. Solutions containing the metal salt of interest having concentration of ca. 5 $\mu g\,l^{-1}$ are peristatically pumped via a nebulizer into the argon plasma. The metal ions emerge from the plasma at a nominal plasma temperature of 5500 K and experience both radiative and collisional electronic-state relaxation before they enter the flow tube. Collisional relaxation may occur with argon as the extracted plasma cools upon sampling and then by collisions with He atoms in the flow tube prior to the reaction region, but the actual extent of electronic relaxation (either radiative or collisional) is not known. The collisions with He ensure that the ions

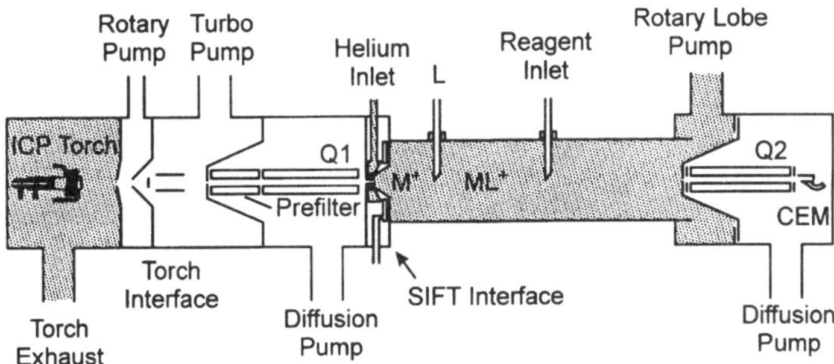

Fig. 10. Schematic view of the Inductively-Coupled Plasma (ICP) Selected-Ion Flow Tube (SIFT) tandem mass spectrometer. The ICP torch produces metal ions M^+ that are mass-selected and injected into the flow tube. The M^+ ions may be ligated with molecules L before they enter the reaction region further downstream

reach a translational temperature equal to the tube temperature of 295±2 K prior to entering the reaction region.

The application of the ICP/SIFT technique was first reported in 2000 and since then has been employed in measurements of reactions of metal ions with O_2 [113, 114] and N_2O [114] and the trapping of metal ions by benzene vapour [111]. As a convenient and highly adaptable source of atomic ions, the ICP clearly has great potential in the exploration of IS metal-ion chemistry.

2.7 Fullerene/Particle Ion Chemistry and the Influence of Charge State

Although there is, as yet, no convincing evidence for the interstellar/circumstellar presence of C_{60}, there are suggestive data [115]. Also, the proposal has been made that C_{60} can serve as a useful model for carbonaceous dust, at least in the limit of spherical carbon structures [110]. Mass spectrometry played a crucial role in the discovery of C_{60} and related fullerenes [116] in the laboratory in 1985, and once C_{60} became available in powder form in 1990 [117], mass spectrometry was used extensively to characterize this molecule with ionization, protonation and fragmentation experiments [118, 119].

Since solid C_{60} is readily sublimed, and because of its high stability, it can be singly and multiply ionized in a conventional MS ion source and can easily be added as a neutral vapour into the reaction region of an appropriate mass spectrometer. The first three charge states of C_{60} have been achieved in the electron impact source and flow tube of a SIFT mass spectrometer and this has led to extensive measurements of the reactivities of C_{60}^{+}, C_{60}^{2+} and C_{60}^{3+} with IS and circumstellar molecules [120]. SIFT experiments with neutral C_{60} have shown that C_{60} is readily ionized by electron transfer to suitable positive ions and by Penning ionization and that reactions with He^+ even lead to two-electron transfer and the formation of doubly charged C_{60}^{2+} [121]. Petrie and Bohme have discussed the formation and reactions of singly and multiply-charged C_{60} cations with a particular emphasis on the chemical derivatization of fullerenes in interstellar and circumstellar environments [122, 123]. Multiple ionization was observed to have some important chemical consequences for chemical derivatization and also for the direct chemical synthesis of neutral IS molecules. It has been shown that dissociative derivatization and the more physical process of electron transfer become increasingly competitive as the charge state increases with electron transfer ultimately being preferred. An example of molecular synthesis assisted by multiple ionization can be found in the SIFT experiments of Milburn et al. [124] which have shown that intramolecular charge separation preceded by bonding to doubly-charged C_{60}, C_{60}^{2+} (and so by inference doubly-charged PAHs or C-containing grains), can lead to new synthetic pathways to cyclic cyanopolyenes. Petrie and Bohme [125] have raised the possible formation of internally cold but kinetically excited ions in interstellar charge separation reactions. Such ions may drive subsequent ion-molecule reactions that are endothermic, or barrier-inhibited, at the low ambient temperatures of IS environments, e.g. H-atom abstraction from H_2.

The formation and reaction of multiply-charged ions may occur more generally with other large molecules in IS and CS regions. For example, naphthalene, a member of the PAH family, has been shown to be doubly-ionized by thermal reaction with He^+ [126]. The chemistry of such multiply-charged species remains to be explored, particularly with regard to their role in the ion-assisted synthesis of neutral molecules.

One of the most interesting applications of mass spectrometry in the past decade has uncovered the processes that may be involved in the formation of fullerenes within astrophysical environments. An ion's diffusion coefficient in, say, a helium bath gas shows some dependence upon its gross geometrical configuration, and the research group of Mike Bowers at Santa Barbara has used this effect in the development of an 'ion chromatography' technique [127] which provides an insight into the shapes of large molecular ions. The groups of Bowers [128, 129] and Jarrold [130, 131] have subsequently studied the evolution of large carbon cluster ions, produced by laser ablation of graphite, from monocycles to tricyclic structures to multiple-ring systems and ultimately to fullerenes through a laser-induced 'annealing' process [132] which may mimic the proposed formation of fullerenes in the atmospheres of highly hydrogen-deficient stars. In a development which holds some relevance for PAH growth mechanisms, the structural influence of hydrogenation has also been explored using this technique [133].

3
Positive Ion/Photon Interactions

While dense IS regions are generally well-shielded against high-energy photons, the interaction between molecular ions and photons is relevant to the question of the survival of such ions in the diffuse interstellar medium, where UV irradiation might be expected to be a powerful destructive force. Such effects are, of course, important also for the fate of neutral molecules in the diffuse interstellar radiation field, but UV photoabsorption by molecules of moderate size is often more likely to lead to photoionization (itself an important topic, but not covered herein) than to photodissociation.

3.1 Interstellar Ion Photochemistry

The particular propensity of high-vacuum MS techniques, for confining ions in an environment almost devoid of collision partners, greatly facilitates the study of ion photochemistry [134]. Studies involving polyatomic ions (most often large aromatic ions) and various modes of irradiation have been an important focus of recent ICR investigations [135]. Several recent studies by Vala and co-workers at Gainesville [136–138], by Grützmacher and associates at Bielefeld [139], by Boissel et al. at Orsay [134], and by Joblin et al. at Toulouse [140] have examined the dehydrogenation of PAH cations ranging in size from 10 to approximately 200 carbon atoms, including fluorene, coronene and perylene. Photolysis in these studies was effected by UV/visible [134, 136–138], sustained off-resonance [139], or laser [140] irradiation of the parent PAH cation, yielding in several instances (see Fig. 11) fully dehydrogenated carbon cluster ions such as C_{18}^+, C_{20}^+ and C_{24}^+. Grützmacher and co-workers have also used multiple-excitation collision activation to generate the C_{24}^+ cation from perchlorocoronene [141]: studies of the subsequent reactivity of these species [141–144], with various aromatic molecules and with dimethyldisulfide, indicates that carbon cluster cations formed by such comparatively 'gentle' techniques tend to retain the polycyclic skeleton of the parent PAH. The astrochemical implications of these

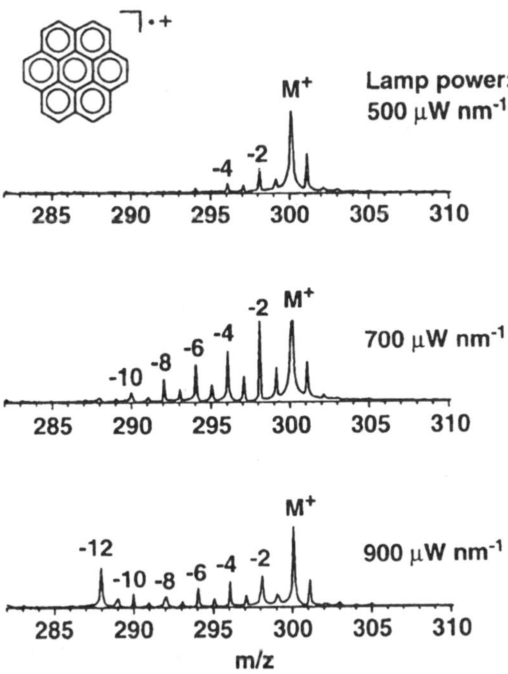

Fig. 11. Mass spectra showing the progression towards complete dehydrogenation of the coronene radical cation as a function of increasing lamp power [138]

results are that UV photolysis of PAHs, in diffuse IS space, will dehydrogenate but not completely destroy the central framework of these molecules: it is likely that mechanisms for their subsequent rehydrogenation will operate in more dense regions (which are shielded from the bulk of the IS radiation field). A startling demonstration of the influence of structure on reactivity is provided by the greater than 1000-fold difference in rate coefficients for reaction of polycyclic and monocyclic C_{24}^+ (produced respectively from sustained off-resonance irradiation of coronene, and from laser ablation of graphite) with benzene [142].

The ion trap technique has also been used to investigate the photochemistry of PAH cations, though in a more restrained fashion than the 'brute force' removal of multiple H atoms which has characterized many of the ICR studies. The von Helden group [145–147] has used the intense IR radiation from a free electron laser to effect excitation (and multiphoton dissociation) of trapped PAH cations ranging in size from $C_{10}H_8^+$ (naphthalene) to $C_{24}H_{12}^+$ (coronene). Such studies, free of the small spectral shifts which influence the spectra of rare-gas-matrix-isolated ions, yield infrared spectra which can then be compared directly with IS spectral features thought to be due to PAH cations [148]. Other MS techniques to have been applied to this topic include ion beam [149–151] and time-of-flight [151–156] approaches.

4
Positive Ion/Electron Interactions

A decade ago, while considerable data had been compiled on the kinetic measurement of dissociative recombination (DR) reactions of small polyatomic ions, laboratory information on the product distributions of such reactions was restricted to the results of a few merged-beam and stationary-afterglow studies on DR of CO_2^+ and of H_3^+ [157, 158], and the first explorations of combined flow tube/Langmuir probe/spectroscopic detector techniques, independently pursued by Rowe and co-workers (at Rennes) [159, 160] and by Adams and co-workers (at Birmingham, and subsequently Atlanta) [161, 162]. Considerable advances have since been made, both in measurement of recombination coefficients (particularly for larger ions) and in the elucidation of product distributions for a still small but growing sample of important IS ions.

4.1
Recombination Reaction Rate Measurements

The Rennes group is currently using the FALP-MS technique [163], a flowing afterglow/Langmuir probe device fitted with a moveable mass spectrometer, as depicted in Fig. 12, to determine recombination kinetics for polyatomic cations. The combination of moveable Langmuir probe and MS detectors permits independent (though not simultaneous) determination of the reactant ion and electron concentration profiles as a function of distance, allowing for a greater degree of clarity in elucidating the reaction chemistry in a plasma where several different reactant ion types might be present. In the studies of Rebrion-Rowe, Mitchell, and co-workers [165–168], helium is employed as a buffer gas, supple-

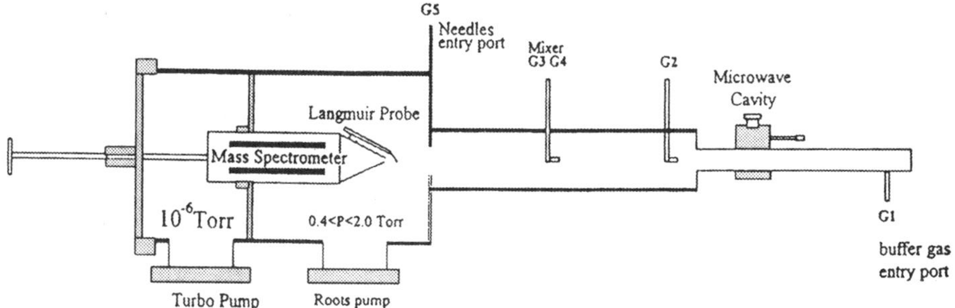

Fig. 12. Schematic view of the Rennes FALP-MS [164]

mented by various 'precursor gases' which are used to generate preferentially the ions of interest from an appropriate reagent gas. Thus, for example, in studies on ionized hydrocarbons, Ar^+, Kr^+, Xe^+ or N^+/N_2^+ would be employed as a precursor ion (from the corresponding rare gas, or from N_2), while to examine the recombination chemistry of a protonated hydrocarbon the preferred precursor gas is an H_2/Ar mixture, yielding H_3^+ which will then protonate the reagent of interest. An important outcome of this work is that, contrary to some earlier theoretical suggestions [169, 170], the recombination coefficients for hydrocarbon cations appear much more dependent on fine details of electronic or molecular structure than on the molecular ion's overall size: their measured recombination coefficient, α, for CH_5^+ (7.0×10^{-7} cm^3 molecule^{-1} s^{-1}) [165] is only marginally less than that for the phenanthrene cation $C_{14}H_{10}^+$ (7.5×10^{-7} cm^3 molecule^{-1} s^{-1}) [171], while the fluoranthene cation $C_{16}H_{10}^+$ apparently has a substantially larger value (α=$2.5\pm1.5\times10^{-6}$ cm^3 molecule^{-1} s^{-1}) [168]. The fluoranthene study [168] identifies several complicating factors arising both from the inability to compensate satisfactorily for problems of mass discrimination in attempted measurements on the wide variety of ion masses (from m/z=14–450) present in the reaction region at any one time, and from the presence of pyrene (which also has the formula $C_{16}H_{10}$) as a contaminant: pyrene undergoes electron attachment, with the result that the reaction plasma contains both a cocktail of reactant cations and a mixture of negative charge carriers. Electron attachment to pyrene (β~10^{-9} cm^3 molecule^{-1} s^{-1}) is not as intrinsically rapid as recombination with a polyatomic ion, but since the reactant neutral number density (even of pyrene as a ~2% impurity in fluoranthene) is many orders of magnitude higher than the positive ion number density, attachment presents a significant impediment to interpreting the recombination kinetics, and this is reflected in the comparatively large uncertainty ascribed to the fluoranthene DR measurement [168]. Such problems are likely to apply also to studies on larger polyatomic ions; nevertheless, extension of these studies to progressively larger species will be very valuable in furthering our understanding of PAH ion chemistry in IS environments. One other result reported from the fluoranthene study [168] is that the recombination coefficient for (dimeric) fluoranthene cluster ions is apparently much larger than the value for ionized fluoranthene itself. Cluster ions typically have larger recombination coefficients than molecular ions of comparable size, as is also borne out by the recent high-pressure flowing afterglow studies, by Glosik, Plasil, and co-workers, of DR with protonated dimers of small molecules [172–175]. Although the types of clusters studied in the latter experiments [172–175] are not in themselves expected to be important constituents of IS clouds, the results do serve to underscore the extremely high efficiency of DR involving weakly-associated complex ions. In an IS context, it may well be that the extent of PAH clustering will significantly influence the interaction of ions with electrons, and hence the degree of ionization, in dense regions.

4.2 Dissociative Recombination Product Channel Determination

Measurement of DR branching ratios is perhaps the most problematic and contentious topic in experimentally-based interstellar chemistry. As the chief means of positive ion neutralization, DR is crucial in determining the eventual outcome of most, if not all, sequences of synthetic ion/molecule steps. Two fundamentally different techniques have been used for DR product analysis. The FALP technique of Smith and Adams, used with considerable success in the study of ion/electron recombination kinetics [171, 176, 177], has been adapted to permit subsequent neutral product detection by LIF (laser-induced fluorescence) or VUV (vacuum ultraviolet) spectroscopy, as shown in Fig. 13. Such studies, first

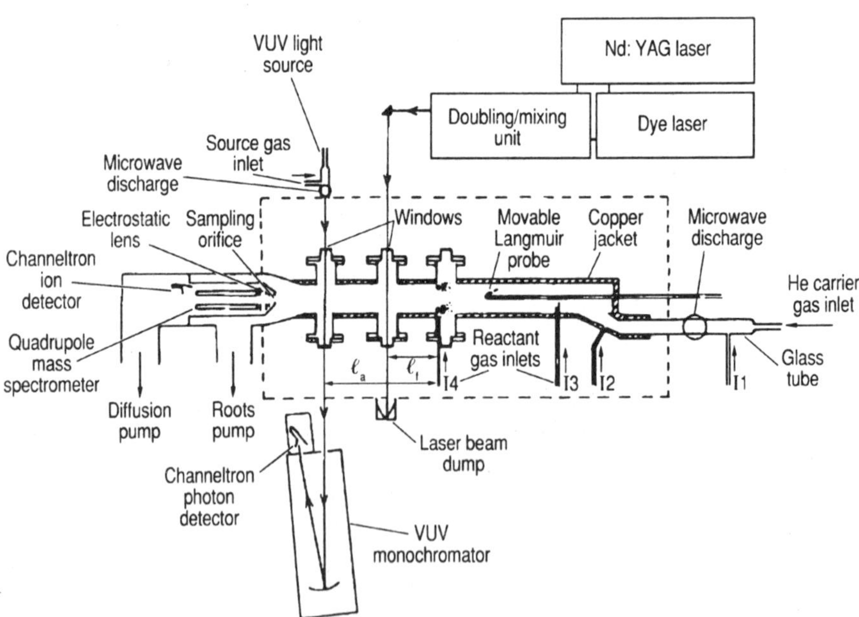

Fig. 13. Schematic diagram of the FALP-LIF/VUV apparatus used, in various configurations, since 1989 by Adams et al. [161] in their characterization of DR neutral products. The *dotted line* encasing the central flow tube represents a vacuum jacket facilitating operation at a broad range of temperatures

conducted in 1989 [161], have been pursued by the research groups of Adams in Atlanta [162, 178], Rowe in Rennes [179] and Johnsen in Pittsburgh [180–182]. In these studies, LIF has been used to characterise the diatomic fragments OH [161, 180, 181, 183], and CO (in the excited $a\ ^3\Pi$ state) [182], as well as indirectly monitoring H atom production from DR, through measurements on the OH yield obtained using the conversion reaction $H+NO_2 \rightarrow OH+NO$ [184]. Direct measurements on H-atom yield, using VUV detection [161, 184] have also been undertaken, as have VUV studies of atomic oxygen [185].

One limiting aspect of the FALP-LIF and FALP-VUV studies is its apparent restriction to detection of highly reactive (in practice, radical) products. Thus, although the DR of protonated species such as H_3^+, HCO_2^+, H_3O^+, H_3S^+ and $HCNH^+$ (all of which play an important role in IS ion chemistry) may very well yield the parent neutrals H_2, CO_2, H_2O, H_2S and HCN, these 'stable' products are not viable candidates for spectroscopic detection in FALP studies since they are necessarily present in the reaction region as precursor gases: remember also that, in a high-pressure laboratory plasma of this type, the concentration of the source gas is routinely many orders of magnitude higher than the concentration of reactant ions. Since only a minuscule fraction of the source gas molecules experience protonation within the reaction region, identification of the yet smaller fraction of parent molecules to re-emerge unscathed from DR is not a tenable proposition. Another important point is that FALP is a flowing-afterglow, and not a selected-ion, approach: once the initial ionization is produced, there is no effective physical constraint or filter to allow preferential selection of a particular reactant ion, and so this selection must be effected chemically. In the first FALP-LIF study reported [161], detection of OH production from the DR of $HOCO^+$ necessitated the sequential occurrence of the following reactions:

$$
\begin{array}{ll}
He^+ + He + He & \rightarrow He_2^+ + He \\
He^* + He + & \rightarrow He_2^+ + e \\
He_2^+ + Ar & \rightarrow Ar^+ + He + He \\
Ar^+ + H_2 & \rightarrow ArH^+ + H \\
 & \rightarrow H_2^+ + Ar \\
XH^+(X{=}H, Ar) + H_2 & \rightarrow H_3^+ + X \\
H_3^+ + CO_2 & \rightarrow HOCO^+ + H_2 \\
HOCO^+ + e & \rightarrow \text{products (including HO + CO).}
\end{array}
$$

Thus, the ionization is initially effected by discharge in helium, while at progressively later stages along the flow tube are added argon, hydrogen and carbon dioxide (and to detect H by LIF, NO_2 must also be added). The intermediate chemistry involving Ar and H_2 is crucial to ensure that the eventual protonation of CO_2 is sufficiently gentle that only ground-state $HOCO^+$, and no contaminating fragment ions, persists into the reaction region [161]. The necessity for the flow-tube chemistry to be 'clean', in the sense of converging, finally, to one specific product ion, has placed further constraints on the DR reactions accessible using this approach, and it is notable, for example, that aside from CH_5^+ [184] none of the important IS hydrocarbon ions – which, as a class, are prone to undergo subsequent chain-growing reactions – have been subjected to this method of product analysis.

The second main approach to DR product identification, using an ion storage ring apparatus [186–190], has recently been used by groups based in Denmark (using the ASTRID storage ring) and Sweden (using the CRYRING installation). A representation of the CRYRING apparatus is shown in Fig. 14. The salient features of this method are, first, acceleration of the reactant ions to very high translational energy (typically several MeV) in the storage ring, second, merging of this fast ion beam with an equal-velocity electron beam and, third, detection of the fast neutral products resulting from individual DR events. Storage times of arbitrary duration (typically several seconds) can ensure that any electronic or vibrational excitation of the reactant ions can satisfactorily be radiatively dissipated prior to reaction, while the very low centre-of-mass collision energies which can be obtained by judicious matching of the ion and electron beam velocities permits occurrence of a 'cold' DR reaction to produce neutral fragments which are, nevertheless, very highly translationally excited in the 'laboratory frame' of the stationary detector. The mode of detection also warrants comment. A dipole magnet immediately downstream from the reaction region deflects the ion beam along the path of the ring: any neutral products are not deflected and travel along the 0° direction to the detector (which is typically an energy-sensitive, ion-implanted silicon detector). Due to the very high translational velocity of the reactants, all neutral products have virtually identical velocity and so all fragments resulting from a single DR event arrive at the detector simultaneously: the detector does not, in fact, register separate products, instead measuring the total collision energy of all fragments from a discrete DR event. If all product fragments from all events reach the detector in separate 'event-packets', then an essentially 'monochromatic' energy signal is registered independent of any fragmentation, yielding no information on the product distribution. If, however, a porous grid is interposed before the detector so that some

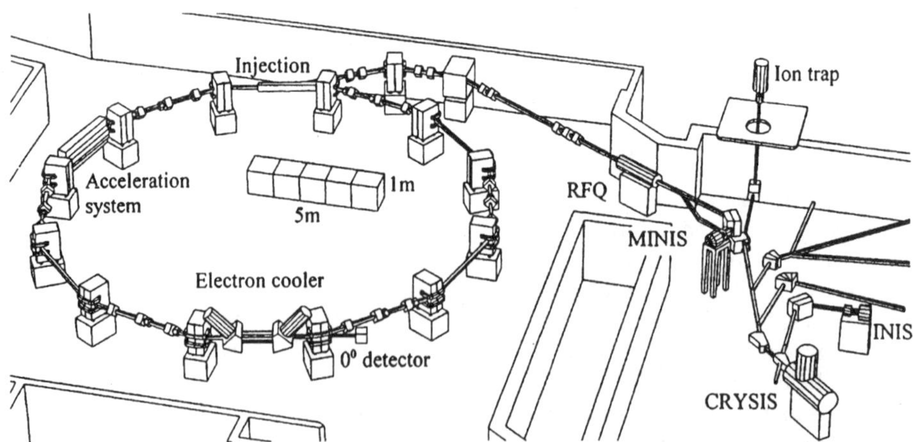

Fig. 14. View of the CRYRING ion storage ring at the Manne Siegbahn Laboratory at Stockholm University [196]. The scale of the instrument can be judged by the cubic-metre blocks depicted near the drawing's centre

fragments are prevented from reaching the detector, then the 'energy spectrum' which builds up from accumulating data from many discrete DR events becomes a 'mass spectrum' of a sort, revealing the distribution of fragment masses (and combination-of-fragment masses) which have successfully reached the detector. Examples of such neutral mass spectra, for DR of H_3O^+ isotopomers, are shown in Fig. 15. Provided that the transmission characteristics of the grid are known, it is then feasible to determine the unique branching ratio of possible product channels which has given rise to the observed energy spectrum. It is, perhaps, ironic that this method relies for its success on the inclusion of a component deliberately designed to block off a part of the signal: in most MS studies, after all, considerable pains are taken to *boost* the signal wherever possible! Storage-ring measurements are capable of yielding information on the dissociative recombination rates, via determination of the electron/ion collision cross-section, but it is undoubtedly their utility for characterizing neutral DR products which has most caught the attention of the interstellar chemistry community. Uncertainties ascribed to the product distributions obtained from this method are very low, generally only 1–2% of the total product yield.

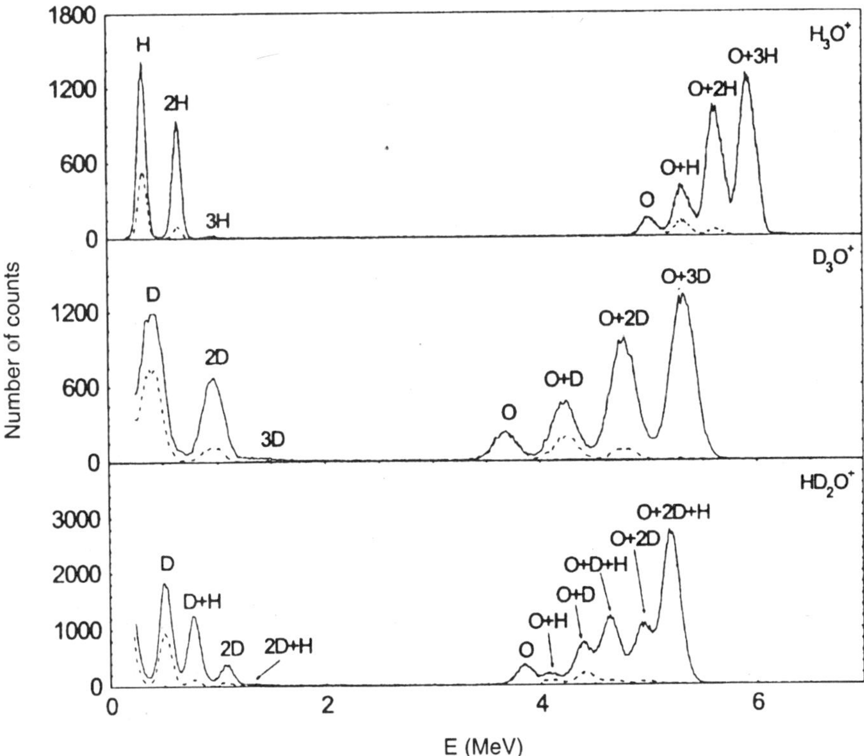

Fig. 15. Neutral product 'mass spectrum' seen for the DR of various H_3O^+ isotopomers, with a 70% transmission grid intersecting the path of the neutral products prior to detection [199]. Analysis of this mass distribution subsequently yields the reaction's product distribution

The first use of an ion storage ring for quantification of relevant DR product channels was reported in 1995: the CRYRING device in Sweden was used to characterize the branching ratios for DR of H_3^+, for collisional energies ranging from 0.001 to 20 eV [192]. Subsequent CRYRING studies have focussed on the DR reactions of H_3O^+, D_3O^+ [193], CH_n^+ (n=2, 5) [191, 194], NH_n^+ (n=2, 4) [195], $HCNH^+$ [196] and $C_2H_2^+$ [197]. The ASTRID device in Denmark has also been employed, with results reported for DR of H_2O^+, H_3O^+ and CH_3^+ [198] and of the variously deuterated isotopomers of ionized and protonated H_2O [199, 200]. Both instruments are currently being used to study the DR of somewhat larger hydrocarbon ions [201, 202]. A common thread emerging from the Scandinavian studies is that three-body dissociative channels, of the form X+H+H or X+H+H_2, are an important and sometimes dominant component of the DR products [203]. This is a phenomenon which was not anticipated from previous experimental (i.e. FALP-LIF) or theoretical studies, and potentially has major ramifications for the chemical evolution of IS clouds [204]. For example, the unexpectedly efficient production of atomic C from the DR of CH_3^+ [198], and of CH from the DR of both CH_3^+ and $C_2H_2^+$ [197, 198] may provide an important route to these highly reactive neutrals, which appear to be surprisingly persistent in comparatively dense IS regions.

Agreement between the ASTRID and CRYRING measurements for H_3O^+ and D_3O^+ – which to date provide the only basis for a comparison of the two instruments – is fair to good [193, 198, 200]. The same cannot be said for some of the ions which have been studied by both FALP-LIF and storage-ring techniques, notably H_3O^+ and $HCNH^+$. For H_3O^+, FALP-LIF measurements of the O atom production yielded a value of 0.30 for the branching fraction of the O+H_2+H channel [185], in contrast with the values of 0.01±0.04 and 0.04±0.06 resulting respectively from ASTRID and CRYRING measurements [193, 198, 200]. For $HCNH^+$, FALP-VUV data showed that CN+H_2 considerably outweighed CN+H+H in efficiency [205], a result very much at odds with the respective branching ratios of 0±0.02 and 0.33±0.03 obtained from CRYRING measurements [196]. Nevertheless, while such discrepancies are very troubling for interstellar modellers, we should also note that the points of broad agreement between FALP and storage-ring DR product distributions are much more numerous than these disagreements. With regard to the resolution of the identified discrepancies, one important distinction [198] is that FALP product distributions effectively result from the compilation of several separately-executed and essentially independent experiments, each of which involves VUV or LIF detection of one particular reaction product such as H, O or OH; in contrast, storage-ring measurements yield a complete product distribution in one experiment (although data accumulation from repeated runs obviously helps to minimize noise in the observed signal). An inherent shortcoming in both techniques is that in some instances the product distribution may not be susceptible to unambiguous determination: for example, the FALP studies on $HOCO^+$ (first begun in 1989) [161, 183–185] have not included attempted detection of a possible HOC product, while the storage-ring measurements on $HCNH^+$ [196] cannot mass-distinguish between HCN and HNC products, nor between any set of isomers. Identification of the exact chemical structure of polyatomic products is technically feasible by FALP-LIF, but appears not to have been reported to date.

It is interesting to emphasize the degree of elegance that has already been demonstrated by the two different techniques. Such elegance is exemplified by FALP-LIF studies which have yielded vibrational product distributions (v=0–6) for CO from the DR of HCO^+ and DCO^+ [206, 207], and by CRYRING measurements which have used a differential absorber (which serves to filter out interference from H-atom products) to quantify the ratio of O (^3P) to O (^1D) produced in DR of H_2O^+ [208]. While population of excited states may not be highly significant for IS cloud chemical evolution (the extremely long timescales between collisions ensure that such excitation will be dissipated away prior to any reactive encounter), the energy distribution from DR certainly plays a role in establishing the thermal equilibrium in these regions. Furthermore, characterization of the dispersal of reaction energy, as well as product fragments, is obviously very valuable in the context of deepening our understanding of the dynamics of the fundamental DR process, and may well lead to an improved theoretical understanding of this important and challenging class of reactions.

4.3 The H_3^++e Reaction Rate

Recombination coefficient measurements are generally less subject to disagreement or discrepancy than are the corresponding product distributions, but there is one very important exception: the reaction of H_3^++e [209]. Over the past several decades, the kinetics of this process have been investigated by methods including the merged-beam [210–212], ion trap [213], stationary afterglow [214–216], flowing afterglow [164, 217–223], infrared absorption [224, 225] and ion storage ring [226–228] techniques. Yet, despite the impressive (and still growing) arsenal of experimental approaches which have been brought to bear, the H_3^+ DR reaction is still the 'John Barleycorn' of ion/electron reactions. Experimental measurements of $\alpha(H_3^+)$ have ranged from $<10^{-11}$ cm^3 molecule^{-1} s^{-1} to $\sim3\times10^{-7}$ cm^3 molecule^{-1} s^{-1}, with several theoretical studies favouring the lower end of this range [229]. The most recent experimental studies have mostly yielded results which are clustered around 10^{-7} cm^3 molecule^{-1} s^{-1}, with the notable exception of an Advanced Integrated Stationary Afterglow (AISA) study which reported α_{eff}~3×10^{-9} cm^3 molecule^{-1} s^{-1} [216]. Attempted explanations of the curious inconsistency in measurements have included the perturbing influence, variously, of H_3^+ vibrational excitation [221, 222], collisional radiative recombination, aerodynamic effects [164], the nuclear spin state [230] and the H_2 partial pressure in the reaction vessel [216]: none of these explanations has proven satisfactory to all of the research groups involved, and debate continues. Some recent theoretical studies suggest [229, 231] that theoreticians may finally be ready to support the comparatively high (~10^{-7} cm^3 molecule^{-1} s^{-1}) values which have predominated in the latest laboratory measurements. It is fortunate for modellers of IS chemistry that most DR rates are not so difficult to pin down. Note also that, although H_3^+ is undoubtedly readily formed by the ion/molecule chemistry which follows cosmic-ray ionization within IS clouds, it is also readily converted (by proton transfer) to HCO^+ which is more generally the dominant ion within dense regions. A conclusive consensus on the H_3^+ DR rate would thus be of greatest value to modellers of diffuse IS clouds, where the higher degree of ionization favours DR over proton transfer as the presumed chief loss process for H_3^+.

5 Interstellar Negative-Ion Chemistry

While positive ions steal all the glory in interstellar ion chemistry, a small undercurrent of continuing research suggests that molecular negative ions may also feature in space chemistry. No interstellar negative ion has yet been incontrovertibly detected [232], and the total abundance of interstellar molecular negative ions is expected to be significantly less than the sum of positive-ion abundances, since free electrons and negatively-charged dust grains are thought to be important negative charge carriers in molecular astrophysical environments. The case for IS negative ions received a significant [233] but temporary [234, 235] boost when it was suggested that the species C_7^- was a carrier of five widely-observed spectral features amongst the 'diffuse interstellar bands' (DIBs). Nonetheless, several types of pathway to IS anions have been outlined [236–238] and various recent mass-spectrometric studies are relevant here. The most obvious pathway to IS negative ions, radiative attachment,

$$X + e \rightarrow X^- + h\nu,$$

has been considered to be generally efficient when X is a polyatomic radical (such species are characterized by large electron affinities) or a large polyaromatic molecule (since the very large number of vibrational modes ensures dissipation of the attachment energy, favouring stabilization). Fullerenes were held as being particularly promising candidates for astrophysically-relevant negative-ion formation, since they possess both a large electron affinity (2.65 eV) [239] and a very large number of vibrational modes. However, FALP studies by Spanel and Smith [240, 241] have indicated that attachment to C_{60} is inhibited by a barrier of ~0.26 eV, effectively ensuring that (except for tunnelling) such attachment at IS cloud temperatures will not occur. The case for attachment to some PAHs is more favourable, with efficient attachment to anthracene ($C_{14}H_{10}$) noted in a CRESU study ranging from 48 to 300 K [242, 243], and the apparent efficiency of a similar process for pyrene ($C_{16}H_{10}$) seen in the FALP-MS [168]. The preceding experimental studies have used 'high-pressure' techniques, but attachment measurements are not exclusively the preserve of the FALP and CRESU methodologies, with 'electron trap' approaches also proving useful for attachment involving comparatively large reactant molecules [244]. Attachment to IS radicals (which may provide the best prospect for the detection of negative ions in IS space) has yet to be probed experimentally by any technique.

The reaction chemistry of smaller IS negative ions is receiving continued attention. Recently, Bierbaum and co-workers have used the Boulder FA-SIFT to monitor the reactivity of various C_n^- and C_nH^- ions with atomic and molecular hydrogen [245]. The absence of any reaction with H_2 is important given the predominance of hydrogen as an IS molecule, while for C_7^-, C_8^-, C_9^- and C_{10}^- association (yielding C_nH^-) was seen to occur in competition with the more customary product channel of associative detachment (e.g. C_7^-+H→C_7H+e) [245]. The C_7H^- species has also featured in a neutralization/reionization mass spectrometry study by Dua et al. [246] which has identified three distinct structural isomers. This C_7H^- study is one of a number [246–254] undertaken by the Adelaide-based group of John Bowie, on structural isomerism in potential IS cumulenes, heterocumulenes and their corresponding anions. Further study in this area, and an increase in our understanding of IS negative-ion chemistry, may well aid in the radioastronomical detection of the first examples of this elusive class of interstellar ions.

6
In-Situ Sampling?

The review this far may have conveyed the sense that MS applications in interstellar chemistry are all somewhat indirect: the primary data on IS chemistry are all obtained through observations using radio-, IR- or optical telescopes, with MS techniques necessarily restricted to informing our understanding of the processes involved in our laboratory attempts to replicate some facet or other of the deep-space chemistry. With the most interesting astrochemical objects all at least several hundred light-years distant, there seems no prospect of in-situ sampling of IS material anytime soon. Or is there? While we can hardly expect the mountain to come to Mohammed (or to his mass spectrometer), some of the material from the mountain may feasibly find a way here, blown on some interstellar wind. Such detritus takes many forms, from the atoms (principally isotopes of H and He) which register as IS 'pickup ions' [255–259] in mass spectrometers on deep-space probes [260, 261], to the microscopic dust grains which reveal their exotic origin in their isotopic distributions [262–265], to the comets which sporadically pass through the solar system. Comets hold a special role in the cosmic hierarchy, as 'pristine', rudimentary bodies whose original (IS) molecular constituents are thought to have undergone comparatively little subsequent chemical processing. This has been borne out by recent observations of several 'interstellar' species such as HNC, HC_3N and CS in cometary comae [266, 267]. While analysis of earlier MS-bearing missions to comets is ongoing [268, 269], the next few years are expected to see a significant upsurge in our understanding of cometary chemistry, with two missions currently underway and a third planned.

One of the cometary probes to have embarked is on a 'sample-return' mission [270], and MS will certainly play a part in the analysis of the material obtained. This will not, however, be the first occasion on which cometary material has reached the Earth. One of the most intriguing, albeit contentious [271–274], reports of the last few years has been the mass-spectroscopic identification of interstellar fullerenes, such as C_{60}, as 'fossil molecules' in the geologic strata associated with the supposed Cretaceous/Tertiary meteoritic [275] ('dinosaur killer') and Permian/Triassic cometary [276] ('the Great Dying') impacts. Similar results are also reported for the Murchison and Allende meteorites, and for material obtained from the ancient (~2 billion-year-old) Sudbury impact feature

[275, 277–279]. This identification has rested on the detection, in gas released from endohedral helium fullerenes, He@C_{60}, upon heating [275], of $^3He/^4He$ ratios which are very much higher than that found naturally in any terrestrial or solar environment, and which are broadly consistent with the helium isotopic composition in diffuse IS regions [280]. The exact conditions under which endohedral fullerenes may have formed, in the depths of space, remain the subject of conjecture, but laboratory experiments have demonstrated several mechanisms for He@C_{60} formation. Mass spectrometry has been used to verify formation of He@C_{60} by heating fullerenes under high helium pressure [281]. Three- and four-sector tandem mass spectrometer experiments reported in 1991 showed that He could be inserted into C_{60}^{+} in high-energy collisions of C_{60}^{+} with He [282–284]. Finally, the observed dissociative insertion of Ne^+ within C_{60} at high Ne^+ energies in an octapole guided-ion beam mass spectrometer [285] has prompted the suggestion that cosmic-ray impact upon interstellar fullerenes may lead to the formation of endohedral complexes of atoms or atomic ions [286].

7 Concluding Remarks

Interstellar chemistry is a dynamic, changing, still-developing field of research. Many of the recent advances, in the mass-spectrometric study of interstellar processes, have been incremental, as the use of pre-existing techniques has matured to tackle more diverse and somewhat more complex chemical problems. In other cases, the development of fundamentally new experimental tools has afforded an opportunity to study processes at a level of detail and under physical conditions that had seemed beyond our grasp. We can extrapolate that many of the most interesting developments in IS chemistry, in the next few years, will be in the areas of dissociative recombination product assignment, of very-low-temperature reaction rate measurement, and in the investigation of detailed chemical networks involving formation and reaction of large IS species such as PAHs, and possibly amino acids, and their corresponding ions. Perhaps some unheralded group of researchers is currently developing an instrument with which they will examine reactions of ions with molecular radicals such as CN, C_2H, OH and CH_3, or which will allow them to identify differences in reactivity between different (rotationally-frozen) conformers of polyatomic ions at temperatures similar to those seen in IS clouds. These and numerous other challenges still lie ahead.

It is also fair to say that, although there are very many research groups with a professed interest in interstellar chemistry, there are probably none that pursue such interest to the exclusion of all else. Most of the mass spectrometric methods outlined in this review can be, and have been, used to study chemical processes beyond the context of IS chemistry. Furthermore, while the cold reaches of space are far removed from our own terrestrial experience, there is very often a tendency for advances, made in the primary hope of elucidating some interstellar phenomenon, to have an effect on the wider world of society. Thus, arguably, efforts to generate interstellar molecules in the laboratory and to reproduce the properties of carbonaceous interstellar dust grains gave rise to the discovery and subsequent production of fullerenes. The development, a quarter-century ago, of the Selected-Ion Flow Tube, which, from its inception, was used extensively to model IS ion chemistry, has produced a technique now being used for medical and analytical purposes. We are not audacious enough to attempt to predict future applications of ion-storage-ring technology, for example, but we confidently anticipate that the actions of researchers using this and other recently-developed mass-spectrometric techniques to study interstellar processes will continue to provide insights of benefit to the wider scientific community.

8 References

1. Solomon PM, Klemperer W (1972) Astrophys J 178:389
2. Herbst E, Klemperer W (1972) Astrophys J 185:505
3. Singh M, Chaturvedi JP (1987) Astrophys Space Sci 135:1
4. Mitchell GF, Ginsburg JL, Kuntz PJ (1978) Astrophys J Suppl 38:39
5. Graedel TE, Langer WD, Frerking MA (1982) Astrophys J Suppl 48:321
6. Millar TJ, Freeman A (1984) Mon Not R Astron Soc 207:405
7. Herbst E, Leung CM (1990) Astron Astrophys 233:177
8. Anicich VG (1993) Astrophys J Suppl 84:215
9. Anicich VG (1993) J Phys Chem Ref Data 22:1469
10. http://astrochem.jpl.nasa.gov/asch/
11. Henis JMS (1972) In: Franklin JL (ed) Ion-molecule reactions, vol 2. Plenum, New York, p 395
12. Marshall AG, Hendrickson CL, Jackson GS (1997) Mass Spectrom Rev 17:1
13. Watson WD, Anicich VG, Huntress WT Jr (1976) Astrophys J 205:L165
14. Huntress WT, Anicich VG (1976) Astrophys J 208:23
15. Huntress WT (1977) Astrophys J Suppl 33:495
16. Adams NG, Smith D (1976) Int J Mass Spectrom Ion Phys 21:349
17. Bohme DK (2000) Int J Mass Spectrom 200:97
18. Smith D (1992) Chem Rev 92:1473
19. Smith D, Spanel P (1992) Acc Chem Res 25:414
20. Bohme DK (1987) In: Ausloos P, Lias SG (eds) Structure, reactivity and thermochemistry of ions. Reidel, p 219
21. Bohme DK (1986) Nature 319:473
22. Bohme DK (1992) In: Adams N, Babcock LM (eds) Advances in gas phase ion chemistry, vol 1. JAI Press, London, p 225
23. Bohme DK, Wlodek S, Wincel H (1991) J Am Chem Soc 113:6396
24. Scott GBI, Fairley DA, Freeman CG, Maclagan RGAR, McEwan MJ (1995) Int J Mass Spectrom Ion Process 150:251
25. Fairley DA, Scott GBI, Freeman CG, Maclagan RGAR, McEwan MJ (1996) J Chem Soc Faraday Trans 92:1305
26. Fairley DA, Scott GBI, Freeman CG, Maclagan RGAR, McEwan MJ (1997) J Phys Chem A 101:2848
27. Milligan DB, Freeman CG, Maclagan RGAR, McEwan MJ, Wilson PF, Anicich VG (2001) J Am Soc Mass Spectrom 12:557
28. McEwan MJ (1992) In: Adams N, Babcock LM (eds) Advances in gas phase ion chemistry, vol 1. JAI Press, London, p 1
29. Adams NG, Smith D, Giles K, Herbst E (1989) Astron Astrophys 220:269
30. Petrie S (1995) Astrophys J 454:L165
31. Bohme DK (1992) Int J Mass Spectrom Ion Process 115:95
32. Cernicharo J, Guelin M (1987) Astron Astrophys 183:L10

33. Squires RR (1992) Int J Mass Spectrom Ion Process 118/119:503
34. Marinelli PJ, Paulino JA, Sunderlin LS, Wenthold PG, Poutsma JC, Squires RR (1994) Int J Mass Spectrom Ion Process 130:89
35. Baranov V, Bohme DK (1996) Int J Mass Spectrom Ion Process 154:71
36. Scott GB, Fairley DA, Freeman CG, McEwan MJ, Spanel P, Smith D (1997) J Chem Phys 106:3982
37. Scott GBI, Fairley DA, Freeman CG, McEwan MJ, Adams NG, Babcock LM (1997) J Phys Chem A 101:4973
38. McEwan MJ, Scott GBI, Adams NG, Babcock LM, Terzieva R, Herbst E (1999) Astrophys J 513:287
39. Scott GBI, Fairley DA, Freeman CG, McEwan MJ (1997) Chem Phys Lett 269:88
40. Scott GBI, Freeman CG, McEwan MJ (1997) Mon Not R Astron Soc 290:636
41. Scott GBI, Fairley DA, Freeman CG, McEwan MJ, Anicich VG (1998) J Chem Phys 109:9010
42. Milligan DB, Fairley DA, Freeman CG, McEwan MJ (2000) Int J Mass Spectrom 202:351
43. Scott GBI, Fairley DA, Milligan DB, Freeman CG, McEwan MJ (1999) J Phys Chem A 103:7470
44. Scott GBI, Fairley DA, Freeman CG, McEwan MJ, Anicich VG (1999) J Phys Chem A 103:1073
45. Milligan DB, McEwan MJ (2000) Chem Phys Lett 319:482
46. Ferguson EE, Fehsenfeld FC, Golden PD, Schmeltekopf AL, Schiff HI (1965) Planet Space Sci 13:823
47. Fehsenfeld FC, Ferguson EE (1971) J Geophys Res 76:8453
48. Sablier M, Rolando C (1993) Mass Spectrom Rev 12:28
49. Le Page V, Keheyan Y, Snow TP, Bierbaum VM (1999) J Am Chem Soc 121:9435
50. Rowe BR, Canosa A, Le Page V (1995 Int J Mass Spectrom Ion Process 149/150:573
51. Smith IWM, Rowe BR (1995) Acc Chem Res 33:261
52. Rowe BR, Rebrion C (1991) Trends Chem Phys 1:367
53. Adams NG, Smith D, Paulson JF (1980) J Chem Phys 72:288
54. Troë J (1985) Chem Phys Lett 122:425
55. Atkinson DB, Smith MA (1995) Rev Sci Instrum 66:4434
56. Hawley M, Mazely TL, Randeniya LK, Smith RS, Zeng XK, Smith MA (1990) Int J Mass Spectrom Ion Process 80:239
57. Smith MA, Hawley M (1992) Adv Gas Phase Ion Chem 1:167
58. Smith MA (1998) Int Rev Phys Chem 17:35
59. Smith MA, Hawley M (1995) Int J Mass Spectrom Ion Process 149/150:199
60. Atkinson DB, Smith MA (1994) J Phys Chem 98:5797
61. Hendrickson CL, Emmett MR (1999) Annu Rev Phys Chem 50:517
62. Marshall AG (2000) Int J Mass Spectrom 200:331
63. Marshall AG, Hendrickson CL (2002) Int J Mass Spectrom 215:59
64. Weddle G, Dunbar RC (1994) Int J Mass Spectrom Ion Process 134:73
65. Cheng Y-W, Dunbar RC (1995) J Phys Chem 99:10,802
66. Ryzhov V, Dunbar RC (1997) Int J Mass Spectrom Ion Process 167:627
67. Ryzhov V, Yang Y-C, Klippenstein SJ, Dunbar RC (1998) J Phys Chem A 102:8865
68. Klippenstein SJ, Yang Y-C, Ryzhov V, Dunbar RC (1996) J Chem Phys 104:4502
69. Dunbar RC (1997) Int J Mass Spectrom Ion Process 160:1
70. Anicich VG, Sen AD, McEwan MJ, Smith SC (1994) J Chem Phys 5696
71. Anicich VG, Sen AD, Huntress WT Jr, McEwan MJ (1995) J Chem Phys 102:3256
72. McEwan MJ, Anicich VG (1995) J Phys Chem 99:12,204
73. McEwan MJ, Fairley DA, Scott GBI, Anicich VG (1996) J Phys Chem 100:4032
74. Filippi A, Occhiucci G, Speranza M (1997) Inorg Chem 36:3936
75. Stöckigt D, Hrusak J, Schwarz H (1995) Int J Mass Spectrom Ion Process 150:1
76. Barlow SE, Dunn GH, Schauer M (1984) Phys Rev Lett 52:902
77. Gerlich D, Kaefer G (1989) Astrophys J 347:849
78. Gerlich D, Horning S (1992) Chem Rev 92:1509
79. Gerlich D (1993) J Chem Soc Faraday Trans 89:2199

80. Sorgenfrei A, Gerlich D (1994) AIP Conf Proc 312:505
81. Tosi P, Dmitriev O, Bassi D, Wick O, Gerlich D (1994) J Chem Phys 100:4300
82. Turner BE (1991) Astrophys J 376:573
83. Kawaguchi K, Kagi E, Hirano T, Takano S, Saito S (1993) Astrophys J 406:L39
84. Ziurys LM, Apponi AJ, Guelin M, Cernicharo J (1995) Astrophys J 445:L47
85. Turner BE, Steimle TC, Meerts L (1994) Astrophys J 426:L97
86. Ziurys LM, Savage C, Highberger JL, Apponi AJ, Guelin M, Cernicharo J (2002) Astrophys J 564:L45
87. Highberger JL, Savage C, Bieging JH, Ziurys LM (2001) Astrophys J 562:790
88. Walmsley CM, Bachiller R, Des Forets GP, Schilke P (2002) Astrophys J 566:L109
89. Petrie S (1996) Mon Not R Astron Soc 282:807
90. Dunbar RC, Petrie S (2002) Astrophys J 564:792
91. Milburn RK, Hopkinson AC, Bohme DK (2002) (in preparation)
92. Petrie S, Becker H, Baranov V, Bohme DK (1997) Astrophys J 476:191
93. Milburn RK, Baranov V, Hopkinson AC, Bohme DK (1998) J Phys Chem 102:9803
94. Petrie S, Dunbar RC (2000) J Phys Chem A 104:4480
95. Boissel P (1994) Astron Astrophys 285:L33
96. Boissel P, Marty P, Klotz A, de Parseval P, Chaudret B, Serra G (1995) Chem Phys Lett 242:157
97. Marty P, de Parseval P, Klotz A, Chaudret B, Serra G, Boissel P (1996) Chem Phys Lett 256:669
98. Marty P, de Parseval P, Klotz A, Serra G, Boissel P (1996) Astron Astrophys 316:270
99. Dunbar RC, Uechi GT, Asamoto B (1994) J Am Chem Soc 116:2466
100. Dunbar RC, Klippenstein SJ, Hrusak J, Stoeckigt D, Schwarz H (1996) J Am Chem Soc 118:5277
101. Lin C-Y, Dunbar RC (1997) Organometallics 16:2691
102. Pozniak B, Dunbar RC (1997) J Am Chem Soc 119:10,439
103. Caraiman D, Koyanagi GK, Scott LT, Preda DV, Bohme DK (2001) J Am Chem Soc 123:8573
104. Frash MV, Hopkinson AC, Bohme DK (2001) J Am Chem Soc 123:6687
105. Serra G, Chaudret B, Saillard Y, Le Beuze A, Rabaa H, Ristorcelli I, Klotz A (1992) Astron Astrophys 260:489
106. Chaudret B, Le Beuze A, Rabaa H, Saillard Y, Serra G (1991) New J Chem 15:791
107. Klotz A, Marty P, Boissel P, Serra G, Chaudret B, Daudey JP (1995) Astron Astrophys 304:520
108. Klotz A, Marty P, Boissel P, de Caro D, Serra G, Mascetti J, de Parseval P, Derouault J, Daudey JP, Chaudret B (1996) Planet Space Sci 44:957
109. Bohme DK, Wlodek S, Wincel H (1991) J Am Chem Soc 113:6396
110. Bohme DK (1992) Chem Rev 92:1487
111. Caraiman D, Bohme DK (2003) Int J Mass Spectrom 223/224:411
112. Koyanagi GK, Lavrov VV, Baranov VI, Bandura D, Tanner SD, McLaren JW, Bohme DK (2000) Int J Mass Spectrom 194:L1
113. Koyanagi GK, Caraiman D, Blagojevic V, Bohme DK (2002) J Phys Chem A 106:4581
114. Koyanagi GK, Bohme DK (2001) J Phys Chem A 105:8964
115. Foing BH, Ehrenfreund P (1997) Astron Astrophys 317:L59
116. Kroto HW, Heath JR, O'Brien SC, Curl RF, Smalley RE (1985) Nature 318:162
117. Krätschmer W, Lamb LD, Fostiropoulos K, Huffman DR (1990) Nature 347:354
118. McElvany SW, Ross MM, Callahan, JH (1992) Acc Chem Res 25:162
119. McElvany SW, Ross MM (1992) J Am Soc Mass Spectrom 3:268
120. Bohme DK (1999) Can J Chem 77:1453
121. Javahery G, Petrie S, Wang J, Bohme DK (1992) Chem Phys Lett 195:7
122. Petrie S, Javahery G, Bohme DK (1993) Astron Astrophys 271:662
123. Petrie S, Bohme DK (2000) Astrophys J 540:869
124. Milburn RK, Hopkinson AC, Sun J, Bohme DK (1999) J Phys Chem 103:7528
125. Petrie S, Bohme DK (1994) Mon Not R Astron Soc 268:103

126. Petrie S, Javahery G, Fox A, Bohme DK (1993) J Phys Chem 97:5607
127. Kemper PR, Bowers MT (1991) J Phys Chem 95:5134
128. Bowers MT, Kemper PR, von Helden G, van Koppen PAM (1993) Science 260:1446
129. von Helden G, Hsu MT, Gotts N, Bowers MT (1993) J Phys Chem 97:8182
130. Hunter J, Fye J, Jarrold MF (1993) Science 260:784
131. Hunter J, Fye J, Jarrold MF (1993) J Chem Phys 99:1785
132. Hunter J, Fye J, Roskamp EJ, Jarrold MF (1994) J Phys Chem 98:1810
133. Lee SH, Gotts N, von Helden G, Bowers MT (1997) J Phys Chem A 101:2096
134. Boissel P, de Parseval P, Marty P, Levere G (1997) J Chem Phys 106:4973
135. Dunbar RC (2000) Int J Mass Spectrom 200:571
136. Ekern SP, Marshall AG, Szczepanski J, Vala M (1997) Astrophys J 488:L39
137. Ekern SP, Marshall AG, Szczepanski J, Vala M (1998) J Phys Chem A 102:3498
138. Dibben MJ, Kage D, Szczepanski J, Eyler JR, Vala M (2001) J Phys Chem A 105:6024
139. Guo X, Sievers HL, Grützmacher HF (1999) Int J Mass Spectrom 185/187:1
140. Joblin C, Masselon C, Boissel P, de Parseval P, Martinovic S, Muller J-F (1997) Rapid Commun Mass Spectrom 11:1619
141. Sun J, Caltapanides S, Grützmacher HF (1998) J Phys Chem A 102:2408
142. Guo X, Grützmacher HF (1999) J Am Chem Soc 121:4485
143. Guo X, Grützmacher HF (2000) J Phys Chem A 104:7811
144. Guo X, Grützmacher HF, Nibbering NMM (2000) Eur J Mass Spectrom 6:357
145. Oomens J, van Roij AJA, Meijer G, von Helden G (2000) Astrophys J 542:404
146. Oomens J, Meijer G, von Helden G (2001) J Phys Chem A 105:8302
147. Oomens J, Sartakov BG, Tielens AGGM, von Helden G (2001) Astrophys J 560:L99
148. Sloan GC, Hayward TL, Allamandola LJ, Bregman JD, DeVito B, Hudgins DM (1999) Astrophys J 513:L65
149. Piest JA, von Helden G, Meijer G (1999) Astrophys J 520:L75
150. Piest JA, Oomens J, Bakker J, von Helden G, Meijer G (2001) Spectrochim Acta A 57:717
151. Kim H-S, Wagner DR, Saykally RJ (2001) Phys Rev Lett 86:5691
152. Brechignac P, Pino T (1999) Astron Astrophys 343:L49
153. Pino T, Boudin N, Brechignac P (1999) J Chem Phys 111:7337
154. Brechignac P, Pino T, Boudin N (2001) Spectrochim Acta A 57:745
155. Pino T, Brechignac P, Dartoid E, Demyk K, d'Hendecourt L (2001) Chem Phys Lett 339:64
156. Boudin N, Pino T, Brechignac P (2001) J Mol Struct 563:209
157. Gutcheck RA, Zipf EC (1973) J Geophys Res 78:542
158. Mitchell JBA, Forand JL, Ng CT, Levac DP, Mitchell RE, Mul PM, Claeys W, Sen A, McGowan JW (1983) Phys Rev Lett 51:885
159. Vallee F, Rowe BR, Gomet JC, Queffelec JL, Morlais M (1986) Chem Phys Lett 124:317
160. Rowe BR, Vallee F, Queffelec JL, Gomet JC, Morlais M (1988) J Chem Phys 88:845
161. Adams NG, Herd CR, Smith D (1989) J Chem Phys 91:963
162. Adams NG (1992) Adv Gas Phase Ion Chem 1:271
163. Abouelaziz H, Gomet JC, Pasquerault D, Rowe BR, Mitchell JBA (1993) J Chem Phys 99:237
164. Laube S, Le Padellec A, Sidko O, Rebrion-Rowe C, Mitchell JBA, Rowe BR (1998) J Phys B 31:2111
165. Lehfaoui L, Rebrion-Rowe C, Laube S, Mitchell JBA, Rowe BR (1997) J Chem Phys 106:5406
166. Rebrion-Rowe C, Lehfaoui L, Rowe BR, Mitchell JBA (1998) J Chem Phys 108:7185
167. Rebrion-Rowe C, Mostefaoui T, Laube S, Mitchell JBA (2000) J Chem Phys 113:3039
168. Rebrion-Rowe C, Le Garrec JL, Hassouna M, Travers D, Rowe BR (2003) Int J Mass Spectrom 223/224:237
169. Biondi MA (1973) Comments At Mol Phys 4:85
170. Bottcher C (1978) J Phys B 11:3887
171. Mitchell JBA, Rebrion-Rowe C (1997) Int Rev Phys Chem 16:201
172. Plasil R, Glosik J, Zakouril P (1999) J Phys B 32:3575
173. Glosik J, Bano G, Plasil R, Luca A, Zakouril P (1999) Int J Mass Spectrom189:103

174. Glosik J, Plasil R (2000) J Phys B 33:4483
175. Glosik J, Plasil R, Zakouril P, Poterya V (2001) J Phys B 34:2781
176. Adams NG, Smith D (1988) Astrophys Space Sci Lib 146:173
177. Spanel P, Smith D (1995) Plasma Sources Sci Tech 4:302
178. Adams NG, Babcock LM (2000) Dissociative recombination: theory, experiment and applications IV. Proceedings of the Conference, Stockholm, Sweden, 16–20 June 1999, p 190
179. Rowe BR, Canosa A, Le Page V (1995) Int J Mass Spectrom Ion Process 149:573
180. Gougousi T, Johnsen R, Golde MF (1997) J Chem Phys 107:2440
181. Gougousi T, Johnsen R, Golde MF (1997) J Chem Phys 107:2430
182. Skrzypkowski M, Gougousi T, Johnsen R, Golde MF (1998) J Chem Phys 108:8400
183. Herd CR, Adams NG, Smith D (1990) Astrophys J 349:388
184. Adams NG, Herd CR, Geoghegan M, Smith D, Canosa A, Gomet JC, Rowe BR, Queffelec JL, Morlais M (1991) J Chem Phys 94:4852
185. Williams TL, Adams NG, Babcock LM, Herd CR, Geoghegan M (1996) Mon Not R Astron Soc 282:413
186. Larsson M (1995) Int J Mass Spectrom Ion Process 149:403
187. Larsson M (1995) Phys Scr T T59:270
188. Larsson M (1997) Annu Rev Phys Chem 48:151
189. Andersen LH (2000) Dissociative recombination: theory, experiment and applications IV. Proceedings of the Conference, Stockholm, Sweden, 16–20 June 1999, p 230
190. Larsson M (2001) Adv Gas Phase Ion Chem 4:179
191. Semaniak J, Larson A, Le Padellec A, Stromholm C, Larsson M, Rosen S, Peverall R, Danared H, Djuric N, Dunn GH, Datz S (1998) Astrophys J 498:886
192. Larsson M, Danared H, Mowat JR, Sigray P, Sundstroem G, Brostroem L, Filevich A, Kaellberg A, Mannervik S (1993) Phys Rev Lett 70:430
193. Neau A, Al Khalili A, Rosen S, Le Padellec A, Derkatch AM, Shi W, Vikor L, Larsson M, Semaniak J, Thomas R, Nagard MB, Andersson K, Danared H, Af Ugglas M (2000) J Chem Phys 113:1762
194. Larson A, Le Padellec A, Semaniak J, Stromholm C, Larsson M, Rosen S, Peverall R, Danared H, Djuric N, Dunn GH, Datz S (1998) Astrophys J 505:459
195. Vikor L, Al-Khalili A, Danared H, Djuric N, Dunn GH, Larsson M, Le Padellec A, Rosen S, Af Ugglas M (1999) Astron Astrophys 344:1027
196. Semaniak J, Minaev BF, Derkatch AM, Hellberg F, Neau A, Rosen S, Thomas R, Larsson M, Danared H, Paal A, Af Ugglas M (2001) Astrophys J Suppl Ser 135:275
197. Derkatch AM, Al-Khalili A, Vikor L, Neau A, Shi W, Danared H, Af Ugglas M, Larsson M (1999) J Phys B 32:3391
198. Vejby-Christensen L, Andersen LH, Heber O, Kella D, Pedersen HB, Schmidt HT, Zajfman D (1997) Astrophys J 483:531
199. Jensen MJ, Bilodeau RC, Heber O, Pedersen HB, Safvan CP, Urbain X, Zajfman D, Andersen LH (1999) Phys Rev A 60:2970
200. Jensen MJ, Bilodeau RC, Safvan CP, Seiersen K, Andersen LH, Pedersen HB, Heber O (2000) Astrophys J 543:764
201. Andersen LH (2002) Personal communication
202. Larsson M (2002) Personal communication
203. Larsson M, Thomas R (2001) Phys Chem Chem Phys 3:4471
204. Herbst E, Lee H-H (1997) Astrophys J 485:689
205. Adams NG, Herd CR, Geoghegan M, Smith D, Canosa A (1991) J Chem Phys 94:4852
206. Adams NG, Babcock LM (1994) Astrophys J 434:184
207. Butler JM, Babcock LM, Adams NG (1997) Mol Phys 91:81
208. Datz S, Thomas R, Rosen S, Larsson M, Derkatch AM, Hellberg F, van der Zande W (2000) Phys Rev Lett 85:5555
209. Larsson M (2000) Philos Trans R Soc London Ser A 358:2433
210. Peart B, Dolder KT (1974) J Phys B 7:1948
211. Auerbach D, Cacak R, Caudano R, Gaily TD, Keyser CJ, McGowan JW, Mitchell JBA, Wilk SFJ (1977) J Phys B 10:3779

212. Mitchell JBA, Forand JL, Ng CT, Levac DP, Mitchell RE, Mul PM, Claeys W, Sen A, McGowan JW (1983) Phys Rev Lett 51:885
213. Mathur D, Khan SU, Hasted JB (1978) J Phys B 11:3615
214. Macdonald JA, Biondi MA, Johnsen R (1984) Planet Space Sci 32:651
215. Glosik J, Plasil R, Poterya V, Kudrna P, Tichy M (2000) Chem Phys Lett 331:209
216. Glosik J, Plasil R, Poterya V, Kudrna P, Tichy M, Pysanenko A (2001) J Phys B 34:L485
217. Smith D, Adams NG (1984) Astrophys J 284:L13
218. Adams NG, Smith D, Alge E (1984) J Chem Phys 81:1778
219. Canosa A, Rowe BR, Mitchell JBA, Gomet JC, Rebrion C (1991) Astron Astrophys 248:L19
220. Canosa A, Gomet JC, Rowe BR, Mitchell JBA, Queffelec JL (1992) J Chem Phys 97:1028
221. Smith D, Spanel P (1993) Int J Mass Spectrom Ion Process 129:163
222. Smith D, Spanel P (1993) Chem Phys Lett 211:454
223. Gougousi T, Johnsen R, Golde MF (1995) Int J Mass Spectrom Ion Process 149:131
224. Amano T (1988) Astrophys J 329:L121
225. Amano T (1990) J ChemPhys 92:6492
226. Larsson M, Danared H, Mowat JR, Sigray P, Sundstroem G, Brostroem L, Filevich A, Kaellberg A, Mannervik S (1993) Phys Rev Lett 70:430
227. Strasser D, Lammich L, Krohn S, Lange M, Kreckel H, Levin J, Schwalm D, Vager Z, Wester R, Wolf A, Zajfman D (2001) Phys Rev Lett 86:779
228. Jensen MJ, Pedersen HB, Safvan CP, Seiersen K, Urbain X, Andersen LH (2001) Phys Rev A 63:052701/1
229. Orel AE, Schneider IF, Weiner AS (2000) Philos Trans R Soc London Ser A 358:2445
230. Cordonnier M, Uy D, Dickson RM, Kerr KE, Zhang Y, Oka T (2000) J Chem Phys 113:3181
231. Kokoouline V, Greene CH, Esry BD (2001) Nature 412:891
232. Aoki K (2000) Chem Phys Lett 323:55
233. Tulej M, Kirkwood DA, Pachkov M, Maier JP (1998) Astrophys J 506:L69
234. Ruffle DP, Bettens RPA, Terzieva R, Herbst E (1999) Astrophys J 523:678
235. McCall BJ, Thorburn J, Hobbs LM, Oka T, York DG (2001) Astrophys J 559:L49
236. Herbst E (1981) Nature 289:656
237. Petrie S (1996) Mon Not R Astron Soc 281:137
238. Petrie S, Herbst E (1997) Astrophys J 491:210
239. Wang L, Pettiette CL, Conceivao J, Chesnovsky O, Smalley RE (1991) Chem Phys Lett 182:5
240. Smith D, Spanel P, Märk TD (1993) Chem Phys Lett 213:202
241. Smith D, Spanel P (1996) J Phys B 29:5199
242. Canosa A, Parent DC, Pasquerault D, Gomet JC, Laube S, Rowe BR (1994) Chem Phys Lett 228:26
243. Moustefaoui Y, Rebrion-Rowe C, Le Garrec JL, Rowe BR, Mitchell JBA (1998) Faraday Discuss 109:71
244. Dunning FB (1995) J Phys B 28:1645
245. Barckholtz C, Snow TP, Bierbaum VM (2001) Astrophys J 547:L171
246. Dua S, Bowie JH, Blanksby SJ (1999) Eur J Mass Spectrom 5:309
247. Blanksby SJ, Dua S, Bowie JH, Schröder D, Schwarz H (1998) J Phys Chem A 102:9949
248. Dua S, Blanksby SJ, Bowie JH (1998) Chem Commun 1767
249. Blanksby SJ, Dua S, Bowie JH (1999) J Phys Chem A 103:5161
250. Blanksby SJ, Bowie JH (1999) Mass Spectrom Rev 18:131
251. Blanksby SJ, Dua S, Bowie JH (1999) Rapid Commun Mass Spectrom 13:2249
252. Dua S, Blanksby SJ, Bowie JH (2000) Rapid Commun Mass Spectrom 14:118
253. Dua S, Blanksby SJ, Bowie JH (2000) J Phys Chem A 104:77
254. Blanksby SJ, McAnoy AM, Dua S, Bowie JH (2001) Mon Not R Astron Soc 328:89
255. Gloeckler G, Geiss J, Balsiger H, Fisk LA, Galvin AB, Ipavich FM, Ogilvie KW, Vonsteiger R, Wilken B (1993) Science 261:70
256. Gloeckler G (1996) Space Sci Rev 78:335
257. Gloeckler G, Geiss J (1998) Space Sci Rev 86:127
258. Kallenbach R, Geiss J, Gloeckler G, von Steiger R (2000) Astrophys Space Sci 274:97
259. Gloeckler G, Geiss J (2001) Space Sci Rev 97:169

260. Boschler P, Fischler J, Wimmer-Schweingruber RF, Geiss J, Kallenbach R (1998) Space Sci Rev 86:497
261. Hilchenbach M (2002) Int J Mass Spectrom 215:113
262. Hoppe P, Amari S, Zinner E, Ireland T, Lewis RS (1994) Astrophys J 430:870
263. Zinner E, Amari S, Wopenka B, Lewis RS (1995) Meteoritics 30:209
264. Zolensky M, Pieters C, Clark B, Papike JJ (2000) Meteorit Planet Sci 35:9
265. Stephan T (2001) Planet Space Sci 49:859
266. Hirota T, Yamamoto S, Kawaguchi K, Sakamoto A, Ukita N (1999) Astrophys J 520:895
267. Bockelee-Morvan D, Lis DC, Wink JE, Despois D, Crovisier J, Bachiller R, Benford DJ, Biver N, Colom P, Davies JK, Gerard E, Germain B, Houde M, Mehringer D, Moreno R, Paubert G, Phillips TG, Rauer H (2000) Astron Astrophys 353:1101
268. Eberhardt P (1999) Space Sci Rev 90:45
269. Altwegg K, Balsiger H, Geiss J (1999) Space Sci Rev 90:3
270. Brownlee DE, Tsou P, Clark B, Hanner MS, Horz F, Kissel J, McDonnell JAM, Newburn RL, Sandford S, Sekanina Z, Tuzzolino AJ, Zolensky M (2000) Meteorit Planet Sci 35:A35
271. Farley KA, Mukhopadhyay S (2001) Science 293:U1
272. Isozaki Y (2001) Science 293:U3
273. Braun T, Osawa E, Detre C, Toth I (2001) Chem Phys Lett 348:361
274. Becker L, Poreda RJ (2001) Science 293:U3
275. Becker L, Poreda RJ, Bunch TE (2000) Proc Natl Acad Sci USA 97:2979
276. Becker L, Poreda RJ, Hunt AG, Bunch TE, Rampino M (2001) Science 291:1530
277. Becker L, Poreda RJ, Bada JL (1996) Science 272:249
278. Becker L, Bunch TE (1997) Meteorit Planet Sci 32:479
279. Becker L, Bunch TE, Allamandola LJ (1999) Nature 400:227
280. Balser DS, Bania TM, Rood RT, Wilson TL (1999) Astrophys J 510:759
281. Saunders M, Jiménez-Vázquez HA, Cross RJ, Mroczkowski S, Gross ML, Giblin DE, Poreda, RJ (1994) J Am Chem Soc 116:2193
282. Weiske T, Bohme DK, Hrusak J, Krätschmer W, Schwarz H (1991) Angew Chem Int Ed Engl 30:884
283. Ross MM, Callahan JH (1991) J Phys Chem 95:5720
284. Caldwell KA, Giblin DE, Hsu C, Cox D, Gross ML (1991) J Am Chem Soc 113:8519
285. Baser Y, Wan Z, Christian JF, Anderson SL (1994) Int J Mass Spectrom Ion Process 138:173
286. Petrie S, Javahery G, Bohme DK (1993) Astron Astrophys 271:662

Top Curr Chem (2003) 225: 77–129
DOI 10.1007/b10477

Transient Intermediates of Chemical Reactions by Neutralization-Reionization Mass Spectrometry

František Tureček

Department of Chemistry, Bagley Hall, Box 351700, University of Washington, Seattle, WA 98195-1700, USA. *E-mail: turecek@chem.washington.edu*

Neutralization-reionization mass spectrometry (NRMS) is a powerful method for the generation of highly reactive molecules, radicals, biradicals, ylids, and organometallics in the gas phase. This chapter brings a brief overview of recent instrumentation developments and discusses the chemical properties of several transient intermediates of relevance to inorganic, organic, and organometallic chemistry and biochemistry. Enols, organic and inorganic acids, carbenes, carbon clusters, heterocumulenes and related unsaturated molecules, and ylids are some of the closed-shell molecules that were generated and studied by NRMS. Hypervalent radicals, oxygenated boron, nitrogen, silicon, phosphorus, and sulfur radicals, heterocyclic and nucleobase radicals, and amide radicals are examples of open shell systems that have been successfully addressed by NRMS.

Keywords. Neutralization-reionization mass spectrometry, Ion chemistry, Collisional electron transfer, Franck-Condon effects, Reactive intermediates

List of Abbreviations

CAD	Collisionally activated dissociation
DMDS	Dimethyldisulfide
EA	Electron affinity
IE	Ionization energy
NCR	Neutralization-collisional activation-reionization
NDMA	*N,N*-Dimethylaniline
NRMS	Neutralization-reionization mass spectrometry
PA	Proton affinity
PES	Potential energy surface
TMA	Trimethylamine
TS	Transition state

1 Introduction

"How does this reaction proceed?" is undoubtedly one of the most tantalizing questions in chemistry. It is fairly accurate to say that most chemical reactions do not proceed as elementary processes involving just the reactant(s), transition state, and product(s), but often involve one or more intermediate chemical species, which are intrinsically stable but highly reactive and hence elusive. The structure and bonding properties of a reactive intermediate can provide important information on the reaction mechanism, and its energy relative to those of the reactants and products represents the lower bound of the potential energy barriers along the reaction path.

Classical approaches to studying reactive intermediates, as used in physical organic and inorganic chemistry, involve running the reaction of interest and attempting to trap the intermediate by another chemical reaction, so that it is converted to a stable species that can be characterized by spectroscopic methods

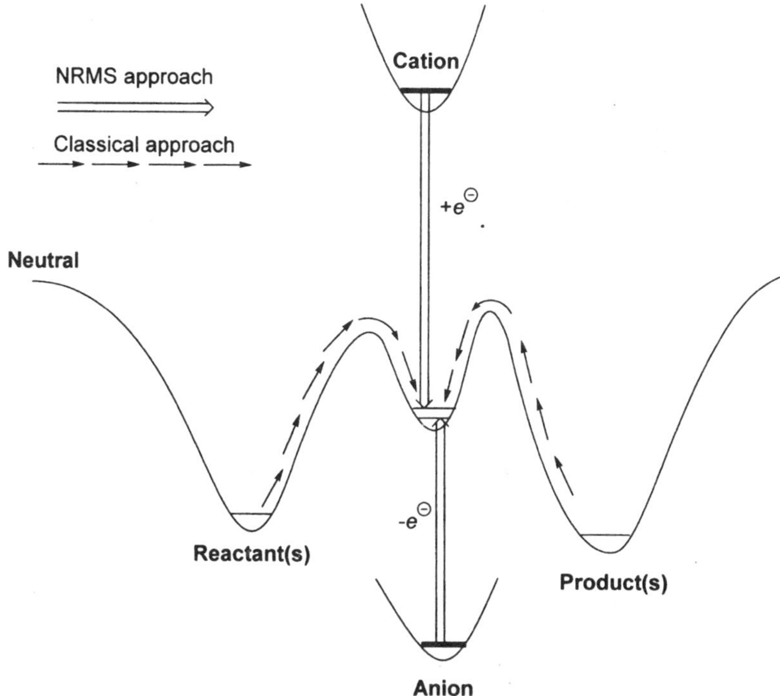

Fig. 1. Schematic potential energy surface for the classical and NR approach to a short-lived intermediate

(Fig. 1). Another approach is to use a fast and selective spectroscopic method that provides specific information about some properties, typically electronic or vibrational, of the intermediate. Alternatively, the intermediate can be trapped in an inert frozen matrix and studied spectroscopically at very low temperatures.

This chapter discusses an entirely different approach to the generation and investigation of highly reactive transient intermediates. The high reactivity is usually due to an unusual electron distribution in the intermediate that was acquired in the course of the chemical reaction. This implies that for an electron rich intermediate there is a corresponding stable cation in which the electron density was lowered by ionization. Likewise, for an electron-deficient intermediate there is a corresponding stable anion in which the electron deficiency was alleviated by electron attachment. Equations (1) and (2) show simple examples of the methoxy and hydroxymethyl radicals, respectively, which are isomeric transient intermediates of hydrogen atom abstraction from methanol:

$$CH_3O^- \rightarrow CH_3O^\bullet \tag{1}$$

$$^+CH_2OH \rightarrow {}^\bullet CH_2OH \tag{2}$$

The corresponding methoxide anion and hydroxymethyl cation are readily prepared by the methods of gas-phase ion chemistry and are very stable as isolated species in the gas phase. Thus, to produce $CH_3O^\bullet$ one needs to neutralize the anion by removing an electron from CH_3O^- [1, 2], whereas to produce CH_2OH requires adding an electron to $^+CH_2OH$ [3]. The electron removal, which is equivalent to *one-electron oxidation*, is accomplished by collisional electron detachment from a fast anion traveling at 10^5 m s^{-1} velocity which corresponds to a kiloelectron volt kinetic energy. Neutralization of cations, which is equivalent to *one-electron reduction*, is carried out by collisional electron transfer from a suitable molecular or atomic donor to the cation having a keV kinetic energy. The fast neutral intermediate is allowed to drift for 100 ns–5 μs when dissociation can occur. The neutral products are ionized to cations or anions that are separated by a mass spectrometer and detected to provide a neutralization-reionization (NR) mass spectrum [4]. The spectrum provides information on the fraction of undissociated neutral intermediates that appear as survivor ions of the same mass to charge ratio (*m/z*). In addition, dissociation products originating from the transient neutral intermediate are detected and used to identify unimolecular reactions that occur on the 100 ns–5 μs time scale following collisional electron transfer or detachment. NR spectra thus provide more or less complete product analysis that can be used to characterize the structure of the neutral intermediate, its electronic state, and vibrational excitation. Neutralization-reionization mass spectrometry was introduced in the late 1970s [5] and the various aspects of the method have been reviewed [6–16]. NR allows for four combinations of ion precursor and detection product polarities that are denoted as $^+NR^+$, $^+NR^-$, $^-NR^+$, and $^-NR^-$ [2] and shown in Scheme 1.

$^+NR^+$ [Cation Precursor]$^+$ → [Neutral]^0 → [Cation Detected]$^+$

$^+NR^-$ [Cation Precursor]$^+$ → [Neutral]^0 → [Anion Detected]$^-$

$^-NR^+$ [Anion Precursor]$^-$ → [Neutral]^0 → [Cation Detected]$^+$

$^-NR^-$ [Anion Precursor]$^-$ → [Neutral]^0 → [Anion Detected]$^-$

Scheme 1

The selection of the cationic or anionic precursor is dictated by the ion stability and accessibility by ion chemistry methods. The selection of the ion polarity for reionization and detection is mainly dictated by the ion stability and ionization efficiency, as discussed in the next section.

2 Collisional Electron Transfer

Collisional electron transfer and detachment are characterized by three parameters, i.e., time scale, energy, and cross section. Consider a precursor ion of 10 Å diameter traveling at 10^5 m s^{-1} velocity that undergoes a glancing collision with a thermal target of 5 Å diameter. The interaction time is given approximately by $(5+10+5)\times10^{-10}/10^5=2\times10^{-14}$ s, taking into account that electron transfer occurs at a close approach of the donor-acceptor pair and its probability decreases exponentially with distance. The interaction time is comparable to the vibrational period (T) of fastest molecular vibrations, e.g., that of an O–H bond (ν=3600 cm^{-1}) which has $T=1/c\nu=9.3\times10^{-15}$ s, but is faster than C–C and C–O bond vibrations of T=2.9 to 3.3×10^{-14} s. This implies that during the collision the heavy nuclei do not have time to change interatomic distances, so that the emerging neutral intermediate is formed with the structure and geometry of the precursor ion. From the point of view of the Fig. 1 energy diagram, this is equivalent to a vertical transfer from the ion potential energy surface to that of the neutral intermediate. Collisional electron detachment from an anion can also occur by fast electron transfer provided the thermal acceptor has a bound state for the incoming electron. This may be the case with oxygen, which has been a popular collision gas used for neutralization of anions. Alternatively, the collision can produce an excited state of the anion that spontaneously detaches an electron in a process analogous to autoionization of highly excited molecules [17]. The time scale for such a spontaneous detachment depends on the electronic properties of the anion and may vary for different species. Little is known about the lifetimes of superexcited anions, and electron detachment is usually viewed as a vertical process.

Recently, a comparison has been made between the neutral formation by laser photodetachment and that by collision-induced electron detachment from the nitromethyl anion, $^{-}CH_2NO_2$ [18]. The nitromethyl radical intermediate was analyzed following collisional ionization to cations. Photoelectron detachment from $^{-}CH_2NO_2$ with 488-nm photons from an argon-ion laser is nearly thermoneutral because of the energy balance between the electron affinity of the radical, EA(CH_2NO_2)=IE($^{-}CH_2NO_2$)=2.475 eV, [19] and the photon energy (2.54 eV), whereas radical excitation through collisional electron detachment is a priori unknown. The difference between the $^{-}NR^{+}$ spectra obtained by laser photodetachment and collisional neutralization (Fig. 2) shows more fragmentation resulting from the collisional process and indicates additional excitation of the neutral intermediate by collisional electron detachment. Hence, photoelectron detachment can be the method of choice for generating extremely fragile radicals from stable anions.

2.1
Neutral Excitation and Franck-Condon Effects

Vertical oxidation or reduction by fast collisions can cause excitation in the neutral intermediate. The excitation energy (E_{exc}) can be expressed as consisting of three terms (Eq. 3):

$$E_{exc} = E_{ion} + E_{FC} + E_{el} \quad (3)$$

where E_{ion} is the precursor ion vibrational energy, E_{FC} is the vibrational excitation caused by Franck-Condon effects, and E_{el} is the electronic excitation due to electron transfer. The source of E_{exc} has been attributed to the energy balance in electron transfer (Eq. 4):

$$E_{exc} \geq -\Delta E = RE_{ion} - IE_{donor} \quad (4)$$

where RE_{ion} is the vertical recombination energy of the precursor ion, taken as a positive value, and IE_{donor} is the vertical ionization energy of the donor atom or

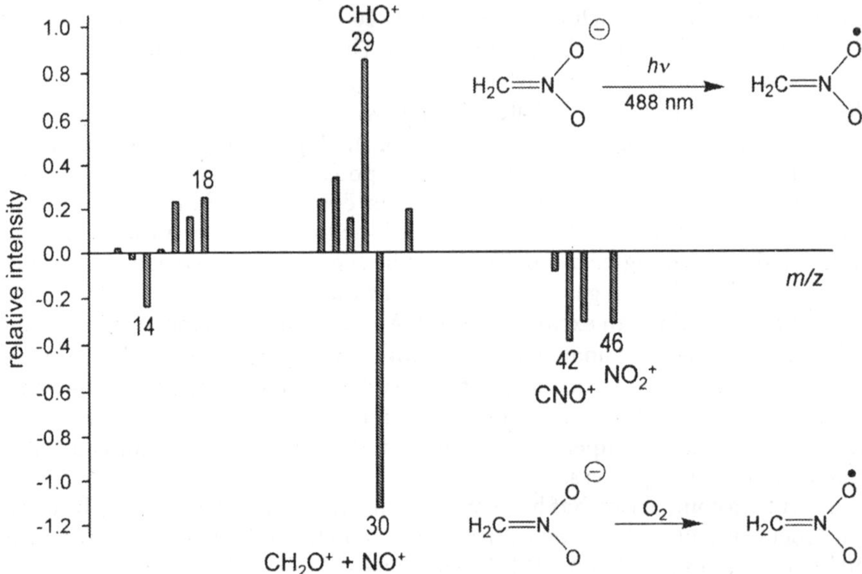

Fig. 2. Laser photodetachment-reionization of $^-CH_2NO_2$ plotted as a difference of laser-on and laser-off measurements. The *upward peaks* are due to dissociations upon photodetachment followed by reionization with O_2 and cation detection. The *downward peaks* are due to dissociations upon collisional detachment due to residual O_2 followed by reionization with O_2 and cation detection

molecule. Equation (4) holds approximately for exothermic electron transfer from atomic donors, e.g., alkali metal atoms, which have low ionization energies, so that $RE_{ion}-IE_{donor}>0$, while the alkali metal cations have high excitation energies and cannot absorb the excess E_{exc}. Equation (4) is less useful for collisional electron transfer from molecular donors, which is often endothermic and yet can result in substantial excitation of the transient neutral formed. E_{exc} in endothermic electron transfer originates from inelastic collisions that result in the conversion of a small fraction of the keV center-of-mass collision energy into the internal energy of the products.

The theory of collisional energy transfer at keV kinetic energies is not well developed [20], and so most of the currently available data on E_{exc} originate from rather crude measurements. Kinetic energy loss in NR was measured to provide the combined energy balance in electron transfer collisions [21, 22]. The distribution of internal energy acquired upon NR has been studied for some model systems using the ion thermometer method introduced by Griffiths et al. [23]. Measurements for NR of $W(CO)_6^+$ [24] and $W(CO)_n^-$ (n=2–5) [25] showed that the deposited energy had a broad distribution that peaked at ca. 4 eV and extended to >15 eV. CH_3OH^+ and $CH_3NH_2^+$ were used as ion thermometers to gauge the energy distribution on NR [26]. This study concluded that, regardless of the energy balance in the neutralizing collision, which was varied from 0.44 eV endothermic to 2.19 eV exothermic, the internal energy distributions in the NR formed ions peaked between 80–200 and 140–320 kJ mol^{-1} for CH_3OH^+ and $CH_3NH_2^+$, respectively, but also showed tails to higher energies [26].

More recently, the internal energy distribution in neutral intermediates was studied using kinetic isotope effects on neutral dissociations [27–30]. It was shown that the internal energy distribution of the ground-electronic state of a neutral intermediate formed by endothermic electron transfer can be expressed by the function shown in Eq. (5):

$$P(E) = \frac{4(E-E)}{W^2} e^{\frac{2(E-E_0)}{W}} \qquad (5)$$

where E_0 is the energy onset and W is a width parameter, such that the most probable internal energy is $E_{max}=E_0+W/2$ and the population mean energy is $<E>=E_0+W$. Importantly, it was found for several systems that $E_{max}\approx E_{ion}+E_{FC}$, which provided a simple formula for estimating the internal energy that the neutral acquired upon collisional electron transfer.

Equation (3) shows the importance of Franck-Condon effects in NR. Franck-Condon effects are due to a mismatch between the ion and neutral potential energy surfaces, such that the equilibrium nuclear configuration (i.e., structure) of the ion, corresponding to its $v'=0$ vibrational states, is different from the equilibrium structure of the neutral. Thus, an ion undergoing vertical transition from the $v'=0$ state lands on the multidimensional wall of the neutral potential energy surface, so that some of the vibrational degrees of freedom in the neutral are excited to $v>0$ states. The reason for the excitation is different bond lengths and angles in the ion and neutral. In particular, changes upon neutralization in hybridization and bond orders often cause large Franck-Condon effects and result in vibrational excitation of the neutral intermediate. The excitation energy can be large enough to exceed the energy barrier for a dissociation and drive fast decomposition of the neutral intermediate. For example, the acetyl radical, CH_3CO, is a bound species that can be produced by gas-phase photolysis of acetone [31]. When generated by collisional neutralization of the stable acetyl cation, CH_3CO^+, the radical undergoes fast dissociation to CH_3 and CO, so that no survivor ions are detected in the $^+$NR$^+$ mass spectrum [32]. In contrast, when CH_3CO is formed as a neutral product of an ion dissociation which does not involve electron transfer, the radical is stable and can be detected as CH_3CO^+ after collisional reionization.

Changes in hybridization and bond orders often occur upon reduction of even-electron cations, such as $CH_3SO_2^+$ [33], $P(OH)_4^+$[34], $(CH_3)_2SiOH^+$ [35], $C(OH)_3^+$ [29], $HC(OH)NH_2^+$ [36], $CH_3C(OH)NHCH_3^+$ [37], etc. These cations show relatively short X–O bonds (X=S, P, Si, C) due to back donation of oxygen p-electrons to the vacant p_z orbital on the central atom X. Upon reduction, the incoming electron enters an antibonding π orbital which in the relaxed radical structure causes pyramidization at X and lengthening of the X–O bonds. However, vertical electron transfer produces the radical with the geometry of the cation, so that the X–O stretching vibrations are excited by bond compression. In addition, planarization about X in the vertically formed radical excites the umbrella vibrational mode which contributes to the overall vibrational energy. Table 1 shows some examples of radical vibrational excitation caused by Franck-Con-

Table 1. Franck-Condon energies in neutralization of cations

Ion	Neutral	Franck-Condon Energy (kJ mol^{-1})
OH, 1.580, Si^+, H_3C, CH_3 — planar	H_3C, Si, 1.664, OH, H_3C — pyramidal	229
OH, 1.285, C^+, HO, OH — planar	HO, C, 1.373, OH, HO — pyramidal	150
HO, OH, 1.563, P^+, HO, OH — D_2	OH, 1.685, HO, P, OH, HO, 1.688 — C_s	129
CH_3, 1.420, S^+, O, O — planar	1.473, O, S, CH_3, O — pyramidal	141
OH, 1.285, C^+, H, NH_2 — planar	H, C, 1.374, OH, H_2N — pyramidal	90-100

don effects, as calculated for vertical neutralization of cations. The changes in bond lengths between the cations and corresponding radicals are on the order of 4–8% and contribute >50% of the overall vibrational excitation energy. Excitation in the deformation modes contributes about 30–45% depending on the particular system [29].

Less is known about Franck-Condon effects on collisional electron detachment from anions. For example, the C–N bond in CH_2NO_2 (1.401 Å) is 4.5% longer than the same bond in the $^-CH_2NO_2$ anion (1.340 Å), and vertical neutralization results in 25–30 kJ mol^{-1} vibrational excitation in the radical, as calculated at several levels of theory [18]. Alkoxy radicals and anions also show different C–O bond lengths, e.g., 1.317 Å and 1.371 Å in the 1-pentoxy anion and radical, respectively [1], although the Franck-Condon energies associated with this bond length change have not been reported. The photoelectron spectrum of $C_2H_5O^-$ shows a major peak for the 0→0′ transition [38], indicating small Franck-Condon effects on vertical electron detachment.

2.2
Excited Electronic States

Another interesting and important feature of collisional reduction of cations is that the reducing electron can enter a high molecular orbital corresponding to an excited state of the neutral molecule, radical, or biradical. The types of excited states are depicted in Fig. 3. Electron capture in a Rydberg-type orbital can give

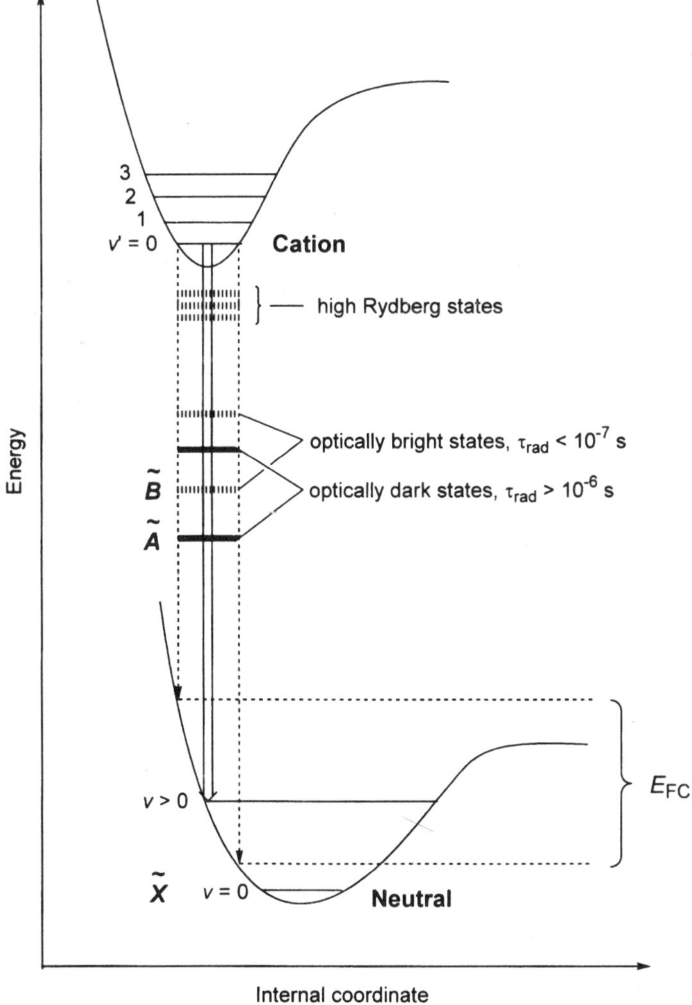

Fig. 3. Schematic energy diagram for electronic states formed by collisional electron transfer

rise to a highly excited state consisting of an ionic core and a loosely bound electron. Rydberg states are atom-like excited electronic states of large principal quantum numbers (n) that converge to the ionization limit of the neutral molecule. Rydberg states have long lifetimes that increase with n^3 [39]. There is experimental evidence for the formation of high Rydberg states of Ar atoms produced by collisional neutralization of Ar^+ [40]. The evidence stems from field ionization measurements where a small fraction of neutralized Ar atoms was ionized in a kV field gradient.

Excited electronic states play a crucial role in the chemistry of hypervalent radicals formed by collisional reduction of organic onium cations, where

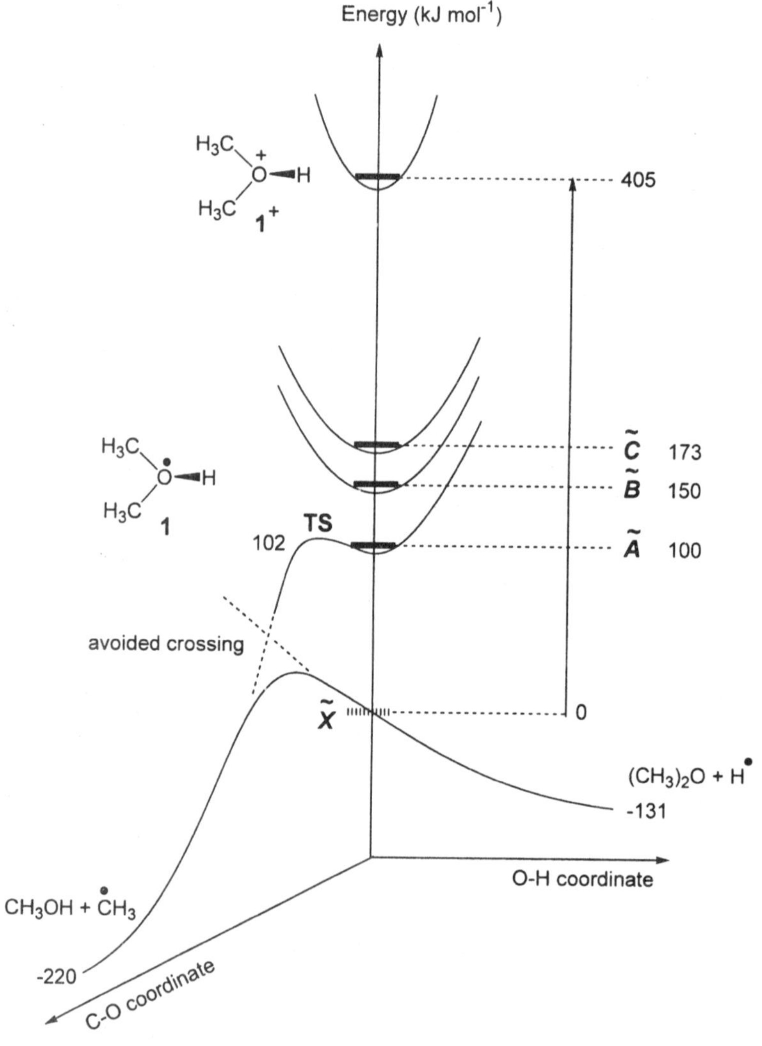

Fig. 4. Excited electronic states in $(CH_3)_2OH$ (**1**)

"onium" stands for ammonium, oxonium, sulfonium, phosphonium, etc. As a rule, these radicals are unbound or only weakly bound in the ground doublet state, but some are surprisingly long-lived [6]. As an example, let us examine the dimethyloxonium radical, $(CH_3)_2O$–H (**1**) which is formed by collisional reduction of protonated dimethyl ether [41–44]. Radical **1** is unbound in the ground $(^2A')X$ electronic state and is predicted to dissociate within one vibration to $(CH_3)_2O$ and H. In contrast, the $^+NR^+$ mass spectrum of the deuterium-labeled radical **1**-OD shows a substantial survivor ion corresponding to undissociated radical, and a fraction of CH_3OD corresponding to loss of CH_3 from **1**-OD. The metastability of **1** and the competitive cleavage of the C–O bond cannot be explained by the properties of the ground electronic state. Examination of the potential energy surfaces of the three lowest excited states, $(^2A')A$, $(^2A'')B$, and $(^2A'')C$ [44], combined with laser photoionization [43], indicated that the metastable (non-dissociating) fraction of **1**-OD was due to the *B* and higher excited electronic states produced by collisional electron transfer. The dissociation of the C–O bond resulting in loss of a methyl starts in the *A* state, overcomes a small energy barrier, and further proceeds by avoided crossing to the repulsive part of the *X* state that occurs at a C–O distance of 1.7 Å (Fig. 4).

Excited electronic states have also been considered to explain NR dissociations of heterocyclic radicals in which low-energy losses of hydrogen atoms compete with high energy ring-cleavage dissociations. The latter reactions were interpreted as starting from excited electronic states of the radicals [27, 28]. Formation of excited electronic states upon collisional electron transfer has also been studied with smaller molecular systems, e.g. CH_n [45], O_2 [46], and H_3 [47, 48].

3 Instrumentation

3.1 Sector Instruments

NRMS instrumentation is basically dictated by the necessity of performing collisions at keV kinetic energies. Both commercial and custom-designed sector mass spectrometers have been used in the majority of NRMS studies, as reviewed [9, 15]. Precursor ions are generated in an ion source that is floated at a kV potential, typically 6–10 kV, accelerated to 6–10 keV kinetic energy, and selected by a magnet mass analyzer. The fast ions are focused into the collision cell which contains a gas at a low pressure for collisional neutralization. A variety of target gases have been used, e.g., alkali metal vapors, Hg, Xe, O_2, NH_3, cyclopropane, NO, trimethylamine (TMA), *N,N*-dimethylaniline (NDMA), dimethyl disulfide (DMDS), and even larger organic molecules. A 1–10% fraction of precursor ions are neutralized and the remaining ions are deflected electrostatically. The neutral beam is allowed to drift to another cell where the neutral products are ionized by collisions with the reionization gas. Molecular oxygen (O_2) is the most common reionization gas, but molecules of high electron affinities such as NO_2 [47], ICl, and $TiCl_4$ have also been used [49]. The ions produced by collisional reionization are separated by kinetic energy or mass. The kinetic energy spectra provide good ion transmission, but suffer from low resolution, especially when the dissociations are accompanied by large kinetic energy release [10]. Mass separation by sector magnet analyzers provides a better mass resolution, but because of the sector instrument design it is available only on multisector instruments.

3.2 Quadrupole Instruments

A different experimental approach to NRMS is embodied in the tandem quadrupole acceleration-deceleration instrument that uses quadrupole mass filters for mass selection and analysis of low-energy (70–80 eV) ions whereas collisional electron transfer is carried out after ion acceleration to 4–8 keV kinetic energies (Fig. 5) [10, 50]. The reionized products are decelerated back to 70–80 eV for mass analysis. The quadrupole instrument achieves unit mass resolution of NR products and it is versatile enough to allow variable-time and photoexcitation experiments described briefly below. Coupling with soft ionization methods

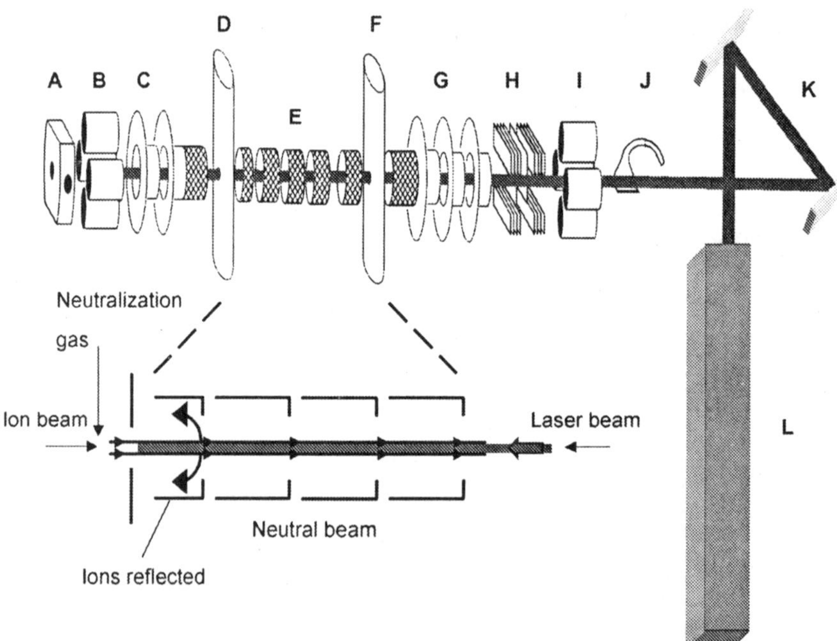

Fig. 5. Tandem quadrupole acceleration-deceleration mass spectrometer for NRMS studies. A – ion source; B – quadrupole mass analyzer; C – ion acceleration lens; D – neutralization cell; E – neutral drift region; F – reionization cell; G – ion deceleration lens; H – energy filter; I – quadrupole mass analyzer; J – off-axis ion detector; K – laser optics; L – Ar-ion laser. *Inset* shows the drift region where residual precursor ions are reflected and the neutral beam overlaps with the laser beam

such as electrospray is also feasible, because the ions have initially low kinetic energies.

4
Methods for Characterization of Neutral Intermediates

The basic NR mass spectrum contains information on the fraction of undissociated (survivor) ions and also allows one to identify dissociation products that are formed by purely unimolecular reactions. NRMS thus provides information on the intrinsic properties of isolated transient molecules that are not affected by interactions with solvent, matrix, surfaces, trace impurities, radical quenchers, etc. However, because collisional ionization is accompanied by ion excitation and dissociation, the products of neutral and post-reionization dissociations overlap in the NR mass spectra. Several methods have been developed to distinguish neutral and ion dissociations and to characterize further short lived neutral intermediates in the fast beam. Moreover, collisionally activated dissociation (CAD) spectra have been used to characterize the ions produced by collisional reionization of transient neutral intermediates [51]. This NR-CAD analysis adds another dimension to the characterization of neutral intermediates, because it allows one to uncover isomerizations that do not result in a change of mass and thus are not apparent from NR mass spectra alone.

4.1
Neutral Collisional Activation and Angle-Resolved NRMS

Collisional activation of the neutral beam is a simple means of inducing dissociations that may provide additional information on the neutral intermediates. Collisional activation is typically carried out with He under multiple collision conditions, and the fraction of ions formed collaterally with neutral activation is removed electrostatically. The acronym for collisional activation of neutral intermediates in NR experiments is NCR [52]. NCR was used to study isomerization of transient vinyl alcohol and CH_3–C–OH to acetaldehyde [53], and isomerization of cyclohexa-2,4-dien 1-one to phenol [54]. It is used rather routinely to probe the stability of neutral intermediates by recording the decrease of survivor ion relative intensity and appearance of new dissociation products. Although NCR is experimentally simple, the interpretation of NCR spectra is not always straightforward. This is because collisional activation is accompanied by scattering, so that instrument factors such as the acceptance angle can skew the data obtained on different sector instruments or even in different field-free regions of the same multi-sector mass spectrometer. Neutral collisional activation does not have to result in dissociation, and the energy deposited in the neutral intermediates can be passed on to ions after reionization and promote their dissociations.

A variation of NCR is the technique of angle-resolved NRMS in which the precursor ion beam is deflected prior to neutralization, so that neutrals back-scattered to the beam axis are reionized, analyzed, and detected [55]. NR mass spectra in angle-resolved measurements have shown that fragment relative intensities increased with the deflection angle, indicating a greater excitation of the transient neutral species at larger scattering angles. A drawback is the loss of ion intensity which can be critical in NR measurements where the overall efficiency is generally low, and the typical ion yields after NR are 0.1–0.001% of the precursor ion intensity [56–58].

4.2 Photoexcitation and Photoionization

As described in Sect. 2.2, radiatively long-lived excited electronic states (dark states) produced upon vertical electron transfer can play an important role in dissociations observed in NRMS. Efforts have been made to characterize such states experimentally in conjunction with theoretical calculations. One such approach is to study photofragmentation in NR mass spectra resulting from laser irradiation. This has been carried out on the tandem quadrupole instrument (Sect. 3.2) using the fixed 488 and 514.5-nm lines from an Ar-ion laser. Since both the neutral and reionized beam are irradiated simultaneously, ion photofragmentation must be studied separately and deconvoluted from the NR data. An obvious shortcoming of the current approach is that only a few discrete photon energies are available from the Ar ion laser. Photoexcitation probes an energy difference between two electronic states, and so at least one reference energy must be known for the transient neutral to interpret photoexcitation data. This information is usually available from time-dependent density functional theory calculations [59] that provide reasonably accurate excitation energies that allow one to calibrate the energy scale for the electronic states and assign the transitions that match the photon energy. The same considerations are valid for laser photoionization, which is experimentally difficult and has been performed for a few readily ionizable radicals only [18, 43, 60]. Photoexcitation to a high Rydberg state has been shown to be a sensitive probe of electronic states in some radicals produced by collisional electron transfer [33, 60]. High Rydberg states have long radiative lifetimes that scale with n^3, where n is the principal quantum number [39], and large collisional cross sections which scale with n^4 [60]. This results in substantially increased ionization cross sections for high Rydberg states in NRMS, so that they can be selectively detected even if the photoexcitation efficiencies are low. Because the Rydberg-series states are limited from above by the ionization limit, it and the photon energy give the upper bound for the energy of the electronic state being photoexcited.

4.3 Variable-Time Kinetics

In NR, dissociations of neutral intermediates and ions formed by collisional reionization are convoluted in the spectrum. An efficient deconvolution is achieved by varying step-wise the time scales for neutral and ion dissociations and using the measured temporal profiles of NR ion intensities to obtain phenomenological rate parameters for both neutral (k_N) and ion dissociations (k_i) [61, 62]. These variable-time measurements not only distinguish neutral and ion dissociations, but the measured rate parameters also provide quantitative information on the contribution of either process and can be used to assess branching ratios for competitive neutral and ion dissociation channels. For neutral species, dissociation of AB to A and B renders both products detectable following collisional ionization to A^+ and B^+. Variable-time measurements for both A and B formations are supposed to give matching rate parameters in each channel if the products arise in a single reaction from a single precursor, such that $k_N(A)=k_N(B)$ [63]. This match, together with the complementary masses (m), such that $m(A)+m(B)=m(AB)$, provides an unequivocal criterion for the occurrence of the particular neutral dissociation. Variable-time measurements are independent of the ion polarity and, in principle, are possible for all four $^{\pm}NR^{\pm}$ combinations.

4.4 Neutral-Ion Dissociation Difference Spectra

A different version of variable-time NRMS relies on qualitative comparison of mass spectra obtained at two neutral lifetimes, but is limited to charge reversal NR, e.g., $^-NR^+$ or $^+NR^-$ [64]. A regular NR mass spectrum measures neutral dissociations over time t_N and ion dissociations over time t_i, as determined by the precursor ion velocity and the neutral and ion path lengths. In contrast, in a charge-reversal (CR) mass spectrum, in which two electrons are added to a cation ($^+CR^-$) or detached from an anion ($^-CR^+$) in a single collision, the time scale for neutral dissociations is considered to be very short ($t_N \approx 0$), while t_i is roughly the same as in NR. Subtraction of normalized ion intensities in NR and CR spectra results in neutral-ion dissociation difference (NIDD) spectra [64], where positive peaks indicate predominant neutral dissociations and negative peaks predominant ion dissociations. Hence, NIDD spectra visualize differences that are often apparent upon comparison of normalized NR and CR mass spectra. NIDD spectra have been used for several systems that provide stable precursor anions allowing one to obtain $^-NR^+$ and $^-CR^+$ spectra. One advantage of NIDD is that the data to be compared can be obtained on commercial sector instruments equipped with two collision cells for NR measurements.

5 Transient Molecules

Generation of unusual molecules by NRMS typically, but not entirely, relies on collisional reduction of cation-radicals produced by dissociative ionization of stable molecular precursors. Owing to the advanced state of gas-phase ion chemistry, ion structures can often be unambiguously assigned to products of ion dissociations and the ions then used to generate unusual neutral molecules.

5.1 Enols

Propen-2-ol (acetone enol) and ethenol (acetaldehyde enol) were some of the first reactive molecules that were generated and characterized by NRMS [65]. Neutral enols are typically less stable than their oxo tautomers and undergo facile isomerization by acid or base catalyzed proton transfer in solution [66] or on active surfaces [67]. In contrast, enol cation-radicals are stable as isolated species in the gas phase [68] and so they provide excellent ion precursors for NR generation and study of very reactive neutral enols. A recent example concerns the generation of 2-hydroxyoxol-2-ene (**2**) which is the enol tautomer of γ-butyrolactone (**3**) [69]. The precursor cation-radical ($2^{+\bullet}$) is generated by the McLafferty rearrangement in 2-acetoxybutyrolactone which proceeds with a good yield (Scheme 2) [70]. Note that the ion dissociation is analogous to the Norrish type II photochemical rearrangement which has also been used to generate neutral enols in solution [66]. Neutralization of ion $2^{+\bullet}$ with dimethyldisulfide is endothermic, $\Delta E = IE_v(CH_3SSCH_3) - RE_v(2^{+\bullet}) = 9.0 - 7.35 = 1.65$ eV, and gives rise to stable **2** that shows a dominant peak of $2^{+\bullet}$ upon reionization.

The high stability of isolated **2** was confirmed by collisional activation ($^+$NCR$^+$) that caused only minor dissociation by elimination of water forming furan. The latter reaction is calculated to be the lowest-energy unimolecular dissociation of **2** that is 4 kJ mol^{-1} exothermic, but is kinetically hampered by an energy barrier to intramolecular hydrogen transfer [69]. The kinetic stability of isolated **2** in the gas phase contrasts its properties in aqueous solution, where the enol is predicted to react rapidly with H_3O^+ or OH^- and isomerizes to the more stable lactone **3**. The equilibrium constant for the isomerization **3**→**2** is calculated to be extremely small in water, $K_{eq} = 5.7 \times 10^{-20}$, so enol **2** would be very difficult to generate and study in solution.

The enol of glycine (**4**) has been generated by neutralization with trimethylamine of the corresponding cation-radical ($4^{+\bullet}$) prepared by dissociative ionization of isoleucine (Scheme 3) [71]. The $^+$NR$^+$ mass spectrum of $4^{+\bullet}$ showed a substantial survivor ion attesting to the stability of isolated **4**. In contrast, the survivor ion from glycine (**5**) is much less stable and appears as a very minor peak in the $^+$NR$^+$ mass spectrum of **5**, in spite of the fact that neutral **5** is thermodynamically more stable than **4**.

The enol of acetamide (**6**) is also quite stable as an isolated species, as documented by the dominant survivor ion in the $^+$NR$^+$ mass spectrum (Fig. 6). Compared to **6**, the thermodynamically more stable neutral acetamide (**7**) shows less

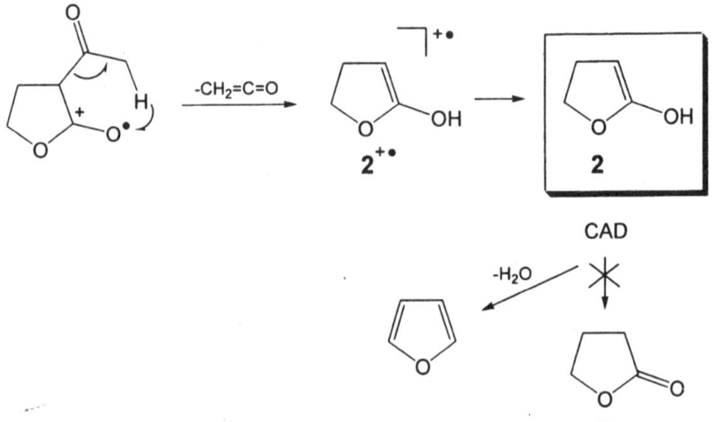

Scheme 2

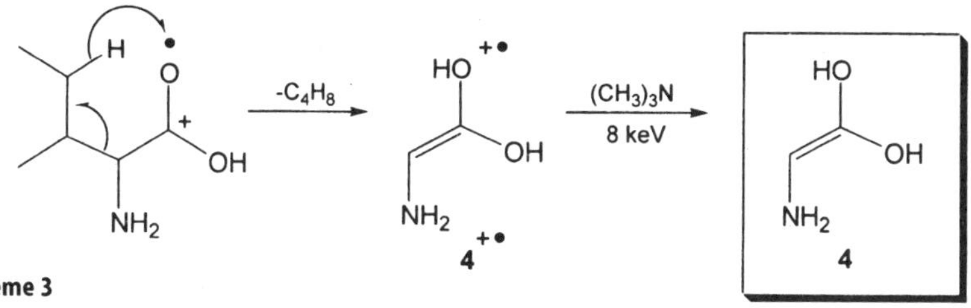

Scheme 3

abundant survivor ion (Fig. 6), which is due to the lower stability of cation-radical $7^{+\bullet}$.

Enol imines have also been generated and characterized by NRMS. For example, *N*-methylformimidic acid, CH_3-N=CH-OH (**8**) was obtained by neutralization of cation-radical $\mathbf{8}^{+\bullet}$, which was generated by dissociative ionization of gyromitrin, CH_3CH=N-N(CH_3)CH=O [72].

The $^+NR^+$ spectrum of **8** showed a small survivor ion, but differed substantially from the spectra of other C_2H_5NO isomers, e.g., **6**, **7**, *N*-methylamino(hydroxy)carbene (**9**), and *N*-methylformamide (**10**). The low intensity of survivor ions in the NR mass spectra of enol imines is due to Franck-Condon effects in collisional reionization that result in vibrational excitation of the resulting cation radical followed by dissociation. Franck-Condon effects were studied for collisional ionization of acetimidic acid, $CH_3C(OH)$=NH, which was one of the neutral dissociation products of 1-hydroxy-1-methylamino-1-ethyl radical, a hydrogen atom adduct to *N*-methylacetamide [37]. The cation-radical dissociates extensively upon reionization, and the dissociation is driven by a 74 kJ mol^{-1} Franck-Condon energy acquired by vertical ionization.

1,1-Dihydroxyethene (**11**, enol of acetic acid) and 1,1,2-trihydroxyethene (**12**, enol of glycolic acid) have been generated as transient intermediates by NR and shown to be stable as isolated molecules in the gas phase [73]. The survivor ion of **12** was shown by NR-CAD to retain the structure of trihydroxyethene cation-radical, which indicated that the enol did not isomerize to the more stable glycolic acid molecule [73]. It may be noted that enols of simple carboxylic acids, esters, and lactones have not been generated in solution and remain elusive. Thus, NRMS currently provides the only experimental approach to these highly reactive molecules. As a further note, hydroxyacetylene (**13**), which is the ynol of ketene, has been generated by collisional reduction of its cation-radical that is formed by dissociative ionization of propiolic acid. Ynol **13** is stable as an isolated molecule in the gas phase and its $^+NR^+$ mass spectrum distinguishes it readily from the more stable ketene [74]. Analogously, stable ethynamine (**14**) is formed by collisional reduction of its cation-radical that can be generated by dissociative ionization of propiolamide (Scheme 4) [75].

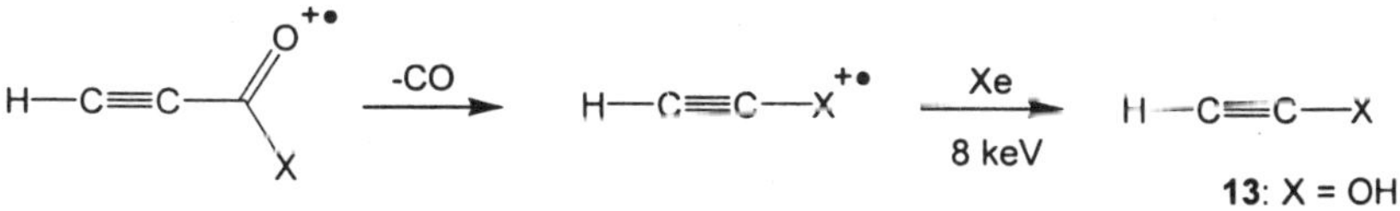

Scheme 4

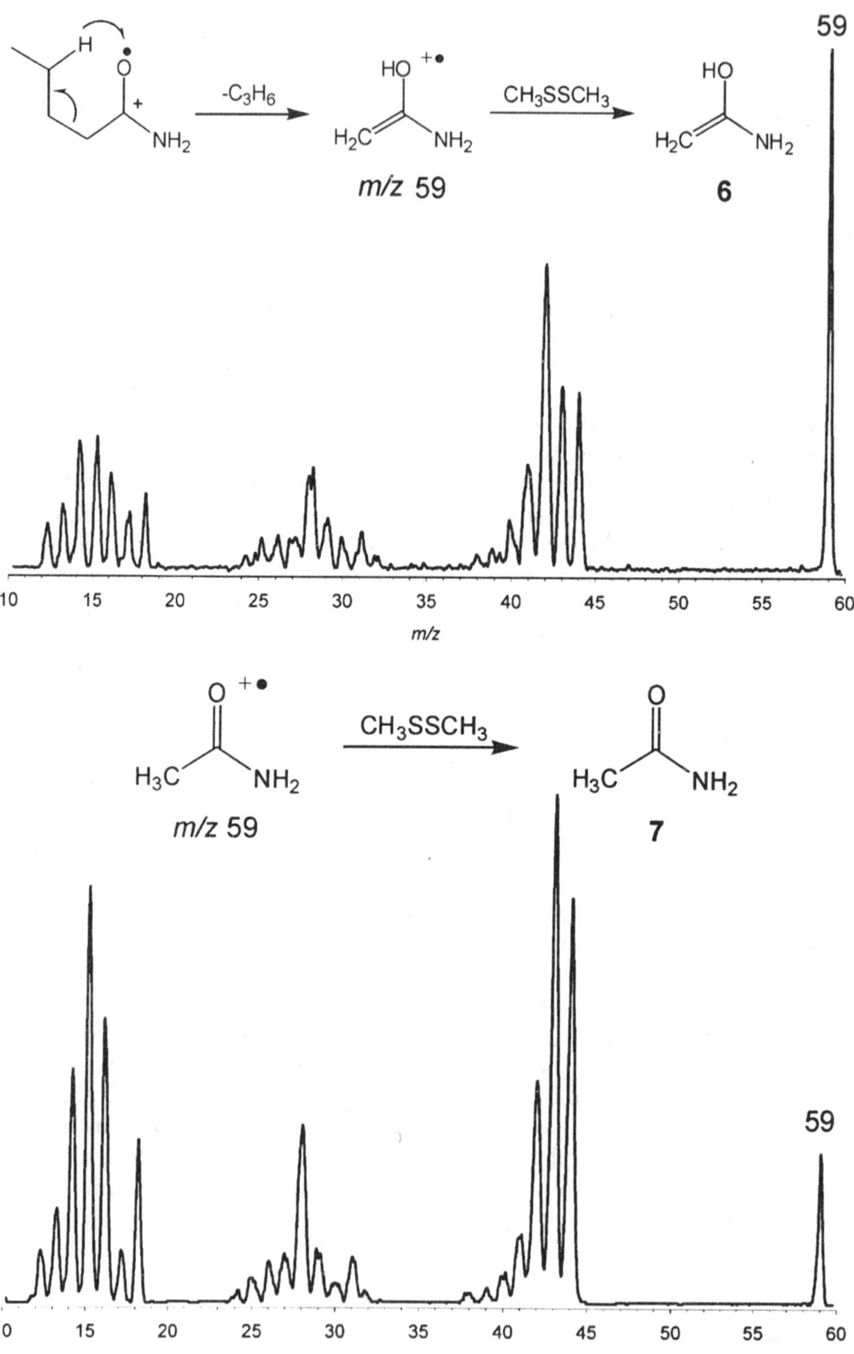

Fig. 6. $^+NR^+$ mass spectra of (*top*) acetamide enol and (*bottom*) acetamide. Neutralization with CH_3SSCH_3 at 70% transmittance, reionization with O_2 at 70% transmittance

5.2 Elusive Acids

NRMS has been used to generate several highly reactive acids that have been elusive in solution or condensed phase studies. Carbonic acid (**15**) [76] and sulfurous acid (**16**) [77] were prepared through the corresponding cation-radicals that were reduced to generate the corresponding neutral molecules. The general route to these and analogous cation-radical precursors involves alkene eliminations from suitable esters, as shown for **16** in Scheme 5.

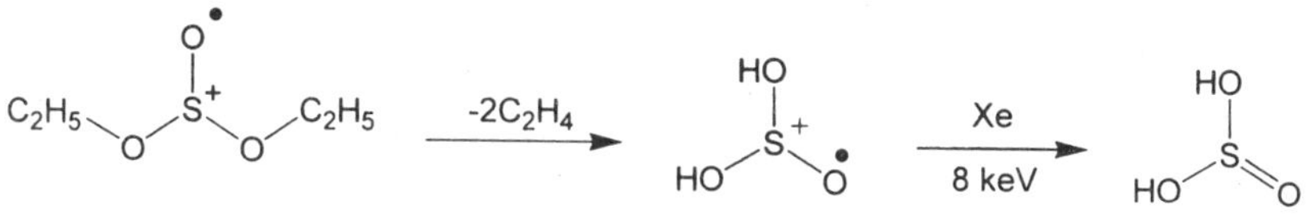

C_2H_5 -2C_2H_4 HO HO Xe 8 keV **16**$^{+\bullet}$ **16**

Scheme 5

Ion **16**$^{+\bullet}$ was distinguished from an isomeric HO-S(H)=O$^+$ structure by CAD of a D-labeled derivative that showed loss of both OH and OD, compatible with the structure having two OH groups. $^+$NR$^+$ of **16** showed a substantial survivor ion that attested to the stability of the isolated molecule, as contrasted to the elusiveness of sulfurous acid in the condensed phase [77].

Dihydroxysulfane, **17**, is another elusive acid that has been generated by NRMS [78]. The precursor cation-radical, **17**$^{+\bullet}$, was obtained by dissociative ionization of dimethylsulfate according to Scheme 6. Upon NR, acid **17** gave an abundant survivor ion showing that the intermediate sulfane was a stable molecule. Collisional activation of neutral **17** caused only minor dissociation by elimination of water, further attesting to the considerable stability of the isolated molecule.

Methanesulfenic acid (**18**) is another highly reactive molecule that was generated in the gas phase and characterized by NRMS [79]. Acid **18** can also be made by flash-vacuum pyrolysis of *tert*-butylmethylsulfoxide, and the neutral molecule had been characterized by microwave spectroscopy [80] and ionization energy measurements [81,82]. In the condensed phase, acid **18** rapidly disproportionates by bimolecular condensation forming $CH_3S(O)SCH_3$ and water [80]. It is worth mentioning that **18** is a neutral byproduct of dissociations of peptide and protein cations containing *S*-oxidized methionine residues, and the loss of CH_3SOH (64 Da) is highly diagnostic of methionine in protein samples [83]. The value of NRMS in studying sulfenic acids is highlighted by the generation of the sulfuranyl isomer of **18** (CH_2=S(H)-OH, **19**), which is hardly at all

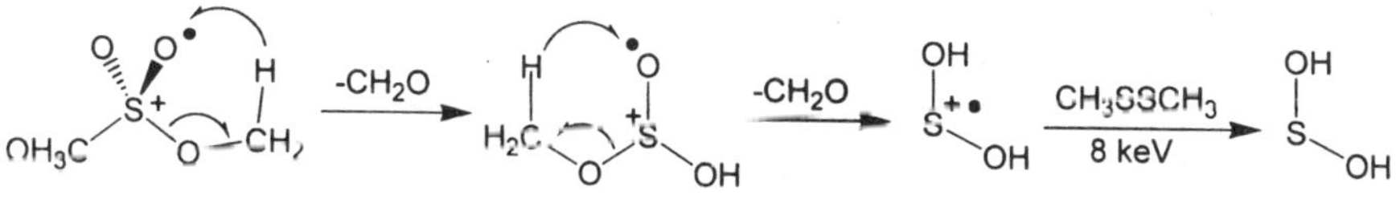

-CH_2O -CH_2O CH_3SSCH_3 8 keV **17**$^{+\bullet}$ **17**

Scheme 6

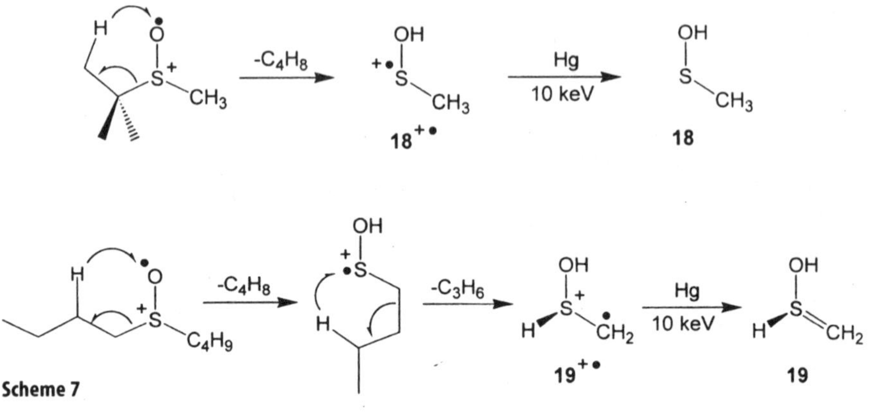

Scheme 7

possible to generate in the condensed phase because of lack of suitable synthetic reactions. Scheme 7 shows that double hydrogen transfer in the cation-radical of di-*n*-butyl sulfoxide gives rise to $\mathbf{19}^{+\bullet}$ which affords the neutral sulfuran upon collisional reduction with Hg atoms.

Ethenesulfenic acid (CH_2=CH-S-OH, **20**) is another transient intermediate that has been generated and characterized by NRMS in conjunction with ab initio calculations. Acid **20** is stable when formed by collisional reduction of its cation-radical and does not isomerize to the more stable ethanethione-*S*-oxide [84].

The *aci*-form of nitromethane, CH_2=N(O)OH (**21**), has been generated by collisional reduction with Xe of the corresponding cation-radical ($\mathbf{21}^{+\bullet}$), which was in turn prepared by elimination of ethylene from the molecular ion of 1-nitropropane (Scheme 8) [85]. An interesting feature of the $^+NR^+$ mass spectrum of **21** is that the elusive *aci*-nitro isomer gives a very abundant survivor ion compared to those produced by NR of its more conventional isomers nitromethane and methyl nitrite. This is due to a combination of two factors. First, although isolated **21** is ca. 90 kJ mol^{-1} less stable than nitromethane, it is separated by a >220 kJ mol^{-1} potential energy barrier to unimolecular isomerization and thus is kinetically very stable [86]. Second, cation-radical $\mathbf{21}^{+\bullet}$ produced from the molecule by collisional reionization is the most stable $[C,H_3,N,O_2]^{+\bullet}$ isomer. The major NR dissociation of **21** is loss of OH, whereas nitromethane dissociates by loss of CH_3, and methyl nitrite dissociates to NO and CH_3O, in analogy with thermal dissociation [87].

Trihydroxyphosphane, $P(OH)_3$ (**22**), is another elusive inorganic acid that has been generated by NRMS [34] and distinguished from the conventional phos-

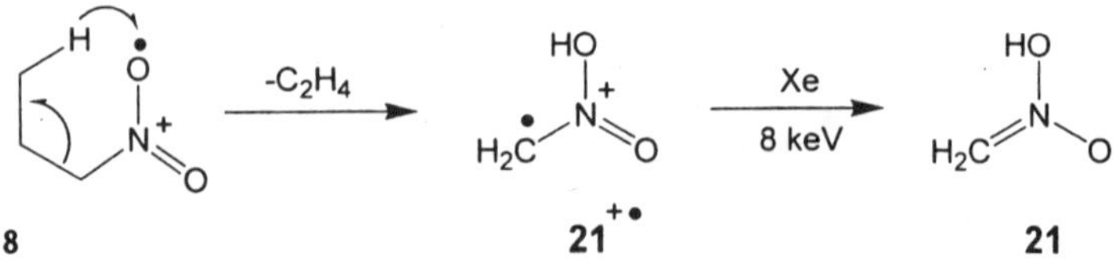

Scheme 8

phorous acid, $HP(=O)(OH)_2$ (**23**) which is the more stable tautomer in the gas phase, $\Delta H_{rxn,298}$(**23**→**22**)=45 kJ mol^{-1}. The relative stabilities are reversed in the cation radicals, where $\mathbf{22}^{+\bullet}$ is 136 kJ mol^{-1} more stable than $\mathbf{23}^{+\bullet}$. Interestingly, neutral **22** partially isomerizes to **23** upon collisional activation, in keeping with the greater stability of the latter tautomer. HTeTeH is another low-valence elusive molecule that was generated by NRMS [88].

Wiedmann and Wesdemiotis used a different strategy to generate the elusive fluoroformic acid, FCOOH (**24**) [89]. Previous attempts at generating neutral **24** by flash-vacuum ester pyrolysis of ethyl and *s*-butyl fluoroformate under molecular flow conditions (10^{-5} Torr) with mass spectrometric detection were unsuccessful and yielded only decomposition products, which were the corresponding alkenes, HF, and CO_2 [90]. Dissociative ionization of ethyl fluoroformate also does not yield $\mathbf{24}^{+\bullet}$, but proceeds by a double hydrogen transfer producing protonated fluoroformic acid, $FC(OH)_2^+$. This ion was mass selected, collisionally activated, and allowed to dissociate by loss of H to form $\mathbf{24}^{+\bullet}$ which was then collisionally reduced with Xe. The resulting $^+NR^+$ mass spectrum showed a survivor ion of $\mathbf{24}^{+\bullet}$ at *m/z* 64, which indicated that a fraction of the intermediate neutral fluoroformic acid was stable on the μs time scale of the measurement. This example nicely illustrates the power of tandem mass spectrometric methods to prepare an ion precursor of the desired structure that is then used to generate elusive neutral species by collisional electron transfer. In particular, the combination of bond making and bond breaking reactions endows gas-phase ion chemistry with a tremendous and unique synthetic potential for generating ion precursors for unusual and otherwise inaccessible neutral species.

5.3
Carbenes, Carbon Clusters, and Heterocumulenes

While there are photochemical methods for the generation of carbenes in solution and gas phase [91–93], carbenes carrying hydroxy and amino groups are not readily available by standard synthetic methods because of lack of suitable precursors. By contrast, carbene cation-radicals are readily accessible by dissociative ionization of stable molecules [94] and serve as precursors for the preparation of neutral carbenes by collisional reduction. For example, the fluorohydroxymethylene cation-radical, $F{-}C{-}OH^{+\bullet}$, is formed by dissociative ionization of methyl fluoroformate, and upon collisional reduction with Xe the ion provides neutral F–C–OH of a microsecond lifetime [95]. Several other monosubstituted carbenes have been prepared and characterized by NRMS, e.g., H–C–OH [96], $H{-}C{-}OCH_3$ [97], $H{-}C{-}OC_2H_5$ [98], $CH_3{-}C{-}OH$ [99], and $H{-}C{-}NH_2$ [100]. Dihydroxycarbene cation-radical, $HO{-}C{-}OH^{+\bullet}$ ($\mathbf{25}^{+\bullet}$), is cogenerated with ionized formic acid by dissociative ionization of oxalic acid [101]. The route to pure $\mathbf{25}^{+\bullet}$ starts with dihydroxyfumaric acid that upon electron ionization dissociates to dihydroxyketene, which in turn eliminates CO forming $\mathbf{25}^{+\bullet}$ (Scheme 9). Collisional neutralization of $\mathbf{25}^{+\bullet}$ with *N*,*N*-dimethylaniline yields a fraction of non-dissociating **25**. It is noteworthy that **25** and formic acid give quite similar $^+NR^+$ mass spectra, and so unambiguous isomer distinction was achieved by CAD of the corresponding survivor ions.

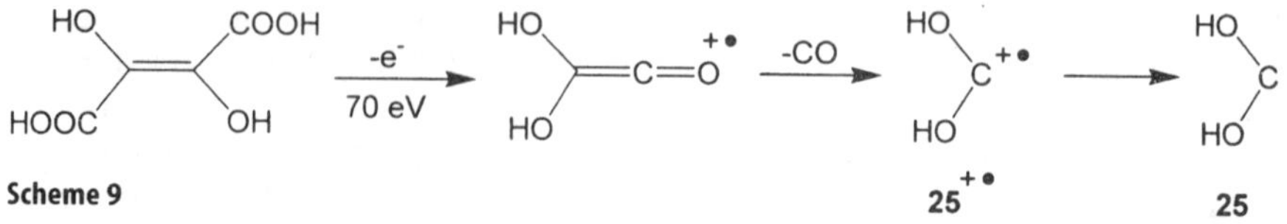

Scheme 9

Diaminocarbene, H_2N–C–NH_2 (**26**), was prepared by collisional reduction of the corresponding cation-radical that was in turn generated by dissociative ionization of aminoguanidine [102]. Carbene **26** gives an abundant survivor ion in the $^{+}NR^{+}$ mass spectrum and is clearly distinguished from its more stable isomer formamidine. Amino(hydroxy)carbene, H_2N–C–OH, has also been prepared by NRMS [103]. Hydroxy-thiohydroxy-carbene cation-radical, HO–C–$SH^{+\bullet}$ ($27^{+\bullet}$), is formed somewhat unexpectedly by ethylene elimination from ionized *S*-ethylthioformate and *O*-ethylthioformate instead of the expected thioformic acid. Carbene ion $27^{+\bullet}$ was characterized by a $^{+}NR^{+}$ mass spectrum that showed a dominant survivor ion of reionized carbene [104]. The energetics of neutral and ionic HO–C–SH have been addressed by ab initio calculations [105]. Di-(thiohydroxy)carbene, HS–C–SH, is also known [106].

In contrast to carbenes carrying σ- and π-electron donating substituents (alkyl, OH, OR, NH_2, SH) that stabilize cation-radical precursors, carbenes having σ- and π-electron accepting groups do not form stable cation-radicals [107] which thus cannot be used as precursors for neutral carbenes. The synthetic route to electron-deficient carbenes involves stable anion-radicals, as exemplified by H–C–NO_2 (**28**) and H–C–C≡N (**29**). The nitrocarbene anion-radical ($28^{-\bullet}$) was produced by electron-capture dissociation of ethyl nitrodiazoacetate and neutralized by collisions with O_2. Reionization to cations gave a weak peak of survivor $[C,H,N,O_2]^{+\bullet}$ in the $^{-}NR^{+}$ mass spectrum that was assigned to reionized **28** [108]. It should be noted here that reionization to anions ($^{-}NR^{-}$) offers a more reliable means of detecting transient species that do not form stable cations, as will be discussed later. Precursor anions for **29** and isocyanocarbene (**30**) were generated by $O^{-\bullet}$ abstraction of H_2 from CH_3CN and CH_3NC, respectively. The $^{-}NR^{+}$ spectra showed very abundant survivor ions that attested to the stability of the intermediate neutral carbenes, **29** and **30**, respectively [109].

Several carbon clusters have been studied by NRMS. Cationic precursors for C_3 and C_4 have been generated from a variety of linear and cyclic precursors and reduced with Xe, Na, K, and Ca to neutral carbon clusters. Unfortunately, ^{13}C labeling has shown that, prior to neutralization, the cations had undergone complete positional scrambling of carbon atoms, resulting in the formation of mixtures of most stable ion isomers [110]. A more successful approach used C_4^- produced by gas-phase desilylation of a linear precursor, 1,4-bis-trimethylsilylbuta-1,3-diyne [111]. Collisional oxidation of C_4^- with O_2, followed by reionization to cations, showed an increased extent of carbon scrambling that was ascribed to linear-rhombic isomerization in neutral C_4. A linear C_5 carbon cluster was also generated from an anionic precursor and shown by ^{13}C labeling to scramble the carbon atoms via skeletal isomerization [112].

Heterocumulenes have been a popular target of NRMS studies and there are now several structures involving O, N, and S atoms and combinations thereof that were generated by collisional neutralization in the gas phase [16]. Considerable attention has been focused on ethenedione, O=C=C=O (**31**), which is the elusive dimer of carbon monoxide. In spite of the fact that the $C_2O_2^{+\bullet}$ cation-radical (**31**$^{+\bullet}$) is readily available from several precursors [113, 114], it does not produce a detectable peak of survivor **31**$^{+\bullet}$ upon NR. The presence in the $^+$NR$^+$ mass spectrum of minor fragments from reionized CO_2 and C_2O was interpreted to suggest that molecule **31** was transiently formed by collisional neutralization [113, 115]. However, another study concluded that **31** dissociates rapidly to CO upon collisional neutralization of the ion. A triplet state of **31** is predicted by calculations to be metastable, but apparently is not formed by electron transfer [116]. Several other oxocarbons were found to be stable in the gas phase, e.g., C=C=O and O=C=C=C=O that were generated from their cation-radicals [113], and C_nO_n (n=3–6), C_2COC_2, and C_4CO [117] that were obtained by collisional oxidation of the corresponding anion-radicals [118]. Of these, carbon suboxide is a well known and relatively stable molecule [119, 120], whereas the others are transient species.

In contrast to **31**, its dithio-analogue ethenedithione, S=C=C=S (**32**), is a very stable molecule when produced by collisional reduction of its cation-radical, which is formed abundantly by dissociative ionization of 1,3,4,6-tetrathiapentalene-2,5-dione (Scheme 10). The high stability of **32** is documented by the fact that following reionization **32**$^{+\bullet}$ represents the dominating base peak of the $^+$NR$^+$ mass spectrum [121]. Molecule **32** was subsequently generated by flash-vacuum pyrolysis of 1,3-dithiane-2-cyclopentylideneketene and characterized by mass spectrometry [122].

A number of stable heterocumulenes have been generated by collisional reduction of the corresponding cation-radicals which are accessible by electron-ionization induced cycloreversion of heterocyclic rings, as reviewed [123]. These included molecules such as SC_nS [124], C_nS (n=2–6) [125], RN=C=C=S [126], S=C=N=C=O [127], Ar-N=C=C=C=N-Ar [128], RN=C=C=O [129], RN=C=C=C=O [130], and radicals S=N=C=C=S [131], HC=C=C=O [132], O=C-N=C=O and O=C=C-N=O [133]. Related to heterocumulenes are vinylidenes, R-C=C:, which have also been generated in the gas phase and characterized by NRMS. Examples of such stable or metastable vinylidenes are H_2C=C: [134], H_2C=C=C=C: [135], NH=C=C: [109], and NC-CH=C: [136].

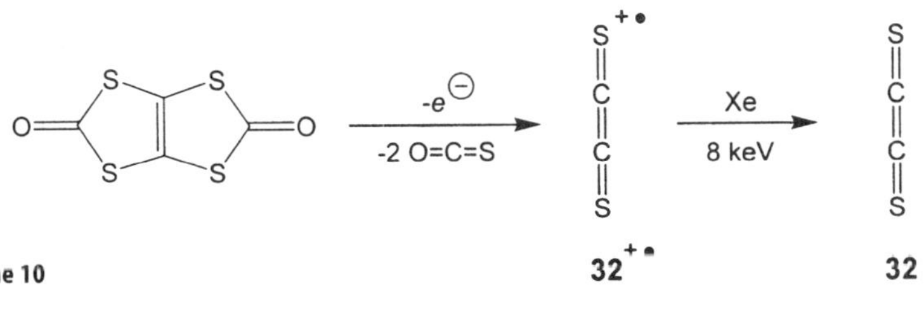

$$\text{1,3,4,6-tetrathiapentalene-2,5-dione} \xrightarrow[-2\ \text{O=C=S}]{-e^{\ominus}} \text{S=C=C=S}^{+\bullet}\ (\mathbf{32}^{+\bullet}) \xrightarrow[8\ \text{keV}]{\text{Xe}} \text{S=C=C=S}\ (\mathbf{32})$$

Scheme 10

5.4 Ylids

Ylid ions and distonic ions are known to be very stable species in the gas-phase, quite unlike their neutral counterparts that are often unstable in the condensed phase, with the exception of phosphorus (e.g., Ph_3P^+–CH_2^-) and sulfur (e.g., $(CH_3)_2(O)S^+$–CH_2^-) ylids, which are the known reactive intermediates of organic reactions in solution and which have been characterized spectroscopically. Ylid ions have structures that can be formally drawn as having the radical and charge sites at adjacent atoms, e.g., $^{\bullet}H_2C$-OH_2^+, in which the carbon and oxygen atom are trivalent. Distonic ions formally have the radical and ionic sites on more distant atoms (hence "distonic") [137], e.g., $^{\bullet}CH_2CH_2CH_2NH_3^+$. A special feature of gas-phase distonic ions is that they are comparably stable to and sometimes more stable than their classical alcohol or amine isomers, and can be generated by hydrogen transfer rearrangements in gas phase ion precursors. Ylid and distonic cation-radicals thus represent convenient precursors of exotic neutral species.

The first attempts to use ylid ions for the generation of neutral ylids reported that stable $^-CH_2OH_2^+$, $^-CH_2FH^+$, and $^-CH_2ClH^+$ were detected [138] upon collisional neutralization with Hg. However, in a subsequent detailed work where Xe was used for collisional neutralization, Hop et al. concluded that the survivor peaks that were previously assigned to ylids were in fact due to impurities from stable classical isomers or isotope interferences, and that the neutral ylids were unstable [51]. In both cases, the precursor cation radicals were prepared by dissociative ionization of acetic acid derivatives as shown for $^{\bullet}CH_2ClH^+$ (**33**$^{+\bullet}$, Scheme 11), and characterized by charge stripping and CAD spectra. The elimination of CO_2 from glycolic acid to yield $^{\bullet}CH_2OH_2^+$ is very inefficient, the main dissociation being the formation of CH_2OH^+ which contributes 1.1% of ^{13}C isotope satellite at *m/z* 32 which overlaps with the very weak peak of $^{\bullet}CH_2OH_2^+$. Similar although less severe interferences are encountered with the formation of $^{\bullet}CH_2FH^+$ from fluoroacetic acid and **33**$^{+\bullet}$ from chloroacetic acid (Scheme 11) [139].

The interpretation of Hop et al. [51] is consistent with ab initio calculations that predict a very shallow potential energy minimum for neutral ylid **33** that was bound by only 17 kJ mol^{-1} against dissociation to CH_2 and HCl and, moreover, vertical electron transfer to **33**$^{+\bullet}$ was predicted to result in 84 kJ mol^{-1} excess vibrational energy in the neutral ylid due to Franck-Condon effects [140]. $^-CH_2FH^+$ and $^-CH_2OH_2^+$ were calculated to be unbound in their ground singlet states [141].

On the other hand, there are now several examples of metastable neutral species produced by collisional electron transfer, for which high-level calculations predict no potential energy minima in the ground electronic state and whose existence is entirely due to the formation of long-lived excited states. Some examples will be discussed in the section dealing with hypervalent radicals. Another recent example is the nitrogen dimer, N_4, which shows a small fraction of metastable neutral species when formed by collisional neutralization of the stable $N_4^{+\bullet}$ cation radical [142]. However, this observation was not confirmed by another research group under very similar experimental conditions [143]. The simple ylids discussed above may belong to the same category, where particular reaction conditions (ion

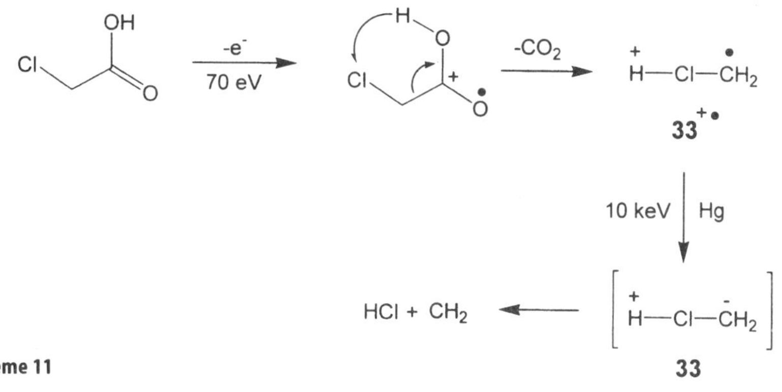

$ClCH_2COOH \xrightarrow[70\ eV]{-e^-}$ [$ClCH_2C(O)OH^{+\bullet}$] $\xrightarrow{-CO_2}$ $H\overset{+}{—}Cl—\overset{\bullet}{C}H_2$ ($33^{+\bullet}$) $\xrightarrow{10\ keV,\ Hg}$ [$\overset{+}{H}—Cl—\overset{-}{C}H_2$] (33) $\longrightarrow HCl + CH_2$

Scheme 11

source pressure, neutralization reagent and its pressure in the collision cell, neutral flight time, collection angle, etc.) may favor the formation and detection of a small fraction of metastable excited states of the transient neutral species of interest. Recent developments in experimental techniques of NRMS, in particular, variable-time photoexcitation, and NR-CAD measurements, bring additional insight into the neutral transient lifetimes and excited state energetics and in some cases allowed for unambiguous interpretation of NRMS data.

Such an unambiguous structure assignment has been reported for the simplest phosphonium methylide, $^{-}CH_2PH_3^{+}$, (**34**) that was generated by collisional neutralization with cyclopropane of the corresponding ylid ion $\mathbf{34}^{+\bullet}$ (Scheme 12) [144]. Reionization with O_2 resulted in an abundant survivor ion, which was selected by kinetic energy and submitted to additional CAD. The CAD spectra of reionized **34** and its classical isomer CH_3PH_2 clearly differed from each other, but resembled closely the CAD spectra of the corresponding precursor cation-radicals. This spectroscopic evidence ruled out isomerization of $^{-}CH_2PH_3^{+}$ to CH_3PH_2 and showed unequivocally that the ylid was stable as an isolated species in the gas phase. The experimental results were in agreement with previous ab initio calculations that predicted **34** to be a bound species, and also the Franck-Condon energy upon vertical reduction of $^{\bullet}CH_2PH_3^{+}$ was calculated to be lower than the threshold energy for dissociation to CH_2 and PH_3 [145].

An interesting class of elusive neutral species is heterocyclic ylids, as represented by the so-called Hammick intermediate that was postulated 65 years ago to explain the accelerated decarboxylation of 2-picolinic acid [146, 147]. An analogous dissociation takes place in ionized 2-picolinic acid in the gas-phase and was employed to generate the pyridine ion isomer $\mathbf{35}^{+\bullet}$ (Scheme 13) [148]. Collisional neutralization of $\mathbf{35}^{+\bullet}$ with *N,N*-dimethylaniline produced neutral ylid **35**, which can also be represented as a singlet α-carbene (Scheme 13). Ylid **35** showed a survivor ion in the $^{+}NR^{+}$ mass spectrum, which was further char-

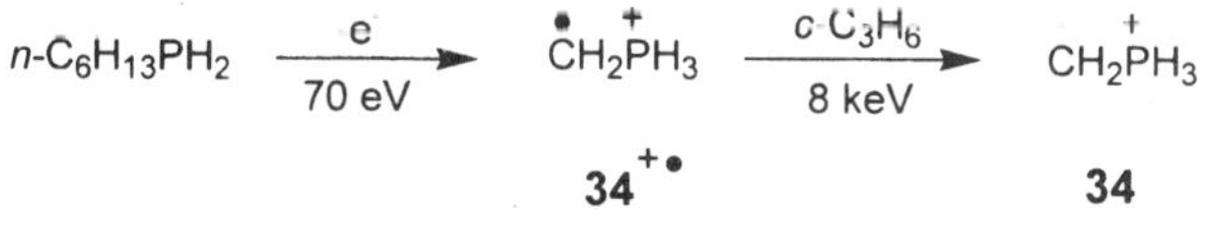

$n\text{-}C_6H_{13}PH_2 \xrightarrow[70\ eV]{e}$ $\overset{\bullet}{C}H_2\overset{+}{P}H_3$ ($\mathbf{34}^{+\bullet}$) $\xrightarrow[8\ keV]{c\text{-}C_3H_6}$ $CH_2\overset{+}{P}H_3$ (**34**)

Scheme 12

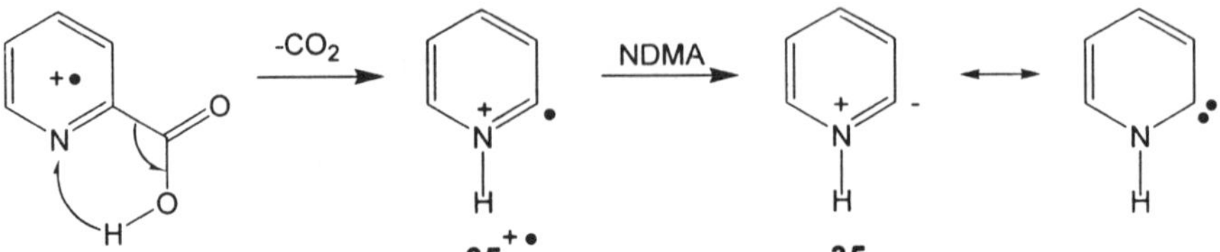

Scheme 13

acterized by NR-CAD as $35^{+\bullet}$, providing supporting evidence for the neutral structure. It may be noted that the CAD and NR-CAD spectra of $35^{+\bullet}$ and its more stable conventional isomer [pyridine]$^+$ were overall quite similar and the isomer distinction relied on the different relative intensity of a minor HC=NH$^+$ fragment which was somewhat higher for $35^{+\bullet}$. Ab initio calculations of the system revealed that while neutral singlet **35** was some 196 kJ mol^{-1} less stable than pyridine, it was separated from the latter by a 356 kJ mol^{-1} energy barrier to H atom migration, in keeping with the kinetic stability of **35** formed by collisional reduction. In addition, cation radical $35^{+\bullet}$ was calculated to be nearly isoenergetic with [pyridine]$^{+\bullet}$, which explained both its facile formation from 2-picolinic acid and stability in the gas phase [148].

Other heterocyclic ylids have been generated in the gas-phase by collisional reduction of stable ylid cation-radicals using synthetic strategies similar to that described for $35^{+\bullet}$, e.g., imidazol-2-ylidene [149], thiazol-2-ylidene [150], pyrazine ylids [151], pyrimidine ylids [152], and pyridinium methylids [153].

A benzenoid ylid has been recently generated in the gas phase by collisional reduction of an ylid isomer of benzonitrile cation-radical [154]. Dissociative ionization of *o*-cyanobenzaldehyde produced a cation-radical ($36^{+\bullet}$) that was distinguished by ion-molecule reactions from its conventional [benzonitrile]$^+$ isomer. In particular, $36^{+\bullet}$ reacts with *tert*-butylnitrite by NO transfer to form 2-nitrosobenzonitrile, while [benzonitrile]$^{+\bullet}$ reacts by proton transfer. Collisional reduction of $36^{+\bullet}$ provided a $^+$NR$^+$ mass spectrum that was distinctly different from that of benzonitrile (Fig. 7).

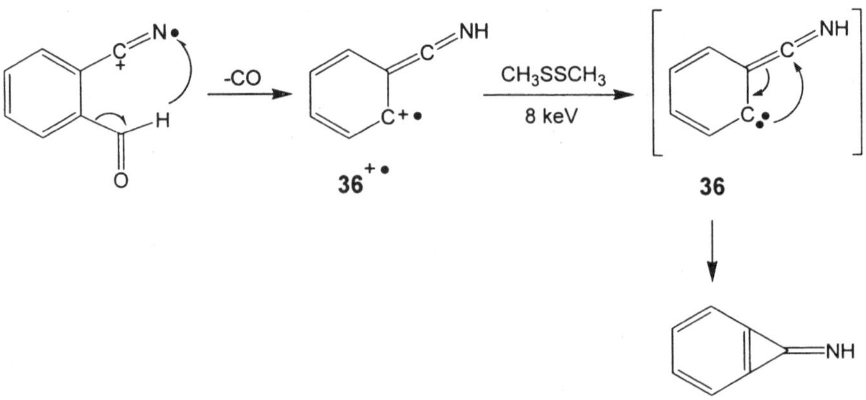

Scheme 14

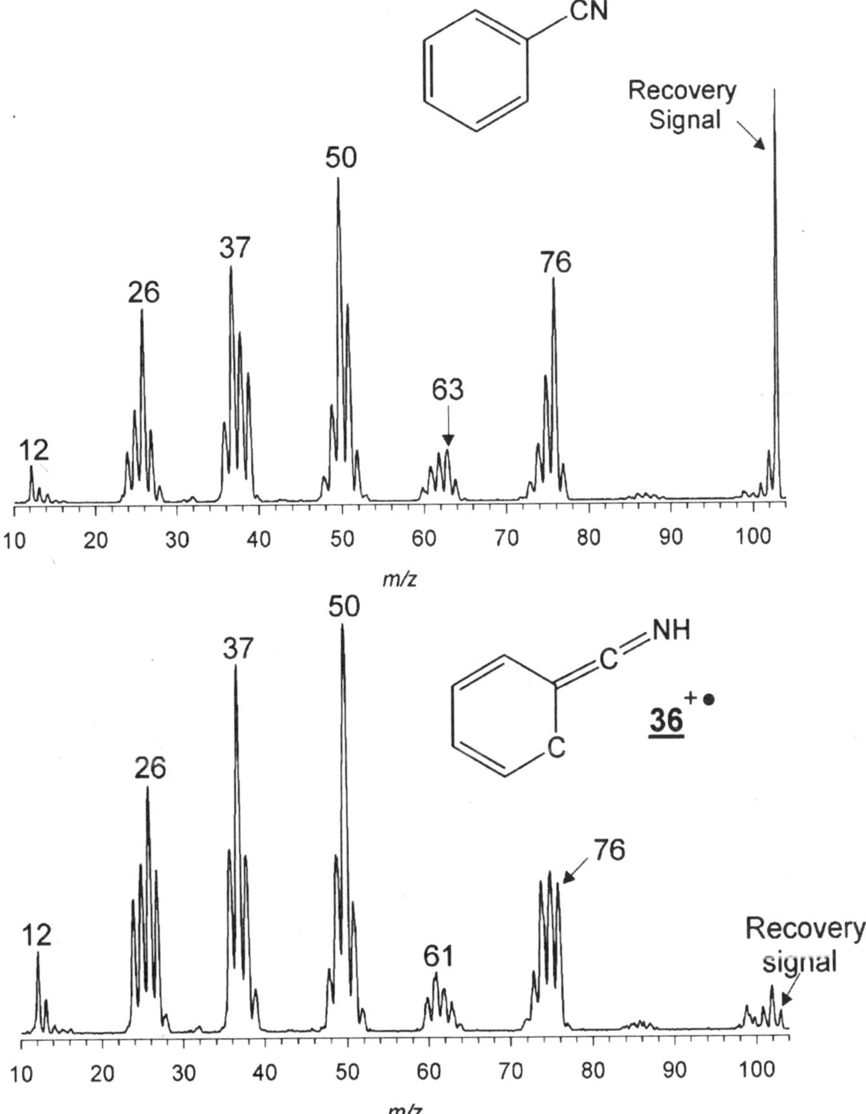

Fig. 7. $^{+}NR^{+}$ mass spectra of (*top*) benzonitrile and (*bottom*) ylid isomer. Neutralization with CH_3SSCH_3 at 70% transmittance, reionization with O_2 at 70% transmittance

In particular, $36^{+\bullet}$ showed a weak survivor ion and dissociation products by multiple losses of H atoms. The low relative abundance of the survivor ion was explained by density-functional theory calculations that showed that neutral **36** was intrinsically unstable and underwent ring closure to form another high-energy intermediate, norcaradiene imine (**37**), which is 260 kJ mol^{-1} less stable than benzonitrile (Scheme 14) [154].

5.5 Other Elusive N, S, P, and Si-Containing Molecules

In addition to short-lived molecules that were assigned to the structure classes discussed above, there are various interesting intermediates that are mentioned here separately. Nitrosomethane (**38**), which is the less stable tautomer of formaldoxime, was generated by collisional reduction of the stable cation-radical and characterized by NRMS [155, 156]. The precursor cation for **38** was produced by three different reactions, e.g., elimination of OH upon exothermic protonation of nitromethane [156], electron-induced loss of O from nitromethane [155, 156], and electron-induced CH_2O extrusion from ethyl nitrite [156] (Scheme 15). Nitrosomethane gives rise to a moderately abundant survivor ion in the $^{+}NR^{+}$ mass spectrum and does not undergo unimolecular isomerization to any of its more stable tautomers.

Formaldonitrone (**39**) is another elusive molecule that has been generated and characterized by NRMS [157]. Although **39** has been predicted by ab initio calculations to be stable as an isolated molecule, its generation in solution by classical methods of nitrone chemistry has been unsuccessful. The gas-phase preparation of **39** consisted of a cycloreversion in 1,2-oxazolidine cation-radical that produced cation-radical $39^{+\bullet}$. The ionization conditions were tuned by charge exchange ionization with $COS^{+\bullet}$ that supplied the 270 kJ mol^{-1} excitation energy necessary for ring cycloreversion in $[oxazolidine]^{+\bullet}$ and elimination of C_2H_4 while preventing side reactions that would produce other, high-energy, isomers (Scheme 16).

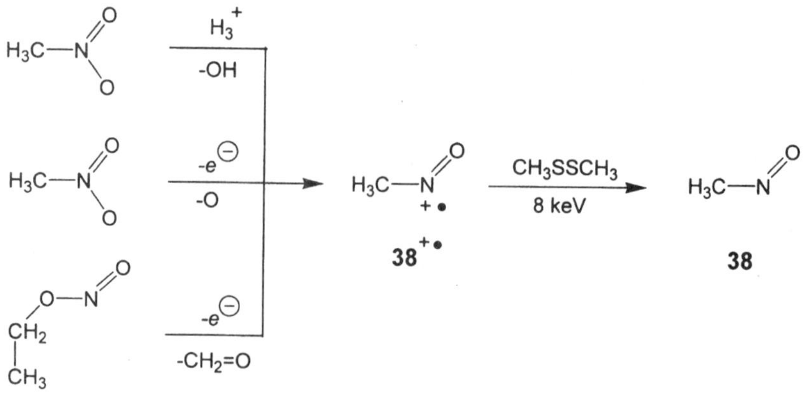

H3C—N(=O)O
H3+
-OH
H3C—N(=O)O
-e⊖
-O
O—N=O
CH2
CH3
-e⊖
-CH2=O
H3C—N=O +•
38+•
CH3SSCH3
8 keV
H3C—N=O
38

Scheme 15

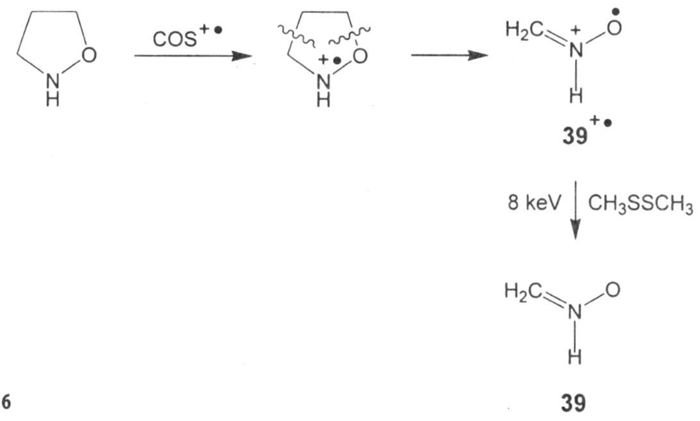

Scheme 16

It is worth noting that, prior to NR measurements, the potential energy surface for the dissociations of [oxazolidine]$^{+\bullet}$ was mapped by Gaussian 2(MP2) calculations and the energy and kinetic data obtained computationally were used successfully to plan the experiments [157].

Nitrone **39** turned out to be a very stable molecule under NR conditions and did not undergo any substantial dissociations, as determined by variable-time NR measurements of unimolecular rate constants.

A similar cycloreversion in 1,2-dithiolan cation-radical was used to prepare $CH_2S_2^{+\bullet}$ ions which were reduced with Xe to generate elusive CH_2S_2 molecules. The structure of neutral CH_2S_2, dithiirane or CH_2=S=S, was not determined from the $^+$NR$^+$ mass spectra [158]. Density functional theory calculations favored both dithiirane and its cation-radical as being more stable than CH_2=S=S and CH_2=S=S$^{+\bullet}$, respectively, and so the authors tentatively assigned cyclic structure to the ion and neutral.

Nitrile-N-sulfides and selenides were synthesized as the corresponding cation-radicals by S^+ or Se^+ transfer in ion molecule reactions in the gas phase and utilized for the generation of transient neutral molecules. As a source of S^+ the authors used CS_3^+ ions that were made from CS_2 in a high-pressure ion source. CS_3^+ was found to transfer S^+ to a variety of nitriles and cyanogen halides forming stable cation-radicals of the R-C=N=S$^{+\bullet}$ type. Collisional reduction with ammonia produced the corresponding neutral nitrile-N-sulfides that were found to give abundant survivor ions in the $^+$NR$^+$ mass spectra, thus attesting to the stability of the neutral intermediates [159]. Analogously, gas-phase NC-CN-Se$^{+\bullet}$ reacts with nitriles by Se^+ transfer producing nitrile-N-selenide cation-radicals that were used to generate neutral nitrile-N-selenides [160, 161]. Se^+ transfer followed by collisional reduction was also used to generate the elusive pyridine-*N*-selenide, which was found to be a stable molecule under NR conditions [160].

Thionitrosylchloride (Cl-N=S) [162] and thionitrosyl cyanide (NC-N=S) [163] were generated in the gas phase and found to be moderately stable on the microsecond time scale. Both molecules and their isomers were characterized by high-level ab initio calculations so that the energetics of their isomerization and dissociations is well understood.

Ammonia-N-oxide, $H_3N\text{–}O$, **40**, is the lowest homologue in the series of organic amine N-oxides. Those derived from tertiary amines are known as stable compounds; however, **40** has been elusive because of lack of synthetic methods for its preparation in solution or gas phase. The precursor ion for **40** was generated by ion-molecule reactions in a mixture of gaseous HN_3 and H_2O [164]. Although the reaction sequence leading to $\mathbf{40}^{+\bullet}$ is somewhat obscure, the $[N,H_3,O]^{+\bullet}$ ion formed that way showed a CAD spectrum that slightly differed from that of NH_2OH^+ and gave rise to an abundant survivor ion peak in the $^+NR^+$ mass spectrum [164]. Structure assignment to the $[N,H_3,O]^{+\bullet}$ cation-radical was mainly based on density functional theory calculations that showed that $H_3N\text{–}O^{+\bullet}$ was a stable structure that was separated from the more stable $NH_2OH^{+\bullet}$ by an energy barrier [164]. The $^+NR^+$ mass spectrum of **40** was very similar to that of NH_2OH and did not permit isomer distinction. That two different isomers were made was inferred from NR-CAD of the survivor ions which showed the same differences as those in the CAD of the precursor ions.

Finally, several elusive phosphorus-containing molecules have been generated and characterized by NRMS. Those included P_3S and P_2S [165], (alkylthio)thioxophosphines, R-S-P=S [166], phosphinedithiol, $HP(SH)_2$ [167], and phenylphosphinidene, Ph-P: [168]. P_6 (**41**), a new allotrope of elemental phosphorus, was generated as a transient neutral fragment by collisionally activated dissociation of an ionized bis-pentamethylcyclopentadienyl adduct (Scheme 17) [169]. The neutral-fragment reionization spectrum showed a peak of reionized P_6 that was taken as evidence for the existence of P_6. Although no structure information could be inferred from the mass spectrum alone, the molecule had been characterized previously by numerous computations [170] and the electron affinity of P_6 was determined by photoelectron detachment measurements of mass-selected P_6^- [171].

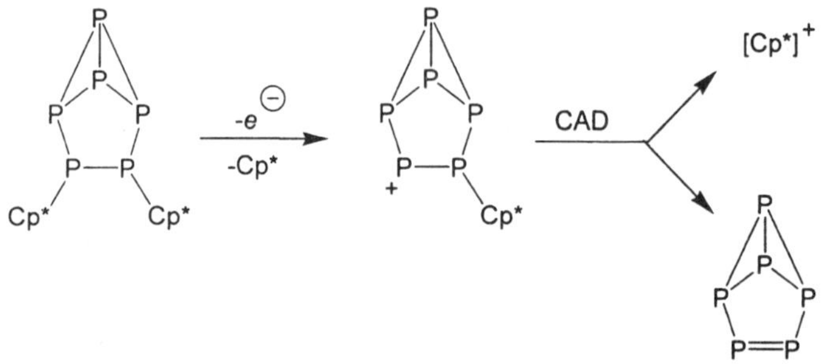

Scheme 17

6 Transient Radicals and Biradicals

While collisional neutralization of open-shell cations and anions produces neutral molecules, neutralization of closed-shell cations or anions provides a general synthetic route to various transient radicals. A great strength of this approach is that by choosing the ion charge one can vary the coordination numbers (*n*) at heteroatoms, e.g., $n=1$ in alkoxide anions, RO^-, and $n=3$ in oxonium cations, e.g., $(CH_3)_2OH^+$, and thus generate various hypovalent or hypervalent radicals, some of which are virtually impossible to synthesize by any other chemical method. The following section illustrates some of the highlights in this area.

6.1 Hypervalent Radicals

Lewis' octet rule dictates that elements of the second period cannot have more than eight electrons in the valence shell in stable molecules [172]. The octet rule is violated in several stable molecules containing elements of the third or higher periods, e.g., SF_6, and such compounds are denoted as hypervalent [173]. Unbeknownst to Lewis, evidence for hypervalent nitrogen molecules had been around since 1872 when Schuster observed emission bands from discharge in gaseous ammonia [174] which, combined with other spectroscopic evidence [175], were much later explained by Herzberg as originating from excited electronic states of the hypervalent ammonium radical $NH_4^\bullet$ [176]. Ammonium has a tetracoordinated nitrogen atom with formally nine electrons in the valence shell and thus can be denoted as a 9-N-4 hypervalent species according to the nomenclature introduced by Perkins et al. [177]. Hypervalent ammonium radicals were the first transient neutral species generated and characterized by neutral beam spectroscopy developed by Porter and coworkers, as reviewed [6]. The general NRMS approach to hypervalent radicals consists of gas-phase protonation of a stable molecule to form an onium cation, where "onium" stands for ammonium, oxonium, sulfonium, phosphonium, etc. Collisional reduction of onium ions generates hypervalent radicals that can be probed by the various methods of NR mass spectrometry. The properties of interest are (1) the radical lifetime, (2) competitive unimolecular dissociations, and (3) intramolecular interactions of the hypervalent group with other functional groups in the radical.

The lifetimes of hypervalent radicals have been found to depend rather dramatically on isotope substitution. For example, dimethyloxonium, $(CH_3)_2OH^\bullet$, dissociates completely on a 1-μs time scale when formed by collisional reduction of the stable cation $(CH_3)_2OH^+$. By contrast, $(CH_3)_2OD^\bullet$ furnishes an abundant survivor ion in the $^+NR^+$ mass spectrum that is evidence that the deuterated hypervalent radical is metastable [178, 179]. From the time scale of the NR measurements and the survivor ion relative intensities one can estimate that $(CH_3)_2OH$ dissociates ≥5 times faster than $(CH_3)_2OD^\bullet$. Similar isotope effects have been observed for $CH_3OH_2^\bullet$ [180], $C_2H_5OH_2^\bullet$ [181], and hypervalent ammonium radicals, e.g., $CH_3NH_3^\bullet$ [182], $(CH_3)_2NH_2^\bullet$ [60], $(CH_3)_3NH^\bullet$ [183], and [pyrrolidinium]$^\bullet$ [184], which are metastable only as deuterated species.

The metastability and bond dissociations in hypervalent radicals have been studied in detail for $(CH_3)_2NH_2^\bullet$ (**41**) [60, 185] and $(CH_3)_2OH^\bullet$ using product analysis, photoionization [186], photoexcitation [60], and theoretical calculations [187]. When formed by collisional reduction, radical **41** dissociates by N–H and N–C bond cleavages that occur in a 2:1 ratio [60]. This experimental finding is incompatible with the potential energy of ground ($^2A_1$) state of **41** that was calculated to have a very low barrier for exothermic N–H bond dissociation [185], which should be greatly preferred according to RRKM calculations [60]. Analysis of excited electronic states of **41** showed that they were strongly bound along the N–H coordinate and should not dissociate by H atom loss. However, the *B* state was weakly bound along the N–C coordinate and, after overcoming a 17 kJ mol^{-1} energy barrier, it can dissociate by avoided crossing to the repulsive part of the ground state potential energy surface (Fig. 8).

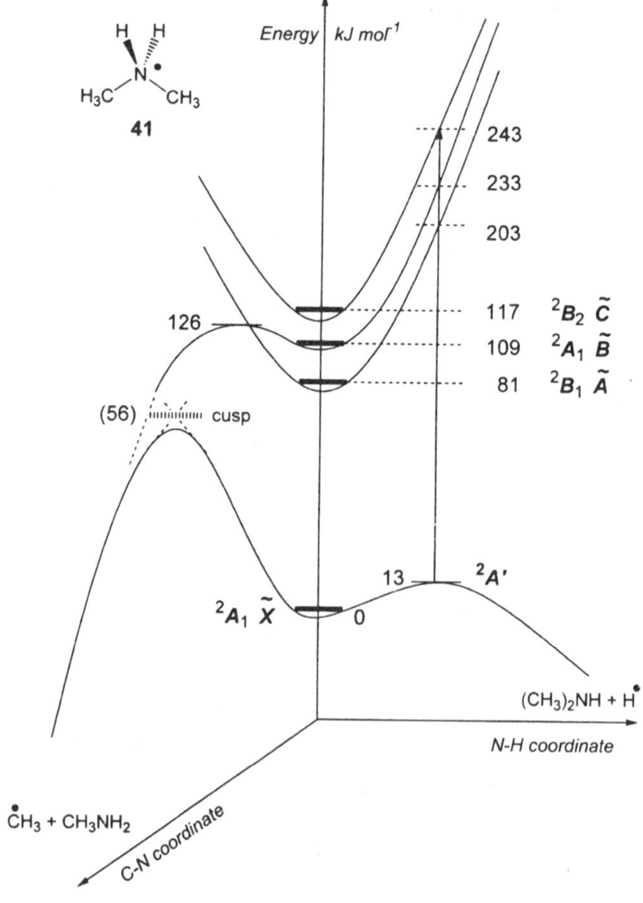

Fig. 8. Potential energy diagram for electronic states of $(CH_3)_2NH$

N–C bond dissociation can also occur from the *A* state by curve crossing to the *B* state potential energy surface, which occurs close to the transition state configuration. The *A* and *B* states thus account for the 33% fraction of **41** that dissociate by loss of a methyl. The *C* and higher excited states are calculated to be bound and can account for the metastable fraction of **41**. The formation of bound excited states of **41** was supported by laser photoionization and photoexcitation experiments. Metastable **41** (in a deuterated form) is photoexcited with 2.41 and 2.54 eV photons to high Rydberg states that show both increased lifetimes and very large cross sections on collisional reionization, as predicted by theory [60]. The isotope effects on metastability have been explained for $(CH_3)_2OH^\bullet$ and its isotopomers by diminished vibronic coupling between the metastable and dissociative electronic states in partially deuterated radicals. The coupling is restricted by symmetry to a few bending vibrational modes, and it is weakened by shifts in vibrational frequencies due to the presence of deuterium. The effect is predicted to disappear in fully deuterated radicals, as indeed observed for $(CD_3)_2ND_2^\bullet$ which dissociates completely on the μs time scale [60], whereas both $(CD_3)_2NH_2^\bullet$ and $(CH_3)_2ND_2^\bullet$ give substantial fractions of metastable radicals [60]. The mode symmetry restriction is consistent with the fact that no hypervalent radical lacking molecular symmetry has been found to be metastable, deuterium substitution notwithstanding.

Hypervalent sulfonium radicals, $H_3S^\bullet$ [188] and $(CH_3)_2SH^\bullet$ [189], have been shown to be metastable on the microsecond time scale when generated as partially deuterated species and studied by variable-time NR mass spectrometry. The metastability was attributed to the formation of excited electronic states, as the ground doublet states were calculated to be unbound and predicted to dissociate exothermically and without a barrier by S–H bond cleavage [188, 189]. Hypervalent sulfonium radicals of the $R_3S^\bullet$ type have been the long sought-after intermediates of radical substitution in sulfides in solution, but have never been detected [190]. The NRMS studies, combined with ab initio calculations, show that such intermediates most likely do not exist on the ground state potential energy surface, but may be of importance in photochemical reactions.

Another mechanism in hypervalent radical chemistry that is acquiring broader implications has been proposed to account for highly endothermic dissociations in hypervalent benzylammonium radicals bearing electronegative (F, NO_2) substituents in the aromatic ring (Scheme 18) [191]. In spite of the presence of weak N–H and C–N bonds in the ammonium group, substantial dissociations occurred in the phenyl group either by ring cleavage or following a transfer of the ammonium proton. The ring cleavage dissociations were explained by electron capture in a vacant π-orbital to form a zwitterion corresponding to an excited electronic state resembling a charge-transfer complex. Ring fragmentation can be induced by the substantial electron energy in such excited states that can either be converted non-radiatively to radical vibrational energy, or drive dissociations on the potential energy surface of the excited state. Rydberg states formed by electron capture have recently been invoked to explain dissociations of peptide cation-radicals in ion-cyclotron resonance mass spectrometry.

Intramolecular hydrogen atom transfer in hypervalent onium radicals is another reaction that has implications in the chemistry of gas-phase peptide and

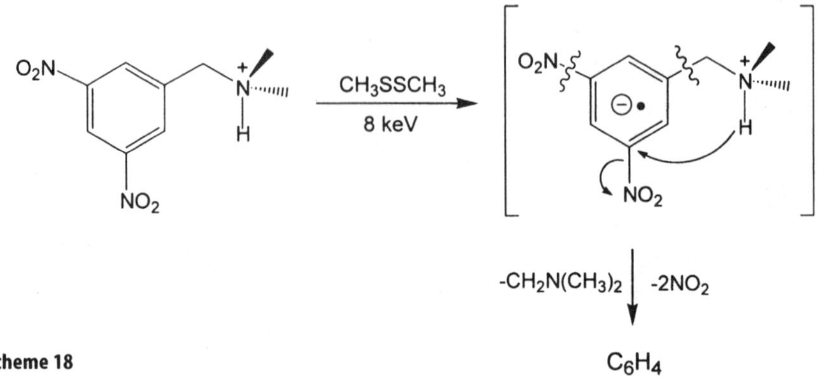

Scheme 18

protein cation-radicals [192]. In hex-3-en-1-ylmethyloxonium (**42**) and hex-3-en-1-ylammonium (**43**) radicals, the proton of the hypervalent group is positioned close to the double bond acceptor group. The proximity is secured by hydrogen bonding in the precursor cations that stabilizes the hydrogen-bonded structures by 18–25 kJ mol^{-1} [193, 194]. Upon collisional reduction, the hydrogen atom of the oxonium group in **42** migrates to the double bond, and the overall exothermicity of the isomerization ($-\Delta H_{rxn}$=250 kJ mol^{-1}) drives further dissociation by loss of methanol (Scheme 19).

The reaction mechanism has been established by deuterium labeling that showed complete retention of the oxonium deuterium in the C_6H_{11} fragment. In contrast, the ammonium hydrogen atoms in **43** do not migrate to the double

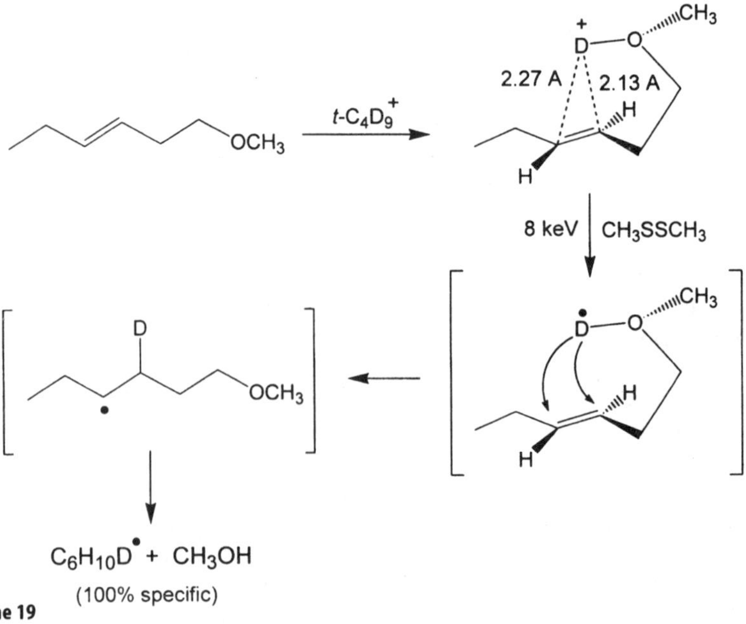

Scheme 19

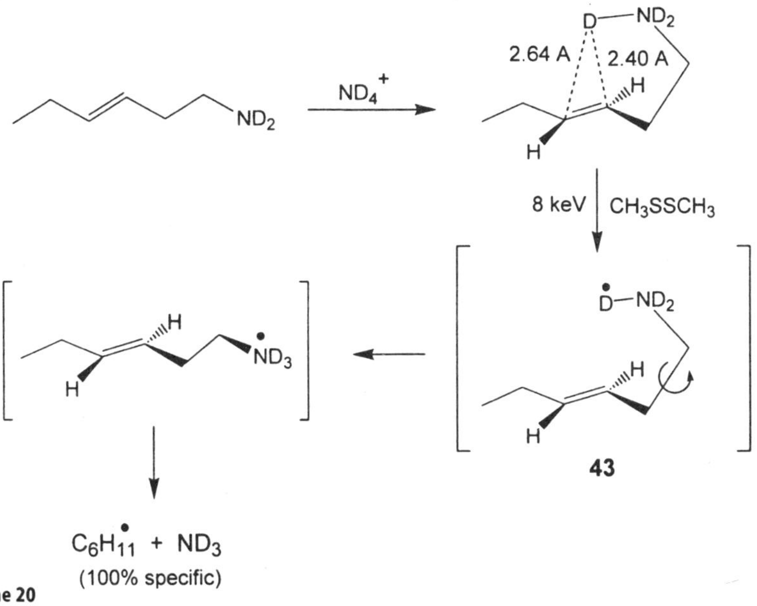

Scheme 20

bond, and the radical dissociates by loss of H and ammonia. Again, deuterium labeling showed complete retention of deuterium atoms in the ND_3 group [194]. The extraordinary difference in the dissociation mechanisms was explained by the stabilities of the hypervalent oxonium and ammonium radicals. The oxonium radical is unbound along the O....H coordinate, so that the O–H bond in vertically formed **42** dissociates exothermically within a single vibration. The hot hydrogen atom released by the dissociation has sufficient kinetic energy to overcome a small barrier for an exothermic addition to the double bond. The hydrogen migration is facilitated by the proximity and orientation of the double bond π-orbital and the oxonium hydrogen atom (Scheme 19). In contrast, **43** is weakly bound along the N–H coordinate. When formed by vertical electron transfer, the radical undergoes spontaneous rotation about the C–N bond to reach the potential energy minimum which has a gauche conformation of the NH_3 group (Scheme 20). The rotation moves the ammonium hydrogen atoms away from the double bond and thus prevents hydrogen transfer. The reason for the repulsive interaction of the hypervalent ammonium group with the double bond π-electrons is mainly electrostatic, as inferred from the calculated population analysis that showed substantial negative charge at the ammonium hydrogen atoms in **43** and also in related ammonium radicals. For the generation and dissociations of other types of cyclic and linear hypervalent radicals see [195–197].

6.2
Oxygenated Sulfur Radicals Relevant to Atmospheric Chemistry

Tropospheric oxidation of the so-called "reduced sulfur", mostly dimethylsulfide, hydrogen sulfide, and dimethyldisulfide, is thought to proceed via several steps that involve partially oxidized sulfur molecules and radicals. Although these molecules are produced naturally on a global scale of 1–50 million tons a year [198, 199], the stationary concentrations are too low to study the mechanisms of atmospheric oxidations in situ, let alone to identify short-lived intermediates. Several transient intermediates have been generated and characterized by NRMS that provided information on the radical stability and identified their unimolecular dissociations. Relevant to oxidation of H_2S, HSO, SOH, $H_2S{=}O$, and HSOH have been generated from the corresponding cations and found to give rise to abundant fractions of stable molecules [200] in keeping with ab initio calculations [201, 202]. The $^+NR^+$ mass spectra of $H_2S{=}O$ and HSOH are very similar and the isomers were distinguished on the basis of ion CAD spectra that showed different relative intensities of very minor OH^+ and $H_2O^{+\bullet}$ fragments [200]. $HOSO^\bullet$ and $HSO_2^\bullet$ radicals are the presumed intermediates of OH radical attack on SO, and H atom attack on SO_2, respectively, and both have been studied intensively by theory [203–206]. The ion precursor for $HOSO^\bullet$ was formed by mildly exothermic gas-phase protonation of SO_2 [207]. Collisional reduction of $HOSO^+$ with TMA yielded stable $HOSO^\bullet$ which was shown by variable-time NRMS to dissociate by loss of H, but not OH. Loss of H is further enhanced by laser photoexcitation of stable $HOSO^\bullet$ [208]. The ion precursor (HSO_2^+) for the much less stable $HSO_2^\bullet$ was generated in a mixture with $HOSO^+$ and the radical was shown to undergo very facile dissociation of the H–S bond forming SO_2, in keeping with ab initio calculations [205, 208].

Both cationic and anionic precursors were used to generate the $O_2SOH^\bullet$ radical (**44**) which is the presumed intermediate of OH radical attack on SO_2 to form SO_3 in the last step of the sulfur oxidation cascade [198]. The hydrogen sulfite cation was produced by dissociative ionization of methanesulfonic acid and used to generate radical **44**, albeit in a low yield [209]. A improved preparation and NR mass spectrometric characterization of **44** relied on an anion precursor, O_2SOH^-, which was made by termolecular associative reaction of SO_2 with OH^- [210]. Neutralization with Xe of O_2SOH^- produced stable radical **44** that gave rise to a dominant survivor cation in the $^-NR^+$ mass spectrum [211].

$(CH_3)_2SOH^\bullet$ (**45**) and $CH_3SO_2^\bullet$ (**46**) are the presumed transient intermediates of tropospheric oxidation of dimethylsulfide, a major natural compound produced by biomass decay in the marine biolayer [212]. An attempt at generating **45** by collisional reduction of the readily available cation, $(CH_3)_2SOH^+$, resulted in complete dissociation to $CH_3SOH{+}CH_3^\bullet$ and $(CH_3)_2S{+}OH^\bullet$ [213]. This result contradicted ab initio calculations that predicted a weakly bound radical with a long S–O bond [214], in keeping with pulse radiolysis studies that reported UV-VIS spectra that were assigned to **45** [215]. A further computational study [216] revealed that vertical neutralization of $(CH_3)_2SOH^+$ is accompanied by large Franck-Condon effects that result in fast dissociation of the nascent radical. Potential energy surface mapping in the vicinity of the landing point for the vertical transition, $(CH_3)_2SOH^+ \rightarrow$ **45**, indicated preferential dissociation to $(CH_3)_2S+$ OH, but also allowed for the formation of $CH_3SOH{+}CH_3^\bullet$ from highly excited radicals [216].

$CH_3SO_2^\bullet$ (**46**) and the isomeric $CH_3OSO^\bullet$ (**47**) were generated by collisional reduction of the corresponding cations with TMA and found to give different

$^+NR^+$ mass spectra [217]. Even though **46** is predicted by calculations to be bound by 59 kJ mol^{-1} against dissociation to CH_3 and SO_2, it acquires >140 kJ mol^{-1} by Franck-Condon effects upon vertical electron transfer and completely dissociates on the time scale of the experiment. By contrast, **47** is stable when formed by collisional reduction. The main dissociation of **47** is loss of CH_3 that proceeds by rate-determining isomerization to **46**, whereas direct cleavage of the O–CH_3 bond must overcome a large potential energy barrier and is kinetically disfavored [217].

6.3
Oxygenated Nitrogen and Phosphorus Radicals

Nitrogen dioxide and organic nitro compounds are important oxidizers and high-energy materials that undergo a variety of reactions in the gas phase, solution, and solid state. NRMS provides an insight into the intrinsic properties of elusive nitro and nitrate radical intermediates that are generated and studied as isolated species. For example, $CH_3NO_2H^\bullet$ (**48**) is the presumed intermediate of hydrogen atom capture by nitromethane in hydrocarbon flames [218]. In spite of extensive kinetic studies [219], radical **48** had been elusive until generated by collisional reduction and characterized by variable-time NRMS [220]. Upon vertical neutralization with DMDS, **48** gives a substantial fraction of undissociated radicals that are detected as the corresponding cations in the $^+NR^+$ mass spectrum. The main dissociations of **48** were elucidated by variable time measurements that revealed formation of CH_3NO and $OH^\bullet$ [220]. Interestingly, dissociation of **48** to $CH_3^\bullet$ and HONO is calculated to compete with the loss of OH but is difficult to recognize from the $^+NR^+$ mass spectra because of fast dissociation of $HONO^{+\bullet}$ after collisional reionization. In related studies, $C_6H_5NO_2H^\bullet$ was found to be stable when generated by collisional reduction of $C_6H_5NO_2H^+$ [221], whereas neutralization of $CH_3O(H)NO_2^+$ and $CH_3ONO_2H^+$ resulted in complete dissociation of the transient nitrate radicals, $CH_3O(H)NO_2^\bullet$ (**49**) and $CH_3ONO_2H^\bullet$ (**50**), respectively, which yielded no survivor ions [222]. The experimental results for the methyl nitrate radicals are in complete agreement with ab initio calculations. Radical **49** was found to be unbound and predicted to dissociate without barrier to methanol and NO_2. In contrast, the various conformers of **50** are weakly bound with respect to dissociation to HONO+$^\bullet OCH_3$ and CH_3ONO+$OH^\bullet$, which both require 35 kJ mol^{-1} in the respective transition states. However, upon vertical reduction radical **50** receives 88–103 kJ mol^{-1} by Franck-Condon effects which drive rapid dissociation, as also confirmed by RRKM kinetic analysis [222].

Tetravalent phosphorus radicals have the geometry of an incomplete trigonal bipyramid in which the p-orbital on phosphorus containing the unpaired electron can occupy an equatorial or axial position, giving rise to conformational isomers. These can interconvert by bending motions of the substituents, as in Berry pseudorotation (Scheme 21), or by pseudoinversion [223]. Although tetravalent 9-P-4 phosphorus radicals have been studied quite in detail by ab initio theory [224], there are only scant experimental data describing their properties.

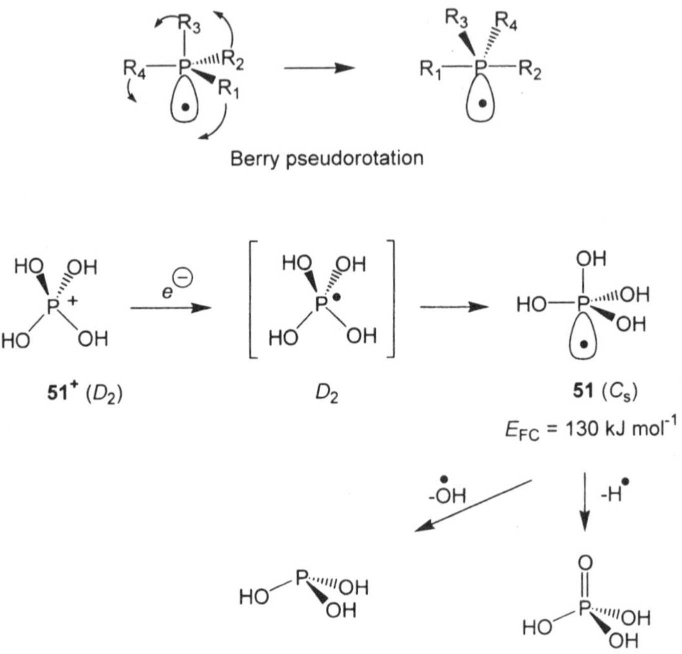

Berry pseudorotation

51^+ (D_2) D_2 **51** (C_s) E_{FC} = 130 kJ mol^{-1}

Scheme 21

The tetrahydroxyphosphoranyl radical, $P(OH)_4^{\bullet}$ (**51**), has been generated transiently by collisional reduction of the stable $P(OH)_4^+$ cation (**51**$^+$). Because of a large difference in the equilibrium geometries of the cation and radical, vertical electron transfer results in large Franck-Condon effects and the radical is formed with 130 kJ mol^{-1} vibrational energy (Scheme 21). This exceeds the transition state energies for both loss of H forming phosphoric acid, and loss of OH forming trihydroxyphosphane [34]. Thus, despite the fact that **51** is both thermodynamically and kinetically stable, it undergoes rapid dissociation when formed by vertical electron transfer, so that no survivor ions are present in the $^+$NR$^+$ mass spectrum.

$P(OH)_2^{\bullet}$ (**52**) is a hypovalent phosphorus radical that has recently been generated and characterized by NRMS [225]. The precursor cation for **52**, $P(OH)_2^+$, was produced by dissociative ionization of triethyl phosphite. $P(OH)_2^{\bullet}$ was found to be stable on the μs time scale. The extent of radical dissociation, judged by the relative intensity of the survivor ion, was shown to strongly depend on the precursor ion internal energy. Precursor ions produced by charge-exchange ionization with only 17 kJ mol^{-1} internal energy gave rise to **52** that showed a dominant fraction of survivor ions in the $^+$NR$^+$ mass spectrum, indicating little or no radical dissociation. More energetic precursor ions, produced by 70-eV electron ionization, gave rise to radicals that showed more extensive neutral dissociations by loss of H, OH, and H_2O that occurred in a 1:1.2:3.2 branching ratio, as established by variable-time measurements. The experimental H:H_2O branching ratio could be reproduced by RRKM calculations that were based on transition state energies from high-level ab initio calculations. However, the absolute rate constants from RRKM calculations ($k=10^8-10^9$ s^{-1}) were much larger than those from the variable-time data (10^5-10^6 s^{-1}). The experimental and theoretical data were reconciled by considering formation of excited electronic states of **52** that were examined by combined configuration interaction singles and time-dependent density functional theory calculations. The first excited (A) state of **52** receives 2.85 eV energy when formed by vertical electron transfer from $P(OH)_2^+$. Internal conversion, $A \rightarrow X$, forms the ground state with sufficient vibrational energy to rapidly dissociate by loss of H and H_2O in a branching ratio that is close to that observed experimentally. However, the overall kinetics for internal conversion $A \rightarrow X$ folowed by dissociation is controlled by the rate-determining first step, which occurs with rate constants in the 10^5-10^6 s^{-1} range. It may be noted that vertical electron transfer in $P(OH)_2^+$ is accompanied by modest Franck-Condon effects that result in a 33 kJ mol^{-1} vibrational excitation in **52** [225].

6.4 Boron and Silicon Radicals

Whereas no hypervalent boron or silicon radicals have been studied by NRMS, there are several hypovalent species that have been successfully generated and characterized. Dihydroxyboron radical, $B(OH)_2^\bullet$ (**53**), dissociates extensively by loss of H when generated by collisional reduction of $B(OH)_2^+$. The dissociation is driven by the >190 kJ mol^{-1} internal energy acquired by Franck-Condon effects, whereas cleavage of the OH bond requires only 79 kJ mol^{-1} at the thermochemical threshold [226]. In a related study, collisional reduction of methoxyborinium (CH_3O-B^+-H) and methyl(hydroxy)borinium (CH_3-B^+-OH) cations was found to give fractions of stable radicals, in spite of sizeable Franck-Condon effects in vertical electron transfer [227]. Large Franck-Condon effects dominate the chemistry of $(CH_3)_2SiOH$ which dissociates extensively when formed by collisional reduction of $(CH_3)_2SiOH^+$ [228]. The generation and characterization of $[CH_3,Si,O]$ radicals met with some difficulties that were due to isomerization of precursor cations, as elucidated by CAD spectra and ab initio calculations. The radical structures that appear to have been assigned unambiguously are sila-acetyl radicals $H_3SiCO^\bullet$ and $CH_3SiO^\bullet$ [229]. The latter species were also made from CH_3SiO^+ that in turn was generated by dissociative ionization of $(CH_3)_3SiOH$ [228]. The homologous $H_2SiOH^\bullet$ radical, which is the sila-analog of the hydroxymethyl radical, is stable when generated by collisional reduction of H_2SiOH^+, as reported for precursor cations prepared from trimethylsilanol [228] and tetramethoxysilane [230]. For other NRMS studies of a variety of transient silicon molecules and radicals see [231–237].

Returning to the introductory sentence of this section, it may be noted that hypervalent silicon hydrides have been prepared as anions in the gas phase and their ion chemistry has been investigated by Squires and co-workers [238]. Thus, the synthetic methods for the precursor anions are known and it is perhaps only a question of time for pentavalent silicon radicals to be generated by NRMS as transient species in the gas phase.

6.5 Heterocyclic and Nucleobase Radicals

Reactions with aromatic and heterocyclic molecules of small radicals, H, OH, OOH, etc., are of considerable interest in several fields of science and technology. For example, combustion involves radical chain reactions that are propagated by H, alkyl, and OH radicals that can add to the aromatic ring or abstract hydrogen atoms. Radical reactions are also important in DNA damage by ionizing radiation and oxidative stress that is linked to aging, cancer, and other processes. The intermediates of such reactions are short-lived radicals that are very difficult to study in situ because of a multitude of components present in the very complex mixtures. By generating transient radicals by specific reactions under well-defined conditions, NRMS provides a surgical tool for studying the intrinsic properties of such elusive species in the gas phase.

The general strategy for generating radical adducts to heterocyclic molecules relies on selective protonation of a suitable neutral precursor to prepare a cation of a well-defined structure. The gas-phase acid is chosen so as to attack only the most basic site in the molecule, or alternatively, non-selective protonation can be used to prepare a mixture of ions. For example, protonation of imidazole with NH_4^+ occurs selectively on the imine nitrogen atom (N-1), which has the highest proton affinity and is the only position that can be protonated by an exothermic reaction (Scheme 22) [239].

The other positions are much less basic than ammonia (PA=853 kJ mol^{-1}) and cannot be protonated with NH_4^+ in the gas phase under conditions of thermal

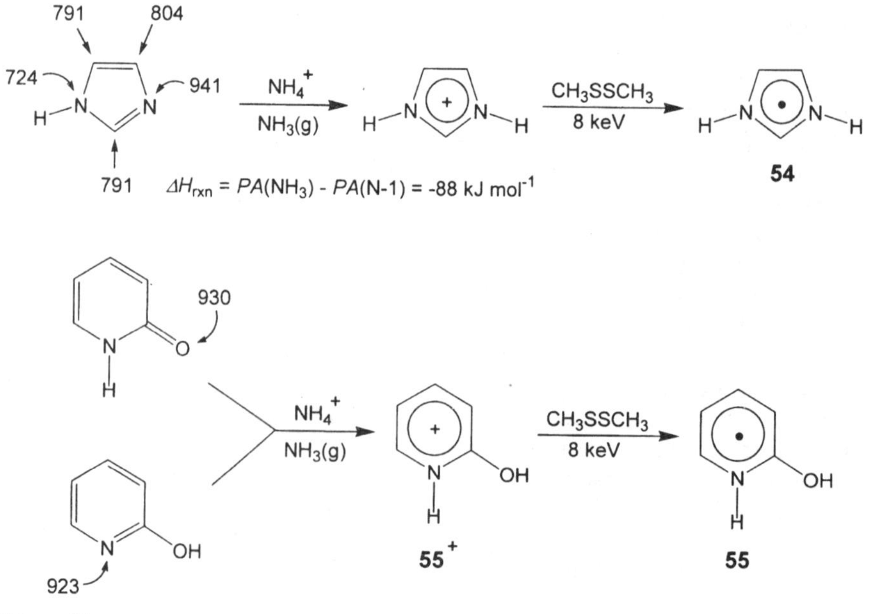

Scheme 22

equilibrium. The topical proton affinities that predict the reaction course are obtained with very good accuracy by ab initio or density functional calculations. Convergent protonation can be used to produce a single ion from a mixture of tautomers, as shown for the 2-hydroxypyridine-2(1*H*)-pyridone system. While the neutral molecules coexist as a mixture of tautomers in the gas phase, protonation with NH_4^+ occurs on O-2 in 2(1*H*)-pyridone and on N-1 in 2-hydroxypyridine, so that a single cation **55**$^+$ is produced (Scheme 22). Collisional reduction of the heterocyclic cations then generates transient radicals **54** and **55**, respectively, that are investigated by NRMS techniques.

Hydrogen atom adducts to heteroaromates are intrinsically stable species, as shown for pyrrole [239], imidazole [239], pyridine [240], 2-hydroxypyridine [241], 3-hydroxypyridine [242], pyrimidine, 2-aminopyrimidine, and 4-aminopyrimidine [243, 244]. The (1*H*,3*H*)-dihydroimidazolyl radical (**54**) undergoes extensive dissociation by hydrogen atom loss which shows a dual mechanism, as established by deuterium labeling. The major pathway is direct cleavage of one of the N–H bonds forming imidazole directly. The other mechanism involves a rate-determining isomerization to the more stable (1*H*,2*H*)dihydroimidazolyl radical (**56**) that eliminates one of the hydrogen atoms from the newly formed C-2 methylene group (Scheme 23). The dissociation is driven by large Franck-Condon effects in vertical electron transfer to the precursor cation that deposits >180 kJ mol^{-1} in radical **54**, whereas isomerization to **56** requires only 132 kJ mol^{-1} in the transition state [239]. Ring-cleavage dissociations were also observed for vertically formed **54** that produced HCN and C_2H_xN fragments.

Competitive dissociations by loss of H and ring cleavage are typical of the unimolecular chemistry of heterocyclic radicals, as observed by NRMS. However, ab initio calculations show that, for the radicals studied so far, the loss of H had substantially lower transition state and product threshold energies and thus should be kinetically preferred. This is shown in the potential energy surface diagram for dissociations of the hydrogen atom adduct to the O-4 position in the RNA nucleobase uracil (**57**, Fig. 9) [245]. Dissociation of the O-4–H bond in **57** requires 111 kJ mol^{-1} in the transition state, which is substantially less than the transition state energy for the most favorable ring opening by cleavage of the N-1–C-2 bond that requires 178 kJ mol^{-1} [246]. In spite of this difference, which is

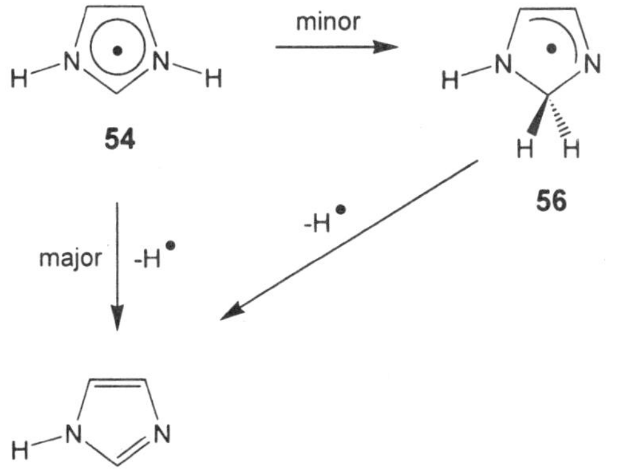

Scheme 23

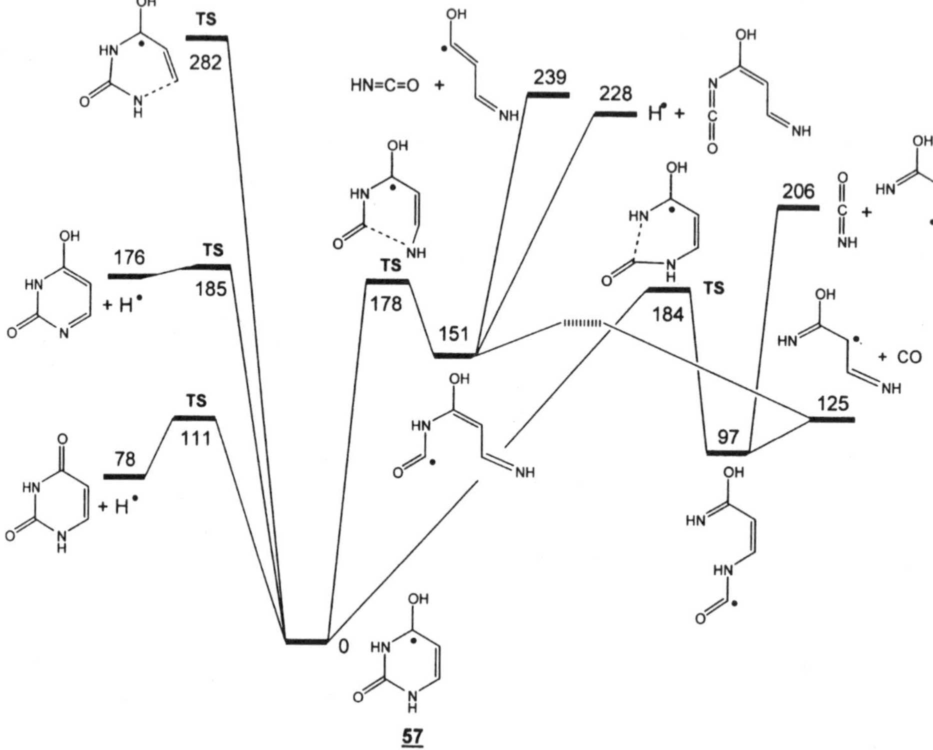

Fig. 9. Potential energy diagram for dissociations of uracil radical 57

also reflected in the relative RRKM rate constants for unimolecular dissociations that favor H loss, the majority of fragments in the $^{+}NR^{+}$ mass spectrum of 57 (80%) arise by ring cleavage. Analysis of excited electronic states in 57 indicated that the *A* and *B* states had both sufficient energy and electronic properties to promote ring dissociations [246].

C-5 is the most reactive position for radical attack in uracil and uridine. The reactive intermediate of H atom addition (58) was generated in the gas phase as shown in Scheme 24. Dissociative ionization of 6-ethyl-5,6-dihydrouracil was accomplished by charge exchange with COS^{+} that delivered the energy needed

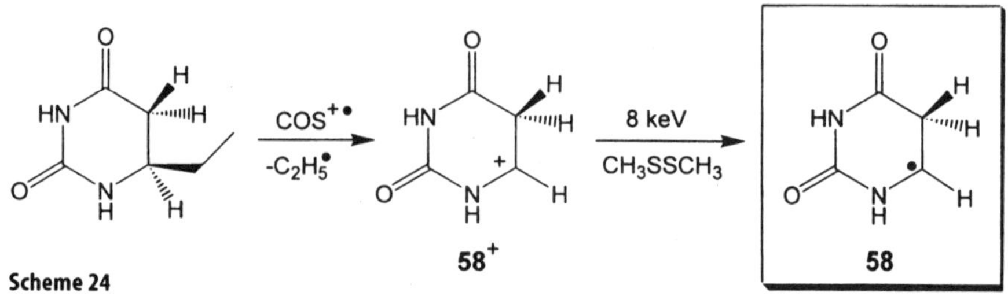

Scheme 24

for ionization and ion dissociation by loss of ethyl radical. COS was chosen on the basis of ab initio calculations that provided the appearance energy for the formation of $\mathbf{58}^+ + C_2H_5^{\bullet}$ which was matched with the recombination energy of the charge-exchange reagent. These precautions were necessary, because $\mathbf{58}^+$ is substantially less stable than five other protonated uracil isomers (e.g., the most stable $\mathbf{57}^+$) and energy-selective formation was warranted to prevent ion isomerization. Collisional neutralization of $\mathbf{58}^+$ gave a fraction of stable radicals **58** that were detected as survivor ions in the $^+NR^+$ mass spectrum [247]. Loss of H from C-5 was a major dissociation of **58** that was shown by deuterium labeling to be >95% regiospecific.

In contrast to H atom adducts, the elusive OH adduct to N-1 in pyridine (**59**) was unstable under NR conditions. The precursor cation ($\mathbf{59}^+$) was prepared by regioselective protonation of pyridine-*N*-oxide where, according to ab initio calculations, the oxygen atom is the most basic site (Scheme 25) [248]. Collisional reduction of $\mathbf{59}^+$ caused extensive dissociations of the intermediate radical **59** by loss of H, OH, and ring fragmentations, so that no survivor ion was detected in the $^+NR^+$ mass spectrum. The low stability of **59** is explained by the potential energy surface, which shows that the radical is metastable with respect to the 49 kJ mol^{-1} exothermic loss of OH, and is kinetically stabilized by a small (15 kJ mol^{-1}) barrier to dissociation [248]. This barrier is readily overcome by the vibrational energy (22 kJ mol^{-1}) the radical acquires upon vertical electron transfer by Franck-Condon effects. In light of the facile dissociation by loss of OH, the concurrent loss of H and ring fragmentations should not be competitive, as they require 124 and 229 kJ mol^{-1} in the respective transition states. The fact that these dissociations are observed was attributed to the formation of excited electronic

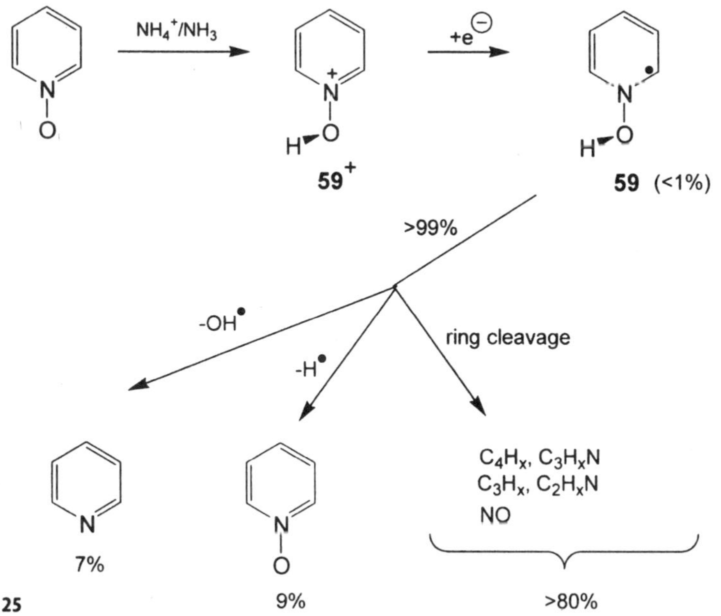

NH$_4^+$/NH$_3$ → $\mathbf{59}^+$ → +e$^-$ → **59** (<1%)

>99%: $-OH^{\bullet}$ → pyridine 7%; $-H^{\bullet}$ → pyridine-N-oxide 9%; ring cleavage → C_4H_x, C_3H_xN, C_3H_x, C_2H_xN, NO >80%

Scheme 25

states of **59**. While the *A* state is isoenergetic with the transition state for loss of H, the *B* state has sufficient energy to lose H, and the *C* and higher states can undergo ring dissociations.

Radicals arising by hydrogen atom abstraction from pyridine have also been generated and studied. NR of isomeric 2-, 3-, and 4-pyridyl cations furnished survivor ions in the $^{+}NR^{+}$ mass spectra that also showed somewhat different fragmentation patterns [249]. It was noted that although the NR spectra alone would not allow unambiguous ion identification, the differences were significant enough to prove the existence of distinct pyridyl radicals.

6.6 Alkoxy Radicals and Related Transient Species

Alkoxy radicals are well known intermediates of radical reactions in the gas phase and solution. The synthetic route to NR generation of alkoxy radicals uses gas-phase alkoxide anions that are readily produced by deprotonation of alcohols [2]. Alkoxy radicals produced by collisional electron detachment have been studied by the NIDD technique and several dissociations have been identified. In particular, alkoxy radicals derived from higher alcohols were found to undergo α-cleavage dissociations and hydrogen transfer rearrangements analogous to the Barton reaction upon nitrite photolysis in solution [1]. Isomeric peroxy radicals, $CH_2OOH^{\bullet}$ and $CH_3OO^{\bullet}$, were generated from the respective cation and anion and analyzed after reionization to cations [250]. $CH_2OOH^{\bullet}$ gives only a very weak survivor cation in the $^{+}NR^{+}$ mass spectrum, possibly due to a combination of extensive radical and ion dissociations that were not resolved in this case. In contrast, $CH_3OO^{\bullet}$ gives an abundant survivor cation attesting to the stability of the radical.

Collisional electron detachment from anion-radicals has been used to generate transient diradicals that were characterized by $^{-}NR^{+}$ mass spectra complemented by density functional theory calculations. Neutralization of $^{\bullet}CH_2COO^{-}$, followed by reionization to cations, resulted in substantial dissociation yielding mostly CO_2 and CO, but also a small peak of the survivor ion. The data were interpreted by the formation of singlet and triplet $^{\bullet}CH_2COO$, where the singlet was postulated to cyclize to *α*-acetolactone [251]. The isomeric methylenecarboxy diradical, $^{\bullet}CH_2OCO^{\bullet}$, was generated by neutralization of the cation-radical and found to give a weak survivor ion upon reionization [252].

Another attempt at generating the elusive oxyallyl biradical, $^{\bullet}CH_2C(O^{\bullet})=CH_2$ (**60**), has been reported recently [253]. While the corresponding cation-radical is unstable and undergoes spontaneous ring closure to [methyleneoxirane]$^{+\bullet}$ [254], the anion-radical is calculated to be a stable species. Anion $CH_2C(O^{-})=CH_2$ was generated from acetone by H_2 abstraction with $O^{-\bullet}$ and clearly distinguished from isomeric carbene anion-radicals such as $[CH_3CO-C-H]^{-\bullet}$ and $[CH_3-C-CH=O]^{-\bullet}$. The $^{-}NR^{+}$ mass spectrum of $^{\bullet}CH_2C(O^{-})=CH_2$ shows a very weak peak at *m/z* 56 that was assigned to ionization of the triplet state of **60**, as the singlet state was ruled out by calculations. The major dissociation in the $^{-}NR^{+}$ mass spectrum of **60** is loss of $^{3}CH_2$ which forms $CH_2=C=O$. Interestingly, density functional theory calculations predict triplet **60** to be a very stable species in the isolated state, with the threshold energy for the loss of $^3CH_2$ being as high as 184 kJ mol^{-1}. The low stability of **60** was explained by triplet-singlet intersystem crossing followed by dissociation on the singlet potential energy surface [253].

6.7
Peptide and Related Amide Radicals

Peptides and proteins are attacked by reactive radicals and cation radicals that are formed by oxidative processes of cellular metabolism or by ionizing radiation. Radicals can attack peptides and proteins by a variety of chemical reactions such as hydrogen abstraction, electron transfer, and addition to multiple bonds, that can cause fragmentation, rearrangements, dimerization, disproportionation, and substitution in the peptide backbone or side chains [255]. Interestingly, some enzymes, e.g., pyruvate formate lyase and anaerobic ribonucleotide reductase, adopt radical reactions in their catalytic portfolio and generate transient intermediates that contain α-glycyl radical residues, $-^{\bullet}CH\text{-}CO\text{-}NH-$ [256]. The prototypical α-glycyl radical (**61**) [257–260], its methyl ester (**62**), and *N*-methylamide (**63**) [260] have been generated in the gas phase by collisional reduction of the corresponding cations (Scheme 26). Radical **61** gives an abundant survivor ion in the $^+NR^+$ mass spectrum that is evidence of a fraction of **61** being stable. This finding is supported by combined ab initio and density functional theory calculations [260] that show that dissociations of **61** are substantially endothermic.

For example, the lowest-energy dissociation of **61** to form $H_2N\text{–}CH\text{–}OH$ and CO requires 129 kJ mol^{-1} threshold energy, but must overcome a 310 kJ mol^{-1} barrier for the rate-determining isomerization to $H_2N\text{–}CH(OH)\text{–}CO$ [260]. α-Radicals derived from methyl glycinate (**62**) and glycine *N*-methylamide (**63**) are somewhat less stable, but still show fractions of undissociated survivor ions in the $^+NR^+$ mass spectra. This increased reactivity is supported by theoretical calculations that show additional dissociation channels, e.g., elimination of CH_3OH from **62** and elimination of CO from **63** [260].

Transient intermediates arising by radical additions to the amide bond have been studied recently in an effort to explain the complex processes occurring in electron-capture dissociation of multiply charged peptide and protein cations [192]. The amino(hydroxy)methyl radical, $HC(OH)NH_2$ (**64**), is a prototypical species that represents the simplest model for radical additions to peptide bonds. Radical **64** was generated from cation $\mathbf{64}^+$ which in turn was prepared by

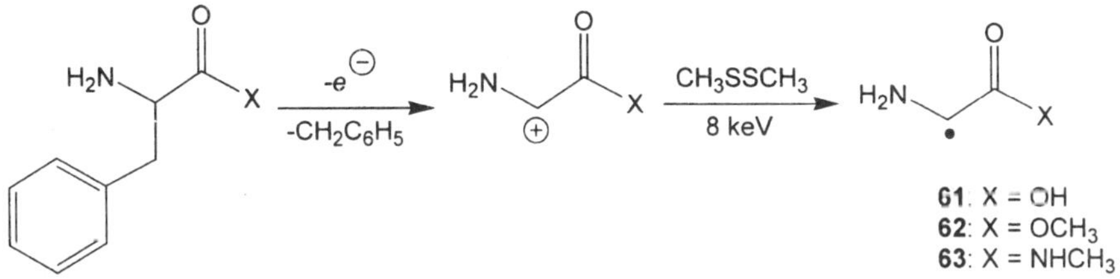

Scheme 26

$-e^{\ominus}$, $-CH_2C_6H_5$; CH_3SSCH_3, 8 keV

61: X = OH
62: X = OCH_3
63: X = $NHCH_3$

gas-phase protonation of formamide. Radical **64** is substantially stable under NR and gives an abundant fraction of reionized **64**$^+$ in the $^+$NR$^+$ mass spectrum. Loss of H from the OH group is the major dissociation of **64** as determined by variable-time spectra and supported by ab initio calculations [261]. The hydrogen atom adduct to *N*-methylacetamide, $CH_3C(OH)NHCH_3$ (**65**), is another model for radical adducts to peptides. Radical **65** dissociates by competing cleavages of the O–H and N–CH_3 bonds that occur in a 1.67:1 branching ratio [262]. This result is supported by Rice-Ramsperger-Kassel-Marcus calculations of unimolecular rate constants for the dissociations that used a potential energy surface from high-level ab initio calculations [262].

7
Transient Metal Compounds

In addition to transient intermediates of organic and inorganic reactions, NRMS has also been used to generate several short-lived species containing main group and transition metal atoms. The topic has been thoroughly reviewed [15, 263] and will not be covered in detail here. An interesting example of transient metal species is hypervalent radicals $Li_2F^\bullet$ [264], Li-$NH_3^\bullet$, and Li-$OH_2^\bullet$ [265] that show substantial stability in the gas phase and present a nice contrast to the properties of hypervalent onium radicals discussed in Sect. 6.1. The Li_2F^+ precursor cations were prepared by fast-atom bombardment (FAB) of CF_3COOLi and were found to be very stable, such that no dissociations of metastable ions were detected. Collisional reduction with TMA produced neutral $Li_2F^\bullet$ that after reionization gave a dominant peak of the survivor ion. $Li_2F^\bullet$ was well characterized by the $^+$NR$^+$ mass spectrum in spite of a minor contamination with isobaric $CH_3OH_2^+$ from the FAB matrix [264]. Neutralization of $CH_3OH_2^+$ produces an unstable methylhydroxonium hypervalent radical which dissociates, and so the neutral beam is effectively purified as a result of collisional reduction [264]. Interestingly, reionization to negative ions also provides an Li_2F^- survivor anion, which is stable in keeping with theoretical predictions [266]. Theory and NRMS are also in qualitative agreement regarding the stability of neutral Li-NH_3 and Li-H_2O complexes [265]. When generated by collisional reduction with TMA of the corresponding cations, both Li-$NH_3^\bullet$ and Li-$H_2O^\bullet$ showed fractions of non-dissociating radicals that were reionized to furnish survivor cations. Vertical electron transfer in Li-NH_3^+ and Li-H_2O^+ was calculated to result in negligible Franck-Condon effects, and the radicals were bound by 50–60 kJ mol^{-1} against Li–O and Li–N bond dissociation [265].

8
Summary and Future Outlook

The extraordinary variety of transient species pertinent to organic, inorganic, and organometallic chemistry and biochemistry that were generated by collisional neutralization of cations and anions and characterized by NRMS is perhaps the best evidence of the power and generality of this approach to solving difficult and tantalizing chemical problems. The future developments in this area will likely focus on overcoming the current drawbacks and deficiencies of the method. Because electron transfer is intrinsically non-selective, the question of the electronic state(s) of the transient neutral species is of great importance, in particular because such states can undergo unusual dissociations or provide routes to superexcited vibrational states. Access to dark excited states, that are difficult or impossible to generate photochemically, is an added benefit of NRMS, as recognized in several studies and investigated by laser photoexcitation [33, 60, 78, 207, 208, 217, 220] and photoionization [18, 43, 60]. Extending NRMS to larger molecules of interest to organometallic chemistry, biochemistry, and biology will require using soft ionization techniques, such as electrospray (ESI), as recently implemented in this author's laboratory [267]. This new ESI-NRMS coupling promises providing access to transient radicals derived from nucleosides, nucleotides, peptides, and other complex molecules.

So far, NRMS has been mostly a qualitative technique that did not provide direct information on the energetics of neutralized and reionized species. However, by combining experiment and theory, progress has been made in determining internal energy distributions in several molecules and radicals [26–30]. The predictive power of high-level theory for designing and planning experiments is also impressive [29, 33, 247]. These trends are likely to continue and be widely adopted, and the progress in technology, both experimental and computational, will undoubtedly bring new exciting data on short-lived intermediates of chemical reactions of increasing complexity.

Several studies have appeared recently that deserve a note. C_3 carbon clusters produced by neutralization of anions were found by ^{13}C labelling not to undergo full scrambling of carbon atoms [268]. This contrasts with the recent finding regarding NRMS of C_3^{+} [110]. Silicon dihydroxide, $Si(OH)_2$ [269] and trihydroxysilyl radical, $Si(OH)_3^{\bullet}$ [270] are stable when formed by collisional neutralization of the corresponding cations.

Acknowledgement. Financial support was provided by the National Science Foundation (Grants CHE-9712570 and CHE-0090930 for experimental work and CHE-9808182 for computations). Sincere thanks are due to all my coworkers whose work is presented and cited in this review.

9
References

1. Hornung G, Schalley CA, Dieterle M, Schröder D, Schwarz H (1997) Chem Eur J 3:1866
2. McMahon AW, Chowdhury SK, Harrison AG (1989) Org Mass Spectrom 24:620
3. Hop CECA, Holmes JL (1991) Org Mass Spectrom 26:476
4. Danis PO, Wesdemiotis C, McLafferty FW (1983) J Am Chem Soc 105:7454
5. Curtis PM, Williams BW, Porter RF (1979) Chem Phys Lett 65:296; Burgers PJ, Holmes JL, Mommers, AA, Terlouw JK (1983) Chem Phys Lett 102:1
6. Gellene GI, Porter RF (1983) Acc Chem Res 16:200
7. Wesdemiotis C, McLafferty FW (1987) Chem Rev 87:485
8. Terlouw JK, Schwarz H (1987) Angew Chem Int Ed Engl 26:805
9. Holmes JL (1989) Mass Spectrom Rev 8:513
10. Tureček F (1992) Org Mass Spectrom 27:1087
11. Goldberg N, Schwarz H (1994) Acc Chem Res 27:347
12. Zagorevski DV, Holmes JL (1994) Mass Spectrom Rev 13:133

13. Schalley CA, Hornung G, Schröder D, Schwarz H (1998) Chem Soc Rev 27:91
14. Tureček F (1998) J Mass Spectrom 33:779
15. Zagorevski DV, Holmes JL (1999) Mass Spectrom Rev 18:87
16. Gerbaux P, Wentrup C, Flamang R (2000) Mass Spectrom Rev 19:367
17. Čermák V (1973) J Electron Spectrosc Relat Phenom 2:97
18. Polášek M, Tureček F (2001) J Phys Chem A 105:1371
19. Metz RB, Cyr DR, Neumark DM (1991) J Phys Chem 95:2900
20. Lorquet JC, Leyh-Nihant B, McLafferty FW (1990) Int J Mass Spectrom Ion Processes 100:465
21. Bordas-Nagy J, Holmes JL, Hop CECA (1989) Int J Mass Spectrom Ion Processes 94:189
22. Bordas-Nagy J, Holmes JL (1988) Int J Mass Spectrom Ion Processes 82:81
23. Griffiths IW, Mukhtar ES, March RE, Harris FM, Beynon JH (1981) Int J Mass Spectrom Ion Phys 39:125
24. Beranová S, Wesdemiotis C (1994) J Am Soc Mass Spectrom 5:1092
25. Hayakawa S (2001) Int J Mass Spectrom 212:229
26. Nguyen VQ, Tureček F (1996) J Mass Spectrom 31:843
27. Wolken JK, Tureček F (1999) J Am Chem Soc 121:6010
28. Wolken JK, Tureček F (1999) J Phys Chem A 103:6268
29. Gerbaux P, Tureček F (2002) J Phys Chem A 106:5938
30. Tureček F (2002) Int J Mass Spectrom (in press)
31. Adachi H, Basco N, James DGL (1978) Chem Phys Lett 59:502
32. Hop CECA, Holmes JL (1991) Int J Mass Spectrom Ion Processes 104:213
33. Frank AJ, Tureček F (1999) J Phys Chem A 103:5348
34. Tureček F, Gu M, Hop CECA (1995) J Phys Chem 99:2278
35. Nguyen VQ, Shaffer SA, Tureček F, Hop CECA (1995) J Phys Chem 99:15,454
36. Syrstad EA, Tureček F (2001) J Phys Chem A 105:11,144
37. Syrstad EA, Tureček F (2002) J Phys Chem A 105 (submitted)
38. Ramond TM, Davico GE, Schwartz RL, Lineberger WC (2000) J Chem Phys 112:1158
39. Chupka WA (1993) J Chem Phys 99:5800
40. Bordas-Nagy J, Holmes JL, Mommers AA (1986) Org Mass Spectrom 21:629
41. Holmes JL, Sirois M (1990) Org Mass Spectrom 25:481
42. Sirois M, George M, Holmes JL (1994) Org Mass Spectrom 29:11
43. Sadílek M, Tureček F (1996) J Phys Chem 100:9610
44. Tureček F, Reid PJ (2002) Int J Mass Spectrom (in press)
45. Gellene GI, Williams BW, Porter RF (1981) J Chem Phys 74:5636
46. Griffith KS, Gellene GI (1993) J Phys Chem 97:9882
47. Gellene GI, Porter RF (1983) J Chem Phys 79:5975
48. Jeon SJ, Raksit AB, Gellene GI, Porter RF (1985) J Chem Phys 82:4916
49. Shaffer SA, Tureček F, Cerny RL (1993) J Am Chem Soc 115:12,117
50. Tureček F, Gu M, Shaffer SA (1992) J Am Soc Mass Spectrom 3:493
51. Hop CECA, Bordas-Nagy J, Holmes JL, Terlouw JK (1988) Org Mass Spectrom 23:155
52. Feng R, Wesdemiotis C, Baldwin MA, McLafferty FW (1988) Int J Mass Spectrom Ion Processes 86:95
53. Wesdemiotis C, McLafferty FW (1987) J Am Chem Soc 109:4760
54. Tureček F, Drinkwater DE, McLafferty FW, Maquestiau A (1989) Org Mass Spectrom 24:669
55. Fura A, Tureček F, McLafferty FW (1991) J Am Soc Mass Spectrom 2:492
56. Wesdemiotis C, Feng R, Williams ER, McLafferty FW (1986) Org Mass Spectrom 21:689
57. Danis PO, Feng R, McLafferty FW (1986) Anal Chem 58:348
58. Danis PO, Feng R, McLafferty FW (1986) Anal Chem 58:355
59. Stratmann RE, Scuseria GE, Frisch MJ (1998) J Chem Phys 109:8218
60. Nguyen VQ, Sadílek M, Frank AJ, Ferrier JG, Tureček F (1997) J Phys Chem A 101:3789
61. Kuhns DW, Shaffer SA, Tran TB, Tureèek F (1994) J Phys Chem 98:4845
62. Kuhns DW, Tureček F (1994) Org Mass Spectrom 29:463
63. Sadílek M, Tureček F (1996) J Phys Chem 100:224

64. Schalley CA, Hornung G, Schröder D, Schwarz H (1998) Int J Mass Spectrom Ion Processes 172:181
65. Danis PO, Wesdemiotis C, McLafferty FW (1983) J Am Chem Soc 105:7454
66. Kresge AJ (1996) Chem Soc Rev 25:275
67. Tureček F (1989) Adv Mass Spectrom 11:1079
68. Holmes JL, Lossing FP (1980) J Am Chem Soc 102:1591
69. Tureček F, Vivekananda S, Sadílek M, Polášek M (2002) J Am Chem Soc 124 (submitted)
70. Tureček F, Vivekananda S, Sadílek M, Polášek M (2002) J Mass Spectrom 37 (in press)
71. Polce MJ, Wesdemiotis C (2000) J Mass Spectrom 35:251
72. McGibbon GA, Burgers PC, Terlouw JK (1994) Int J Mass Spectrom Ion Processes 136:191
73. Suh D, Francis JT, Terlouw JK, Burgers PC, Bowen RD (1995) Eur Mass Spectrom 1:545
74. van Baar BLM, Weiske T, Terlouw JK, Schwarz H (1986) Angew Chem 98:275
75. Buschek JM, Holmes JL, Lossing FP (1986) Org Mass Spectrom 21:729
76. Sülzle D, Verhoeven M, Terlouw JK, Schwarz H (1988) Angew Chem 100:1591
77. Terlouw JK (1989) Adv Mass Spectrom 11:984
78. Frank AJ, Sadílek M, Ferrier JG, Tureček F (1997) J Am Chem Soc 119:12,343
79. Tureček F, Drinkwater DE, McLafferty FW (1989) J Am Chem Soc 111:7696
80. Block E, O'Connor J (1974) J Am Chem Soc 96:3929
81. Tureček F, Brabec L, Vondrák T, Hanuš V, Hájíček J, Havlas Z (1988) Collect Czech Chem Commun 53:2140
82. Lacombe S, Loudet M, Banchereau E, Simon M, PfisterGuillouzo G (1996) J Am Chem Soc 118:1131
83. Langewerf FM, vandeWeert M, Heerma W, Haverkamp J (1996) Rapid Commun Mass Spectrom 10:1905
84. Tureček F, McLafferty FW, Smith BJ, Radom L (1990) Int J Mass Spectrom Ion Processes 101:283
85. Egsgaard H, Carlsen L, Florencio H, Drewello T, Schwarz H (1989) Ber Bunsenges Phys Chem 93:76
86. McKee ML (1986) J Am Chem Soc 108:5784
87. (a) Stacie EW, Show GT (1934) Proc Roy Soc Ser A, A146:388; (b) Fernandez-Ramos A, Marinez-Nunez E, Rios MA, Rodriguez-Otero J, Vazquez SA, Estevez CM (1998) J Am Chem Soc 120:7594
88. Hop CECA (1994) J Am Chem Soc 116:3163
89. Wiedmann FA, Wesdemiotis C (1994) J Am Chem Soc 116:2481
90. Tureček F (1987) Unpublished results
91. Arduengo AJ (1999) Acc Chem Res 913
92. Moss RA, Jones M Jr (eds) (1973) Carbenes. Wiley, New York
93. Moss RA (1989) Acc Chem Res 22:15
94. Flammang R, Nguyen MT, Bouchoux G, Gerbaux P (2000) Int J Mass Spectrom 202:A8
95. Sülzle D, Drewello T, van Baar BLM, Schwarz H (1988) J Am Chem Soc 110:8330
96. Feng R, Wesdemiotis C, McLafferty FW (1987) J Am Chem Soc 109:6521
97. Wesdemiotis C, Leyh B, Fura A, McLafferty FW (1990) J Am Chem Soc 112:8655
98. Polce MJ, Wesdemiotis C (1993) J Am Chem Soc 115:10,849
99. Wesdemiotis C, McLafferty FW (1987) J Am Chem Soc 109:4760
100. Polce MJ, Kim Y, Wesdemiotis C (1997) Int J Mass Spectrom Ion Processes 167/168:309
101. Burgers PC, McGibbon GA, Terlouw JK (1994) Chem Phys Lett 224:539
102. McGibbon GA, Kingsmill CA, Terlouw JK (1994) Chem Phys Lett 222:129
103. Hop CECA, Chen H, Ruttink PJA, Holmes JL (1991) Org Mass Spectrom 26:679
104. Lahem D, Flammang R, Nguyen MT (1997) Chem Phys Lett 270:93
105. Thanh Le H, Lam Nguyen T, Lahem D, Flammang R, Nguyen MT (1999) Phys Chem ChemPhys 1:755
106. Vivekananda S, Srinivas R, Manoharan M, Jemmis ED (1999) J Phys Chem A 103:5123
107. Tureček F, Drinkwater DE, McLafferty FW (1991) J Am Chem Soc 113:5958
108. O'Bannon PE, Sülzle D, Dailey WP, Schwarz H (1992) J Am Chem Soc 114:344
109. Goldberg N, Fiedler A, Schwarz H (1995) J Phys Chem 99:15,327

110. Fura A, Tureček F, McLafferty FW (2002) Int J Mass Spectrom 217:81
111. Blanksby SJ, Schröder D, Dua S, Bowie JH, Schwarz H (2000) J Am Chem Soc 122:7105
112. Dua S, Bowie JH (2002) J Phys Chem A 106:1374
113. Chen H, Holmes JL (1994) Int J Mass Spectrom Ion Processes 133:111
114. Sülzle D, Weiske T, Schwarz H (1993) Int J Mass Spectrom Ion Processes 125:75
115. Dawson D, Chen H, Holmes JL (1996) Eur Mass Spectrom 2:373
116. Schröder D, Heinemann C, Schwarz H, Harvey JN, Dua S, Blanksby SJ, Bowie JH (1998) Chem Eur J 4:2550
117. Dua S, Blanksby SJ, Bowie JH (2000) Int J Mass Spectrom 195/196:45
118. Schröder D, Schwarz H, Dua S, Blanksby SJ, Bowie JH (1999) Int J Mass Spectrom 188:17
119. Diels O, Wolf B (1906) Chem Ber 39:689
120. Reyerson LH, Kobe K (1930) Chem Rev 7:479
121. Sülzle D, Schwarz H (1988) Angew Chem Int Ed Engl 27:1337
122. Wentrup C, Kambouris P, Evans RA, Owen D, Macfarlane G, Chuche J, Pommelet JC, Cheikh AB, Plisnier M, Flammang R (1991) J Am Chem Soc 113:3130
123. Flammang R, Wentrup C (1999) Sulfur Rep 21:357
124. Wong MW, Wentrup C, Flammang R (1995) J Phys Chem 99:16,836
125. Flammang R, Van Haverbeke Y, Wong MW, Wentrup C (1995) Rapid Commun Mass Spectrom 9:203
126. Flammang R, Landu D, Laurent S, Barbieux-Flammang M, Kappe CO, Wong MW, Wentrup C (1994) J Am Chem Soc 116:2005
127. Vivekananda S, Srinivas R, Terlouw JK (1997) Int J Mass Spectrom Ion Processes 171:L13
128. Flammang R, Laurent S, Barbieux-Flammang M, Van Haverbeke Y, Wentrup C (1994) Rapid Commun Mass Spectrom 8:329
129. Flammang R, Van Haverbeke Y, Laurent S, Barbieux-Flammang M, Wong MW, Wentrup C (1994) J Phys Chem 98:5801
130. Flammang R, Laurent S, Barbieux-Flammang M, Wentrup C (1992) Rapid Commun Mass Spectrom 6:667
131. Wong MW, Wentrup C, Morkved EH, Flammang R (1996) J Phys Chem 100:10,536
132. Peppe S, Blanksby SJ, Dua S, Bowie JH (2000) J Phys Chem A 104:5817
133. Sülzle D, O'Bannon PE, Schwarz H (1992) Chem Ber 125:279
134. Sülzle D, Schwarz H (1989) Chem Phys Lett 156:397
135. Goldberg N, Sülzle D, Schwarz H (1993) Chem Phys Lett 213:593
136. Goldberg N, Schwarz H (1994) J Phys Chem 98:3080
137. Radom L, Bouma WJ, Nobes RH, Yates BF (1986) Pure Appl Chem 56:1831
138. Wesdemiotis C, Feng R, Danis PO, Williams ER, McLafferty FW (1986) J Am Chem Soc 108:5847
139. Linstrom PJ, Mallard WG (eds) (2001) IR and mass spectra. In: NIST Chemistry WebBook, NIST Standard Reference Database Number 69NIST Mass Spec Data Center, Stein SE, director, July 2001, National Institute of Standards and Technology, Gaithersburg MD, 20899 (http://webbook.nist.gov)
140. Yates BF, Bouma WJ, Radom L (1987) J Am Chem Soc 109:2250
141. Pople JA (1986) Chem Phys Lett 132:144
142. Cacace F, de Petris G, Troiani A (2002) Science (Washington D.C.) 295:480
143. Flammang R, Gerbaux P, Nguyen MT (2002) (submitted)
144. Keck H, Kuchen W, Tommes P, Terlouw JK, Wong T (1992) Angew Chem Int Ed Engl 31:86
145. Yates BF, Bouma WJ, Radom L (1984) J Am Chem Soc 106:5805
146. Dyson P, Hammick DL (1937) J Chem Soc 1724
147. Ashworth MRF, Daffern RP, Hammick DL (1937) J Chem Soc 809
148. Lavorato DJ, Terlouw JK, Dargel TK, Koch W, McGibbon GA, Schwarz H (1996) J Am Chem Soc 118:11,898
149. McGibbon GA, Heinemann C, Lavorato DJ, Schwarz H (1997) Angew Chem Int Ed Engl 26:1478
150. McGibbon GA, Hrušák J, Lavorato DJ, Schwarz H, Terlouw JK (1997) Chem Eur J 3:232

151. Dargel TK, Koch W, Lavorato DJ, McGibbon GA, Terlouw JK, Schwarz H (1999) Int J Mass Spectrom 185/187:925
152. Lavorato DJ, Dargel TK, Koch W, McGibbon GA, Schwarz H, Terlouw JK (2001) Int J Mass Spectrom 210/211:43
153. Lavorato DJ, Fel LM, McGibbon GA, Sen S, Terlouw JK, Schwarz H (2000) Int J Mass Spectrom 195/196:71
154. Flammang R, Barbieux-Flammang M, Gualano E, Gerbaux P, Hung TL, Nguyen MT, Tureček F, Vivekananda S (2001) J Phys Chem A 105:8579
155. Hop CECA, Chen H, Ruttink PJA, Holmes JL (1991) Org Mass Spectrom 26:679
156. Polášek M, Sadílek M, Tureček F (2000) Int J Mass Spectrom 195/196:101
157. Polášek M, Tureček F (2000) J Am Chem Soc 122:525
158. Vivekananda S, Srinivas R, Manoharan M, Jemmis ED (1999) J Phys Chem A 103:5123
159. Gerbaux P, Van Haverbeke Y, Flammang R, Wong MW, Wentrup C (1997) J Phys Chem A 101:6970
160. Gerbaux P, Flammang R, Morkved EH, Wong MW, Wentrup C (1998) Tetrahedron Lett 39:533
161. Gerbaux P, Flammang R, Morkved EH, Wong MW, Wentrup C (1998) J Phys Chem A 102:9021
162. Nguyen MT, Flammang R (1996) Chem Ber 129:1379
163. Nguyen MT, Allaf AW, Flammang R, Van Haverbeke Y (1997) THEOCHEM 418:209
164. Brönstrup M, Schröder D, Kretzschmar I, Schalley CA, Schwarz H (1998) Eur J Inorg Chem 1529
165. Keck H, Kuchen W, Tommes P, Terlouw JK, Wong T (1991) Phosphorus Sulfur Silicon Relat Elem 63:307
166. Keck H, Kuchen W, Renneberg H, Terlouw JK, Visser HC (1991) Angew Chem 102:331
167. Keck H, Kuchen W, Renneberg H, Terlouw JK, Visser HC (1990) Anorg Allg Chem 580:181
168. Keck H, Kuchen W, Terlouw JK, Tommes P (1999) Phosphorus Sulfur Silicon Relat Elem 149:23
169. Schröder D, Schwarz H, Wulf M, Sievers H, Jutzi P, Reiher M (1999) Angew Chem Int Ed Engl 38:3513
170. Häser M, Treutler O (1995) J Chem Phys 102:3703
171. Jones RO, Ganteför G, Hunsicker S, Pieperhoff P (1995) J Chem Phys 103:9549
172. Lewis GB (1916) J Am Chem Soc 38:762
173. Akiba K (ed) (1999) Chemistry of hypervalent compounds. Wiley-VCH, New York
174. Schuster A (1872) Br Assoc Adv Sci Rep 38
175. Schüler H, Michel A, Grün AE (1955) Naturforsch 10a:1
176. Herzberg G (1981) Discuss Faraday Soc 71:165
177. Perkins CW, Martin JC, Arduengo AJ, Lau W, Alegria A, Kochi JK (1980) J Am Chem Soc 102:7753
178. Holmes JL, Sirois M (1990) Org Mass Spectrom 25:481
179. Sirois M, George M, Holmes JL (1994) Org Mass Spectrom 29:11
180. Raksit AB, Porter RF (1987) J Chem Soc Chem Commun 500
181. Wesdemiotis C, Fura A, McLafferty FW (1991) J Am Soc Mass Spectrom 2:459
182. Gellene GI, Cleary DA, Porter RF (1982) J Chem Phys 77:3471
183. Shaffer SA, Tureček F (1994) J Am Chem Soc 116:8647
184. Frøsig L, Tureček F (1998) J Am Soc Mass Spectrom 9:242
185. Boldyrev AI, Simons J (1992) J Phys Chem 96:8840
186. Sadílek M, Tureček F (1996) J Phys Chem 100:9610
187. Tureèek F, Reid PJ (2002) Int J Mass Spectrom (in press)
188. Sadílek M, Tureček F (1996) J Phys Chem 100:15,027
189. Sadílek M, Tureček F (1999) Int J Mass Spectrom 185/187:639
190. Schiesser CH, Wild LM (1996) Tetrahedron 52:13,265
191. Shaffer SA, Sadílek M, Tureček F (1996) J Org Chem 61:5234
192. Zubarev RA, Kelleher NL, McLafferty FW (1998) J Am Chem Soc 120:3265

193. Shaffer SA, Sadílek M, Tureček F, Hop CECA (1997) Int J Mass Spectrom Ion Processes 160:137
194. Shaffer SA, Wolken JK, Tureček F (1997) J Am Soc Mass Spectrom 8:1111
195. Shaffer SA, Tureček F (1995) J Am Soc Mass Spectrom 6:1004
196. Wolken JK, Nguyen VQ, Tureček F (1997) J Mass Spectrom 32:1162
197. Tureček F, Polášek M, Frank AJ, Sadílek M (2000) J Am Chem Soc 122:2361
198. Andrae MO (1985) In: Galloway JN et al. (eds) The biogeochemical cycling of sulfur and nitrogen in the remote atmosphere. Reiel, New York, p 5
199. Calvert JG, Stockwell WR (1984) SO_2, NO and NO_2 oxidation mechanism: atmospheric considerations. Butterworth, Boston, chap 1
200. Iraqi M, Schwarz H (1994) Chem Phys Lett 221:359
201. O'Hair RAJ, DePuy CH, Bierbaum VM (1993) J Phys Chem 97:7955
202. Otto AH, Steudel R (2000) Eur J Inorg Chem 617
203. Laakso D, Smith CE, Goumri A, Rocha JDR, Marshall P (1994) Chem Phys Lett 227: 377
204. Qi JX, Deng WQ, Han KL, He GZ (1997) J Chem Soc Faraday Trans 93:25
205. Goumri A, Rocha JDR, Laakso D, Smith CE, Marshall P (1999) J Phys Chem A 103:11,328
206. Morris VR, Jackson WM (1994) Chem Phys Lett 223:445
207. Frank AJ, Sadílek M, Ferrier JG, Tureček F (1996) J Am Chem Soc 118:11,321
208. Frank AJ, Sadílek M, Ferrier JG, Tureček F (1997) J Am Chem Soc 119:12,343
209. Egsgaard H, Carlsen L, Florencio H, Drewello T, Schwarz H (1988) Chem Phys Lett 148:537
210. Fehsenfeld FC, Ferguson EE (1974) J Chem Phys 61:3181
211. Iraqi M, Goldberg N, Schwarz H (1993) Int J Mass Spectrom Ion Processes 124:R7
212. Turnipseed AA, Ravishankara AR (1993) In: Restelli G, Angeletti G (eds) Oceans, atmosphere and climate. Kluwer, Dordrecht, p 185
213. Gu M, Tureček F (1992) J Am Chem Soc 114:7146
214. McKee ML (1993) J Phys Chem 97:10,971
215. Merenyi G, Lind J, Engman L (1996) J Phys Chem 100:8875
216. Tureček F (2000) Collect Czech Chem Commun 65:455
217. Frank AJ, Tureèek F (1999) J Phys Chem A 103:5348
218. Egsgaard H (1993) Ion chemistry of the flame. Risø National Laboratory: Roskilde, Denmark
219. Ko T, Flaherty F, Fontijn A (1991) J Phys Chem 95:6967
220. Polášek M, Tureček F (1999) J Phys Chem A 103:9241
221. Polášek M, Tureček F (2000) J Am Chem Soc 122:9511
222. Polášek M, Tureček F (2000) J Am Soc Mass Spectrom 11:380
223. Berry RS (1960) J Chem Phys 33:933
224. Gustafson SM, Cramer CJ (1995) J Phys Chem 99:2267
225. Srikanth R, Srinivas R, Bhanuprakash K, Vivekananda S, Syrstad EA, Tureček F (2002) J Am Soc Mass Spectrom 13:250
226. Srinivas R, Vivekananda S, Blanksby SJ, Schröder D, Schwarz H, Fell LM, Terlouw JK (2000) Int J Mass Spectrom 197:105
227. Srinivas R, Vivekananda S, Blanksby SJ, Schröder D, Trikoupis MA, Terlouw JK, Schwarz H (2000) Int J Mass Spectrom 202:315
228. Nguyen VQ, Shaffer SA, Tureček F, Hop CECA (1995) J Phys Chem 99:15,454
229. Holthausen MC, Schröder D, Zummack W, Koch W, Schwarz H (1996) J Chem Soc Perkin 2 2389
230. Srinivas R, Böhme DK, Sülzle D, Schwarz H (1991) J Phys Chem 95:9836
231. Iraqi M, Schwarz H (1993) Chem Phys Lett 205:183
232. Goldberg N, Hrušák, J, Iraqi M, Schwarz H (1993) J Phys Chem 97:10,687
233. Srinivas R, Böhme DK, Schwarz H (1993) J Phys Chem 97:13,643
234. Srinivas R, Hrušák J, Sülzle D, Böhme, DK, Schwarz H (1992) J Am Chem Soc 114:2802
235. Goldberg N, Iraqi M, Hrušák J, Schwarz H (1993) Int J Mass Spectrom Ion Processes 125:267

236. Karni M, Apeloig Y, Schröder D, Zummack W, Rabezzana R, Schwarz H (1999) Angew Chem Int Ed Engl 38:332
237. Srinivas R, Sülzle D, Weiske T, Schwarz H (1991) Int J Mass Spectrom Ion Processes 107:369
238. Hajdasz DJ, Ho Y, Squires RR (1994) J Am Chem Soc 116:10,751
239. Nguyen VQ, Tureček F (1996) J Mass Spectrom 31:1173
240. Nguyen VQ, Tureček F (1997) J Mass Spectrom 32:55
241. Wolken JK, Tureček F (1999) J Phys Chem A 103:6268
242. Wolken JK, Tureček F (1999) J Am Chem Soc 121:6010
243. Nguyen VQ, Tureček F (1997) J Am Chem Soc 119:2280
244. Tureèek F, Wolken JK (1999) J Phys Chem A 103:1905
245. Wolken JK, Syrstad EA, Vivekananda S, Tureèek F (2001) J Am Chem Soc 123:5804
246. Wolken JK, Tureček F (2001) J Phys Chem A 105:8352
247. Syrstad EA, Vivekananda S, Tureček F (2001) J Phys Chem A 105:8339
248. Vivekananda S, Wolken JK, Tureček F (2001) J Phys Chem A 105:9130
249. Tureček F, Wolken JK, Sadílek M (1998) Eur Mass Spectrom 4:321
250. Schalley CA, Schröder D, Schwarz H (1996) Int J Mass Spectrom Ion Processes 153:173
251. Schröder D, Goldberg N, Zummack W, Schwarz H, Poustma JC, Squires RR (1997) Int J Mass Spectrom Ion Processes 165/166:71
252. Polce MJ, Song W, Cerda BA, Wesdemiotis C (2000) Eur Mass Spectrom 6:121
253. Schalley CA, Blanksby SJ, Harvey JN, Schröder D, Zummack W, Bowie JH, Schwarz H (1998) Eur J Org Chem 987
254. Tureček F, Drinkwater DE, McLafferty FW (1991) J Am Chem Soc 113:2950
255. Hawkins CL, Davies MJ (2001) Biochim Biophys Acta 1504:196
256. Stubbe J, van der Donk W (1998) Chem Rev 98:705
257. O'Hair RAJ, Blanksby SJ, Styles M, Bowie JH (1999) Int J Mass Spectrom 182/183:203
258. Tureček F, Carpenter FH, Polce MJ, Wesdemiotis C (1999) J Am Chem Soc 121:7955
259. Polce MJ, Wesdemiotis C (1999) J Am Soc Mass Spectrom 10:1241
260. Tureček F, Carpenter FH (1999) J Chem Soc Perkin Trans 2 2315
261. Syrstad EA, Tureček F (2001) J Phys Chem A 105:11,144
262. Syrstad EA, Stephens DD, Tureček F (2003) J Phys Chem A 107:115
263. Zagorevski DV (2002) Coord Chem Rev 225:5
264. Polce MJ, Wesdemiotis C (1999) Int J Mass Spectrom 182/183:45
265. Wu J, Polce MJ, Wesdemiotis C (2001) Int J Mass Spectrom 204:125
266. Gutowski M, Simons J (1994) J Chem Phys 100:1308
267. Langley CL, Syrstad EA, Seymour JL, Tureček F (2002) Proceedings of the 50th ASMS Conference on Mass Spectrometry and Allied Topics, Orlando, FL
268. McAnoy AM, Dua S, Schröder D, Bowie JH, Schwarz H (2002) J Chem Soc Perkin Trans 2 1647
269. Srikanth R, Bhanuprakash K, Srinivas R (2002) Chem Phys Lett 360:294
270. Dimopoulos G, Srikanth R, Srinivas R, Terlouw JK (2002) Int J Mass Spectrom 221:219

II
Metalorganic Chemistry

Top Curr Chem (2003) 225: 133–152
DOI 10.1007/b10474

Diastereoselective Effects in Gas-Phase Ion Chemistry

Detlef Schröder · Helmut Schwarz

Institut für Chemie, Technische Universität Berlin, 10623 Berlin, Germany
E-mail: Helmut.Schwarz@www.chem.tu-berlin.de

A method for the investigation of diastereoselective effects in reactions of gaseous ions bearing flexible backbones is described. It relies on the study of diastereoselectively labeled substrates that are studied by contemporary mass spectrometric means. While most examples discussed here deal with bond activations mediated by bare transition-metal ions in the gas phase, it is shown that the method is not restricted to this particular chemistry. Interestingly, the reason why transition-metal ions show relatively pronounced diastereoselective effects is that the activation barriers associated with their reactions are sufficiently small to be influenced by even subtle steric modifications of the substrate. Provided that suitable isotopically labeled precursors are investigated, even complex mechanistic schemes including competing side reactions can be analyzed by kinetic modeling of the isotope distributions in the reaction products.

Keywords. Chirality, Diastereoselectivity, Isotope labelling, Mass spectrometry, Stereochemistry

1 Introduction

Although mass spectrometry is a powerful and particularly sensitive tool in structural analysis, stereochemical aspects are more difficult to resolve by mass spectrometric means [1]. Of course, there do exist some time-honored examples for stereochemical effects showing up in mass spectra, and as a good example may serve the pronounced differences in the Retro-Diels-Alder (RDA) reactions of polycyclic compounds such as *syn*-**1** and *anti*-**1** upon electron ionization (EI) as shown in Scheme 1 [2].

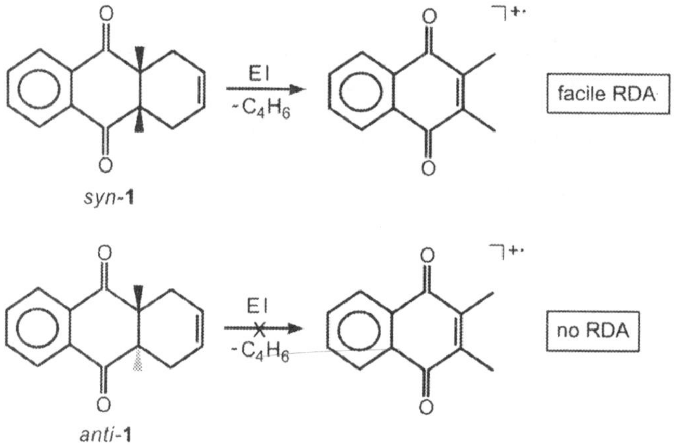

EI
$-C_4H_6$
facile RDA
syn-**1**
EI
$-C_4H_6$
no RDA
anti-**1**

Scheme 1

In general, however, detailed protocols for the direct analysis of stereochemical features by mass spectrometric means have not been achieved so far. Instead, other analytical methods, such as X-ray analysis or nuclear magnetic resonance (NMR), elegantly address these issues [3]. However, with respect to the analysis of trace compounds and in the context of high-throughput analytical methods, it would be beneficial to develop tools for rapid stereochemical assays by mass spectrometric means or hyphenated techniques [4, 5].

Numerous examples of stereospecific reactions in the gas phase are reported in the mass spectrometric literature [1, 2]. Many if not most of them, however, deal with relatively rigid systems, e.g., **1**, or polyfunctional molecules such as derivatives of tartaric acid [6]; the latter gave rise to the first "chirality effect" observed in mass spectrometry [7]. For stereogenic centers linked by flexible alkyl chains, however, diastereoisomeric differentiation in ion fragmentation is often poor. Two epimers of the aminoalkanol **2**, for example, show quite small differences in their mass spectra whereas these differences increase if the two centers are linked by cyclization upon formation of **3** as indicated in Scheme 2 with the epimeric center being marked by an asterisk [8].

Here, we report on methods aimed at obtaining stereochemical information about flexible molecules bearing a single functional group, by a combination of

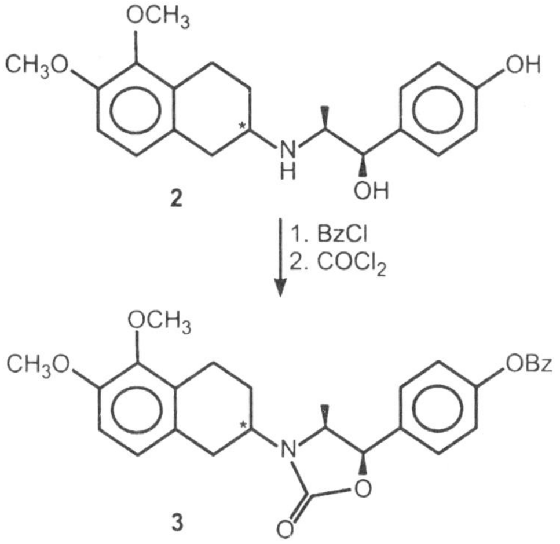

Scheme 2

diastereospecific synthesis with isotopic labeling. A review of the mass spectrometric literature reveals that examples exhibiting significant stereoselective effects (*SE*s) in molecules, where the asymmetric centers are part of a flexible backbone, are quite scarce [9]. However, already back in 1965 Audier et al. reported significant *SE*s in the McLafferty-type rearrangements of substituted but-3-en-1-ols [10]. A notable example for the application of diastereoselective isotopic labeling are the secondary alkylphenolates investigated by Morton and coworkers. Upon electron ionization, these compounds reveal distinct *SE*s in the 1,2-elimination of ionized phenol [11]. In this account we describe the utility of the gas-phase chemistry of bare transition-metal ions for the investigation of diastereoselective effects in molecules bearing a flexible side chain.

2
Arene Hydrogenation

Before illustrating the concept for a mass spectrometric distinction of stereoisomers, let us refer to some of the problems associated with the stereoselective hydrogenation of benzene to cyclohexane; this may serve as a simple and illustrative example. Metal-catalyzed hydrogenation is known to proceed with a large *syn*-selectivity, and the reaction involves at least three separate steps in the case of benzene→cyclohexane (Scheme 3).

Thus, when at a certain stage of the hydrogenation sequence the substrate leaves the metal catalyst [M], with [M] being a homo- or a heterogeneous catalyst, it may coordinate with either side upon return to [M] (e.g., **4**⇌**4a**). Thereby, a mixture of diastereoisomers (e.g., **5** and **5a**) is obtained, even if each individual hydrogenation step were entirely *syn*-specific.

Stereochemistry can, of course, not be monitored in the C_6H_6/H_2 system shown in Scheme 3 because **4/4a** and **5/5a** are indistinguishable. When isotopic labeling is applied, however, different diastereoisomers may result. Let us therefore consider the catalytic hydrogenation of C_6D_6 in some more detail. If the sub-

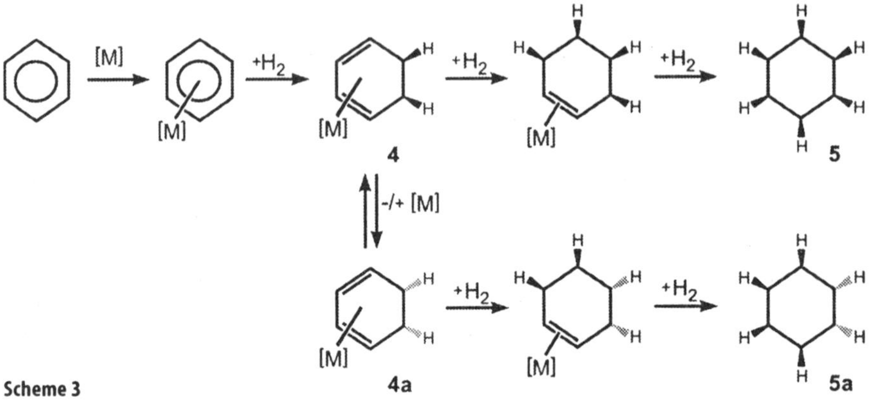

Scheme 3

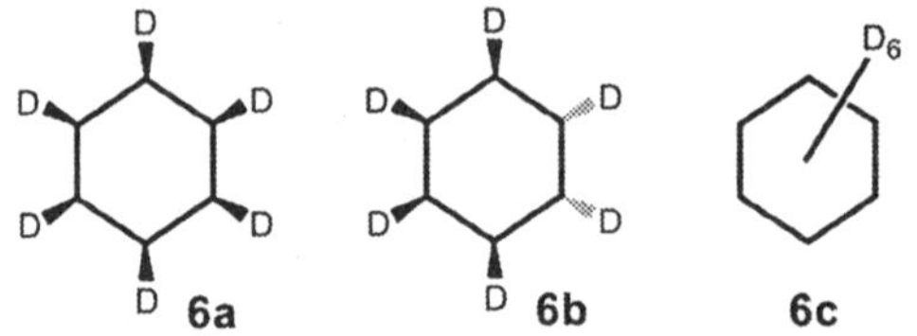

Scheme 4

strate remains coordinated to the catalyst throughout the whole hydrogenation sequence, the all-*syn*-compound **6a** is formed exclusively (Scheme 4). Partial dissociation and re-coordination of the diene- or ene-intermediates inter alia leads to **6b** and further stereoisomers bearing the *syn*-relationship of the actual hydrogenation step. Reversible metal-mediated C–H bond activations, most likely involving allylic positions, may even afford a random distribution of the label, e.g., **6c**. For compounds such as **6a–6c**, the elucidation of the relative stereochemistry by means of ^{1}H- and ^{2}H-NMR encounters severe problems because the spectra of **6a–6c** overlap strongly. In this respect, the gas-phase chemistry of transition-metal ions offers a viable alternative using mass-spectrometric techniques. Particularly the bare atomic ions of early transition metals triply dehydrogenate cyclohexane to afford the corresponding $M(C_6H_6)^+$ complexes [12]. Because the metal ion initially attacks cyclohexane at one particular side and remains bound to it throughout, triple dehydrogenation may occur, reflecting to a large extent the original stereochemical features of $C_6H_6D_6$. As an example, we refer to the reaction of Ti^+ with $C_6H_6D_6$ generated by hydrogenation of C_6D_6 with $(\eta^3\text{-}C_3H_5)Co[OP(OCH_3)_3]_3$ as a catalyst [13]. The gas-phase reaction of mass-selected Ti^+ with the substrate $C_6H_6D_6$ almost exclusively yields $Ti(C_6H_6)^+$ and $Ti(C_6D_6)^+$, respectively (Fig. 1).

Careful analysis of the isotope pattern reveals a >99.5% all-*syn*-hydrogenation of C_6D_6 by the cobalt catalyst. As far as analytical aspects of the stereochemical assignment of $C_6H_6D_6$ are concerned, such a high degree of sensitivity is hard to achieve by any other techniques in this particular case. In fact, Freiser and coworkers have used analogous reactions of bare Co^+ ion to screen the

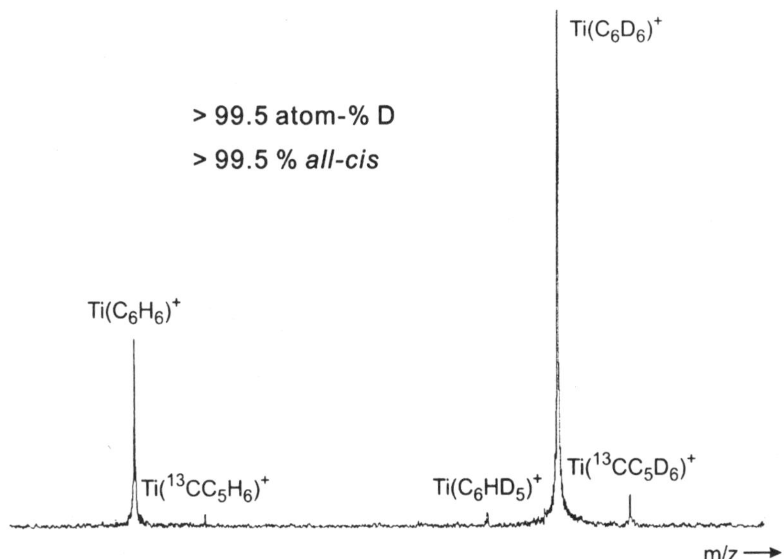

Fig. 1. Partial Fourier-transform ion cyclotron resonance mass spectrum of the triple-dehydrogenation region for the reaction of **6a** with bare Ti^+ cations at 30% conversion

stereochemical performance of various hydrogenation catalysts by mass spectrometric assays [14, 15]. Further, Fig. 1 also proves that the triple dehydrogenation of $C_6H_6D_6$ occurs with >99.5% diastereoselectivity. While in the present example the analysis is unambiguous, in case of mixed eliminations of H_2, HD, and D_2 analogous conclusions would resemble to some extent a self-fulfilling prophecy because no straightforward method for an independent analysis of the stereochemistry in the labeled cyclohexane sample is available; we will return to this aspect further below.

In the following, we will address strategies applicable to flexible molecules mostly bearing a single functional group as well as stereospecific labeling. A priori, there are three scenarios conceivable for the involvement of stereocenters in the gas-phase ion chemistry of diastereoisomers: (i) none, (ii) one, or (iii) both of the stereocenters are involved in the bond-breaking/bond-formation processes in the reactions under investigation. While the first case could play a role in polycyclic or polyfunctional molecules, only cases (ii) and (iii) are of interest for flexible, acyclic systems bearing a single functional group.

3 Activation of One Stereocenter

Bare cations of late transition-metal ions have been shown to be capable of activating C–H and C–C bonds in monofunctionalized alkanes in the gas phase. Because these bond activations occur quite far away from the functional group, these processes have been termed remote functionalization in the gas phase [16],

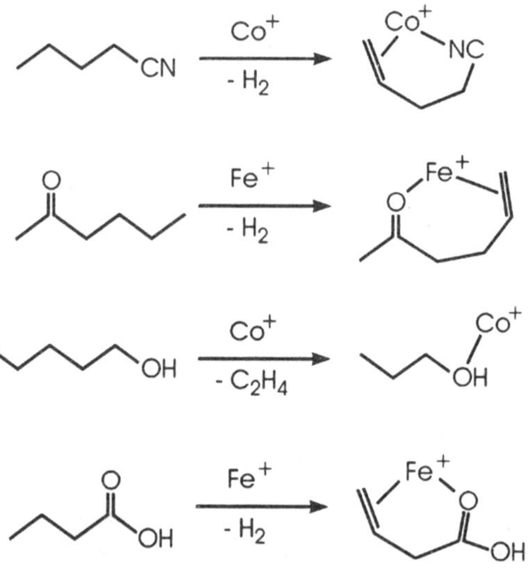

Scheme 5

in analogy to Breslow's seminal concept and studies on steroids [17]. In the gas-phase variant of remote functionalization, a coordinatively unsaturated transition-metal ion undergoes initial complexation ("docking") at a functional group (e.g., CN, COR, OH, COOH, etc.). Then, the alkyl chain of the substrate is subject to a recoil (internal solvation) with the consequence that certain positions of the backbone are exposed to interact with M^+ and to bring about selective C–H as well as C–C bond activations. Typical examples are given in Scheme 5.

The recoil of the alkyl chain crucially depends on the nature of the functional group, the metal atom, as well as additional constraints imposed by the backbone. Accordingly, recoil is quite likely to be influenced by introducing stereogenic centers along the alkyl chain via diastereoselective substitution. In order to probe such effects by means of mass spectrometry, however, a detailed stereochemical protocol needs to be developed which allows one to translate the occurrence of stereoselective processes into measurable mass differences. To this end, diastereoselective isotopic labeling is applied. Before continuing, it is to be pointed out that mass spectrometry in general cannot differentiate enantiomers; consequently, only diastereoselective, rather than enantioselective labeling is required. In fact, most of the substrates discussed below were prepared as racemic mixtures of pure diastereoisomers. For the sake of simplicity, only one of the two enantiomers is shown in the schemes, and in several cases even just a single diastereoisomer is presented. Further, we use the more simple prefixes *syn* and *anti* instead of the correct CIP nomenclature.

The first example of this kind concerns stereochemical aspects of the regiospecific 4,5-dehydrogenation of 3-methyl-pentan-2-one **7** mediated by Fe^+ cations [18]. For the labeled variants **7a**–**7c**, dehydrogenation leads to losses of H_2 and HD, respectively, with remarkably different ratios (Scheme 6). Obviously, a particular advantage of this approach is that compounds **7a** and **7b** only differ in their relative stereochemistry at C(4). Thus, whatever kinetic isotope effect

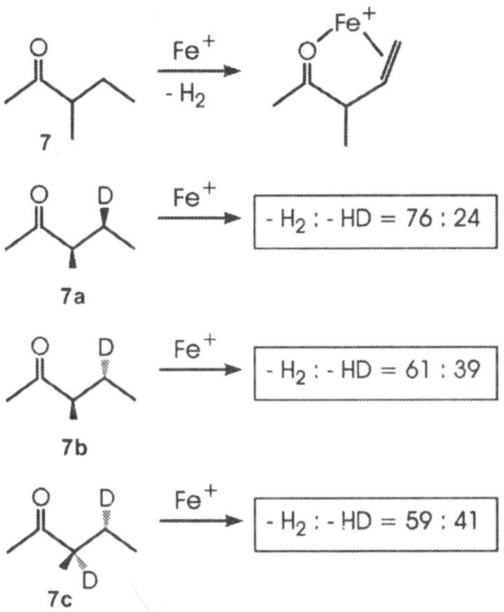

7 $\xrightarrow[-H_2]{Fe^+}$

7a $\xrightarrow{Fe^+}$ $-H_2 : -HD = 76 : 24$

7b $\xrightarrow{Fe^+}$ $-H_2 : -HD = 61 : 39$

7c $\xrightarrow{Fe^+}$ $-H_2 : -HD = 59 : 41$

Scheme 6

(*KIE*) is associated with deuteration, any difference observed between the two diastereoisomers can uniquely be attributed to the *SE*. Provided that the fragmentation occurs regiospecifically, as is the case for **7**/Fe^+, the complementary character of the pair of diastereoisomers **7a, 7b** further allows the determination of the relevant parameters *KIE* and *SE* from the measured ratios of H_2 and HD losses using a simple algebraic formalism (Eqs. 1 and 2), where I stands for the measured intensities, k_H/k_D is representative of the integrated *KIE*s operative in the reaction, and k_{syn}/k_{anti} describes the steric effect:

$$I_{H_2}/I_{HD}(\mathbf{7a}/Fe^+) = k_H/k_D \cdot k_{syn}/k_{anti} = KIE \cdot SE \quad (1)$$

$$I_{H_2}/I_{HD}(\mathbf{7b}/Fe^+) = k_H/k_D \cdot k_{syn}/k_{anti} = KIE/SE \quad (2)$$

Inspection of Eqs. (1) and (2) reveals that both effects accumulate for one of the two diastereoisomers while they attenuate each other for the second. The direction and the magnitude of the *SE* bear information about the steric requirements of the rate-determining step. Note that a differentiation of the *KIE*s associated with the transfer of the diastereotopic H(D) atoms is not required in Eqs. (1) and (2) because any difference between these *KIE*s would itself represent a steric effect and is thus already included in the parameter *SE*. Further, only one of the two stereocenters is actively involved in the bond cleavage, while the second, i.e., the methyl group at C(3), just serves as a spectator. As a consequence, both diastereoisomers lead to the same products. This distinction is quite important in comparison to systems in which the activation of both stereocenters in a diastereoisomer takes place (see below).

For the remote functionalization of **7**/Fe^+, we arrive at values of *KIE*=2.1 and *SE*=1.5. Effects of similar magnitudes (Table 1) were found for the Fe^+-mediated

Table 1. Kinetic isotope and steric effects observed in the remote bond activation of metal-ion complexes of several functionalized alkanes[a]

Substrate/M^+	Process	*KIE*	*SE*
3-Methyl-pentan-2-one/Fe^+ (**7**/Fe^+) [18]	4,5-Dehydrogenation	2.1	1.5
2-Methyl-butyric acid/Fe^+ (**8**/Fe^+) [19]	3,4-Dehydrogenation	1.8	1.6
3-Methyl-valeramide/Fe^+ (**9**/Fe^+) [20]	4,5-Dehydrogenation	2.3	2.0
6-Methyl-octanenitrile/Co^+ (**14**/Co^+) [22]	ω-C-C-cleavage	1.4	1.7
7-Methyl-nonanenitrile/Co^+ (**17**/Co^+) [22]	ω-C-C-cleavage	1.3	1.6
5-Methyl-7-silyl-heptanenitrile/Mn^+ (**18**/Mn^+) [21]	6,7-Dehydrogenation	2.2	2.1
5-Methyl-7-silyl-heptanenitrile/Fe^+ (**18**/Fe^+) [21]	6,7-Dehydrogenation	2.4	1.5
6-Methyl-8-silyl-octanenitrile/Fe^+ (**19**/Mn^+) [21]	7,8-Dehydrogenation	2.2	1.9
6-Methyl-8-silyl-octanenitrile/Fe^+ (**19**/Fe^+) [21]	7,8-Dehydrogenation	2.1	1.4
5-Methyl-7-silyl-heptanenitrile/Fe^+ (**18**/Co^+) [24]	6,7-Dehydrosilylation	1.6	4.1
6-Methyl-8-silyl-octanenitrile/Fe^+ (**18**/Co^+) [24]	7,8-Dehydrosilylation	1.5	3.0

[a] Monitored by unimolecular decomposition of the mass-selected metastable ions generated by chemical ionization.

dehydrogenations of the terminal positions in 2-methyl-butyric acid **8** [19] and 3-methylvaleramide **9** [20] (Scheme 7).

There exist two simple rationales to explain the observed direction of the diastereoselective bond activation in **7**/Fe^+, i.e., more pronounced loss of H_2 from **7a**/Fe^+ in comparison to **7b**/Fe^+. At first, one can safely assume that the reaction proceeds via insertion of the 'docked' Fe^+ in a terminal C–H bond to form a six-membered ring. Depending on the relative stereochemistry at C(3) and C(4), the eliminations of H_2 and HD, respectively, therefore involve quasi-axial or quasi-equatorial orientations of the methyl substituents in the intermediates *eq*- and *ax*-**10a**; of course, similar considerations apply to the associated transition structures (TSs). By analogy to conventional arguments of conformational analysis, an equatorial position of the methyl group is assumed to be preferred, thereby accounting for the experimentally observed H_2/HD ratios. Thus, for the stereoisomer shown in Scheme 8, both the *KIE* and the equatorial position of the methyl substituent favor loss of H_2, whereas the *SE* favors loss of HD from **7b**/Fe^+. However, for the latter this path is impeded by the operation of a kinetic isotope effect that slows down activation of a C–D bond. Secondly, one arrives at

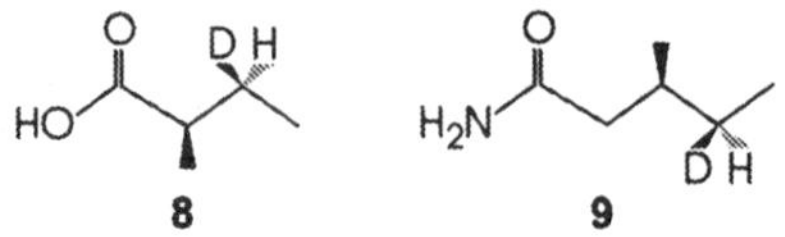

O DH HO **8** O H_2N DH **9**

Scheme 7

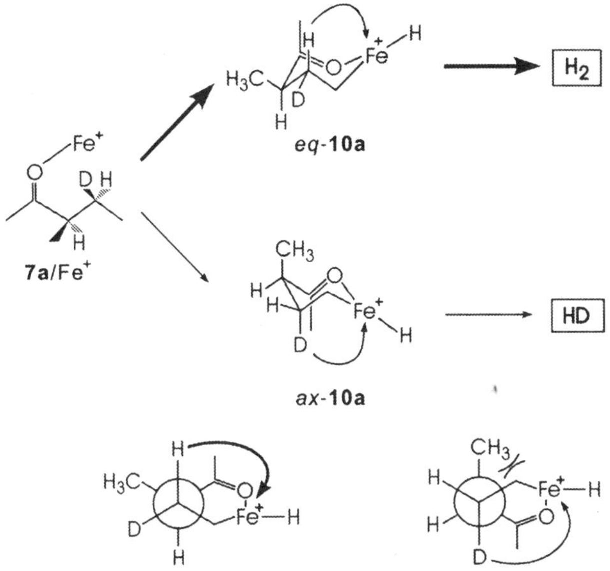

Scheme 8

the same conclusion by consideration of the corresponding Newman projections, where loss of HD from **10a** is disfavored by a *gauche* interaction (Scheme 8b). Note that the latter approach is more general because it is not confined to a particular ring size.

The concept of remote functionalization has been applied to the reactions of the metal ions Mn^+, Fe^+, and Co^+ with a series of substrates bearing longer alkyl backbones, e.g., **11–19** (Scheme 9) [21]. While significant diastereoselectivities are observed in the dehydrogenations of several of these substrates with these metal ions, e.g., **13**/Co^+, **14**/Mn^+, and **16**/Co^+, only few data can be analyzed quantitatively in terms of extracting *KIE* and *SE* on the ground that the M^+-mediated C–H bond activations bear limited regioselectivities with these substrates. Interestingly, however, the activation of the (ω–2) methyl group in the alkanenitriles **14** and **17** by gaseous Co^+ is associated with pronounced diastereo-

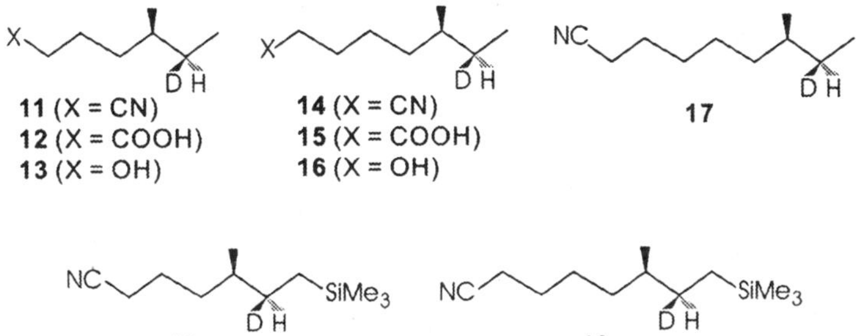

Scheme 9

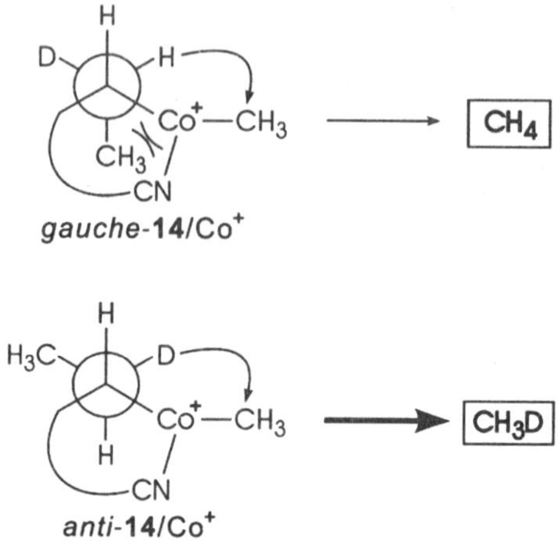

gauche-**14**/Co^+

anti-**14**/Co^+

Scheme 10

selective effects [22]. Again, 1,2-repulsions in the corresponding Newman projections can account for the observed results, e.g., preferred loss of deuterated methane from **14**/Co^+ due to unfavorable steric demands associated with a H transfer (Scheme 10 via *gauche*-**14**/Co^+). Not surprising for C–C bond activations, *SE* exceeds *KIE* in these cases (Table 1). It is important to point out, however, that only a few of the substrates and metals examined exhibit significant differences in the metal-mediated bond activations of the diastereoisomeric pairs. Negligible *SE*s can either be ascribed to increased flexibility or to increased steric strain in the cyclic reaction intermediates. The former case is expected for enlarged ring sizes in that the energy differences of the diastereotopic TSs are small, e.g., medium rings. Decreasing ring-sizes of the insertion intermediates in the latter case imply that the overall activation barriers rise significantly and hence the subtle energetic differences of two competing pathways due to a change in the relative stereochemistry vanish.

As mentioned above, the stereochemical aspects of the metal-mediated dehydrogenation of most substrates shown in Scheme 9 cannot be analyzed quantitatively due to limited regioselectivities of C–H bond activation for these molecules. This obstacle can be bypassed by taking advantage of the directing effect exerted by silyl groups in organic chemistry in general and metal-mediated remote bond functionalization in the gas phase in particular [23]. In fact, ω-silylation of alkanenitriles (i.e., **18** and **19**) affords exclusive $\omega/(\omega-1)$-dehydrogenations by Mn^+ and Fe^+ cations with notable diastereoselectivities (Table 1). Interestingly, the corresponding Co^+ complexes trade-off dehydrogenation and instead promote C–Si bond cleavage, again with considerable diastereoselectivities [24]. With regard to the bulkiness of the trimethylsilyl group, it is not at all surprising that the *SE*s are particularly pronounced for silyl substitution and exceed the *KIE*s more than twice.

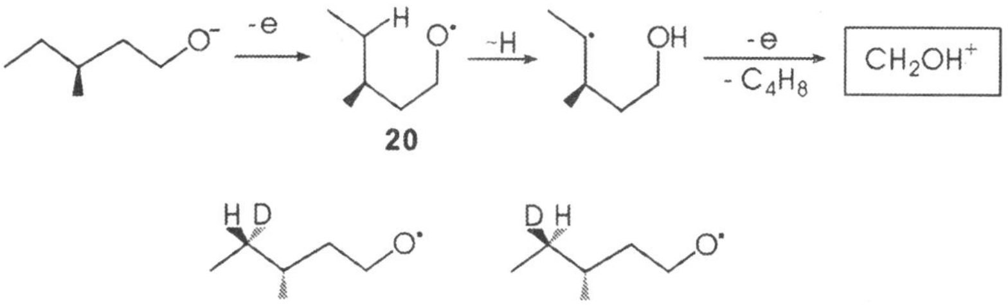

Scheme 11

20 20a 20b

4
Barton Rearrangement

In an extension of the concept to other reactions, we have found that the Barton rearrangement of the neutral alkoxy radical **20** (Scheme 11) shows a notable *SE*. Thus, analysis of neutralization-reionization and charge-reversal experiments of alkoxide ions by means of the NIDD scheme (NIDD=neutral and ion decomposition difference [25, 26]) reveals the occurrence of hydrogen migrations in transient neutral alkoxy radicals generated upon collisional electron detachment of the corresponding anions (Scheme 11). The rearrangement is probed by monitoring the hydroxy methyl cation CH_2OH^+ formed upon collisional reionization of the (rearranged) radical, thus involving even another electron-transfer step in these mass spectrometric experiments. For the diastereoselectively labeled radicals **20a** and **20b**, produced by electron detachment of the corresponding anions, the NIDD analysis of the CH_2OH^+/CH_2OD^+ ratios implies *SE*=3.6±1.3 and *KIE*=1.8±0.3 for the Barton-type reaction of neutral **20** in the gas phase [27]. It is worth mentioning that, despite the harsh conditions of vertical electron transfer in keV collisions, application of the NIDD scheme reveals the operation of a significant *SE* in **20** which even exceeds the *KIE* associated with H (D) atom migration. The direction of the effect (more CH_2OD^+ from **20b**) implies that the reaction proceeds via a chair-like transition structure. The latter finding is not particularly spectacular for a well-known process like the Barton reaction. Yet it is to be pointed out that this mass-spectrometric protocol allows decisive stereochemical insight into the transition structures involved in gas-phase reactions, thus probing the truly intrinsic features of an isolated radical. Moreover, this example demonstrates that the use of diastereospecific labeling as a probe for stereoselective processes is not restricted to transition-metal complexes and not even confined to charged species.

5
Activation of Both Stereocenters

A somewhat different situation evolves if both stereogenic centers of a diastereoisomer take part in the bond activation. Although the actual *SE* is small and the precise mechanistic course may be more complicated than usually anticipated [28], let us address the Mn^+-mediated dehydration of 2-butanol (**21**) and some of

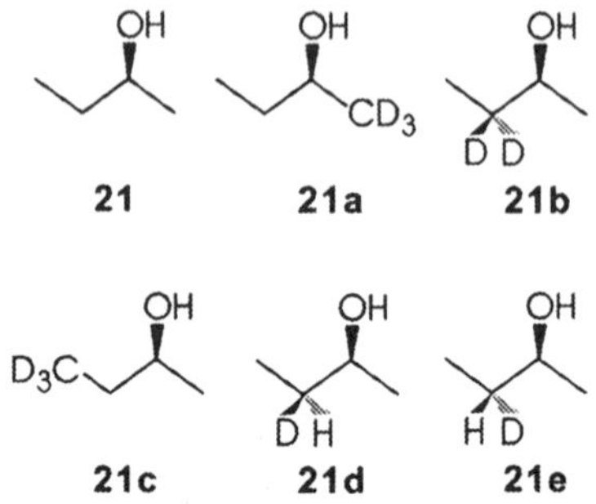

Scheme 12

its isotopologues (Scheme 12) in more detail. **21** is an interesting candidate for it constitutes an almost minimal model for probing diastereospecific processes in which both stereocenters are involved in the actual bond activation. In addition, it may serve to develop possible strategies to treat competing reactions.

Elimination of water prevails in the unimolecular dissociation of metastable 2-butanol/Mn^+ [21]. Inspection of the isotopologues **21a**/Mn^+ and **21b**/Mn^+ (Table 2) indicates reactions involving preferentially the 1,2- and 2,3-positions, respectively (Scheme 13). To a first approximation, the H/D equilibration which is indicated by the small amount of HDO loss from **21c**/Mn^+ might be considered as a minor perturbation. The diastereoisomers **21d**/Mn^+ and **21e**/Mn^+ show a small, but significant difference in the H_2O/HDO ratios. Because of the competition of 1,2- and 2,3-eliminations, however, the simple formalism using Eqs. (1) and (2) is not applicable. Specifically, for any possible isotopic labeling, the competing reaction(s) affect the H_2O/HDO ratio due to the operation of isotopically sensitive branching [29]. Thus, an eventual *KIE* associated with deuteration of a certain position inevitably pushes the reaction towards the competing channel(s). This is qualitatively reflected in the experimental data, in that the HDO loss from **21a**/Mn^+ implies 15% of 1,2-elimination whereas the loss of H_2O from **21b**/Mn^+ suggests twice as much. In such situations one often refrains from an explicit quantitative analysis of the labeling data. However, provided that a sufficiently large set of labeled substrates is available, a simple phenomenologi-

Table 2. Losses of H_2O and HDO from metastable complexes of 2-butanols with Mn^+ cations in the gas phase [21] [a,b]

Substrate		H_2O	HDO
$CH_3CHOHCH_2CH_3$	(**21**)	100	
$CD_3CHOHCH_2CH_3$	(**21a**)	85	15
$CH_3CHOHCD_2CH_3$	(**21b**)	32	68
$CH_3CHOHCH_2CD_3$	(**21c**)	98	2
syn-$CH_3CHOHCHDCH_3$	(**21d**)	67	33
anti-$CH_3CHOHCHDCH_3$	(**21e**)	75	25

[a] Normalized to $\Sigma=100$ for the water losses.

[b] Other reactions are losses of butene and evaporation of the entire butanol ligand. For the unlabeled ions the branching ratio for the eliminations of H_2O, C_4H_8, and C_4H_9OH is 65:25:10.

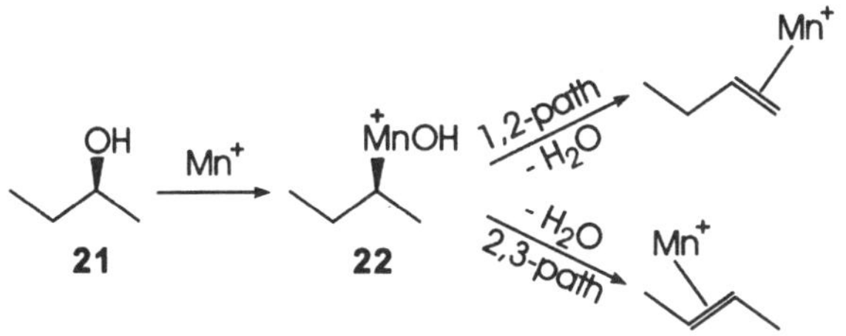

Scheme 13

cal kinetic modeling using some reasonable assumptions can resolve the situation to a considerable extent.

In such a kinetic modeling, the level of sophistication is determined by the number of isotopologues examined and the experimental errors associated with the mass spectrometric measurements [28, 30]. In the present example, there are five H_2O/HDO ratios for isotopologues of **21**/Mn^+ such that five parameters can be determined. Although the actual mechanism may be more complex [28], let us employ a simple formalism here. Thus, the regioselectivity of dehydration is phenomenologically described by the rate constants k_1–k_4 with $\Sigma k_i=1$, where the subscripts stand for the position in the alkyl backbone from which the H(D) atom is delivered; the OH group of the alcohol is assumed to stay intact. Further, a common *KIE* is attributed to all HDO losses, and the 2,3-dehydration of the diastereoisomers is attributed to bear an *SE*. Accordingly, Eqs. (3)–(7) are obtained:

$$I_{H_2O}/I_{HDO}(\mathbf{21a}/Mn^+) = (k_2 + k_3 + k_4)/(k_3/KIE) \quad (3)$$

$$I_{H_2O}/I_{HDO}(\mathbf{21b}/Mn^+) = (k_1 + k_2 + k_4)/(k_3/KIE) \quad (4)$$

$$I_{H_2O}/I_{HDO}(\mathbf{21c}/Mn^+) = (k_1 + k_2 + k_3)/(k_4/KIE) \quad (5)$$

$$I_{H_2O}/I_{HDO}(\mathbf{21d}/Mn^+) = (k_1 + k_2 + {}^1/_2\, k_3/SE + k_4)/({}^1/_2\, k_3/KIE) \quad (6)$$

$$I_{H_2O}/I_{HDO}(\mathbf{21e}/Mn^+) = (k_1 + k_2 + {}^1/_2\, k_3/SE + k_4)/({}^1/_2\, k_3/KIE/SE) \quad (7)$$

Equations (3)–(5) provide information about the branching ratios and the *KIE* whereas only Eqs. (6) and (7) also comprise the *SE*. Within the experimental error margins, a perfect match of experimental and modeled data is obtained for the following set of parameters: k_1=0.21±0.02, k_2=0.00±0.01, k_3=0.76±0.02, k_4=0.03±0.01, *SE*=1.13±0.04, and *KIE*=1.51±0.07. Except for the *SE*, however, these parameters are of limited relevance because a qualitative inspection of the experimental data already reveals, for example, that the branching of 2,3-elimination (k_3) must be between 0.67 (**21b**/Mn^+) and 0.85 (**21a**/Mn^+). Nevertheless, only quantitative analysis permits a reliable evaluation the of *SE*. In particular, neglect of the competing reactions may lead to a substantial overestimation of the *SE*. For example, a mere comparison of the H_2O/HDO ratios of **21d**/Mn^+ and **21e**/Mn^+ might be used to extract *SE*=(75:25)/(67:33)≈1.5, whereas the modeling reveals a significantly smaller effect. The neglect of explicit consideration of competing processes may also explain why Morizur et al. arrived at the somehow disappointing conclusion that "it is not possible to establish a clear-cut correlation between regioselectivity and stereoselectivity" in the dissociation of ionized alkylphenolates [11].

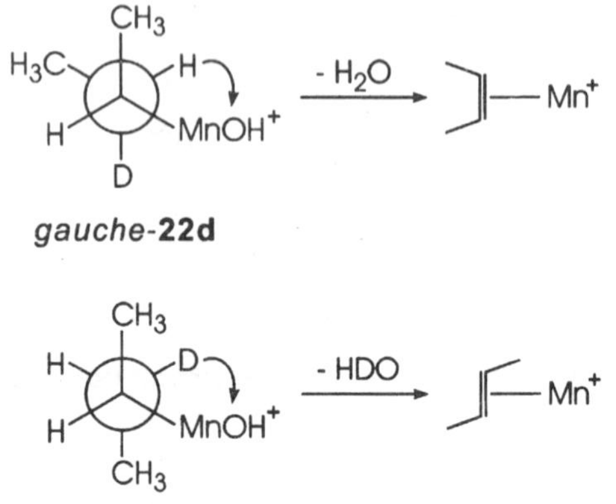

gauche-**22d**

anti-**22d**

Scheme 14

Before proceeding to a more general discussion, let us briefly return to the title question of this section, i.e., the particular case in which both stereocenters take part in bond activation. To this end, consider the Newman projections of the putative insertion intermediate **22** in the Mn^+ mediated dehydration of 2-butanol (Scheme 14). Thus, a *syn*-elimination of H_2O is not only disfavored by *gauche* interactions but also leads to a different product, i.e., *Z*-butene/Mn^+ from *gauche*-**22d** and *E*-butene/Mn^+ from *anti*-**22d**. Accordingly, the observed *SE*s may not only be traced back to the asymmetry of the associated transition structures, but may also arise from mere product stabilization, thus reflecting a thermodynamic rather than a kinetic control. While the stability differences between *E*- and *Z*-alkenes are small, the very same applies to steric constraints in diastereoisomeric transition structures. Accordingly, the formation of products with different stabilities is likely to modulate the *SE*s compared to the above examples with only one stereogenic center being activated and in which only the stereochemistry of the TS matters. Therefore, we generally conclude that activation of both stereocenters is less insightful in terms of uncovering intrinsic details of reaction mechanisms compared to a situation in which one stereocenter acts as a mere spectator, although having a decisive influence on the *SE*. Of course, one way to alleviate the problem of thermodynamic vs kinetic control is to apply the Hammond postulate, and by doing so to integrate a thermochemical description in a kinetic one.

Nevertheless, some notable effects have been observed in the metal-mediated activations of alkanols with larger alkyl backbones, where dehydration still persists but activation of internal C–H bonds via remote functionalization also becomes accessible. Thus, regiospecific 3,4-dehydrogenation of 1,6-hexandiol/Fe^+ [31] is associated with a considerable diastereoselectivity (an *SE* of about 3.2 can be derived from the experimental data). Similarly, 3,4-dehydrogenation of 1,8-octandiol/Fe^+ occurs diastereoselectively [32]. Note, however, that most complexes of the monofunctional alkanols **13** and **16** with Mn^+–Co^+ show negligible *SE*s.

6
General Considerations

The above examples demonstrate that diastereoselectively labeled substrates can be used to probe stereochemical features in gas-phase reactions of flexible molecules. In addition, this approach is not confined to metal-ion chemistry and not even to charged species, cf. the Barton rearrangement of neutral alkoxy radicals mentioned above. Instead, the decisive factor for the observation of a significant *SE* is a rate-determining activation barrier in an appropriate energy regime.

Consider the hypothetical potential-energy surfaces shown in Scheme 15 to illustrate this aspect where we deliberately only shift the relative energy of the central reaction intermediate. In case of a large activation barrier (Scheme 15a), the energy difference between the diastereoisomeric transition structures TS and TS* is small compared to the overall energy demand ($E_a \gg \Delta E_a$) such that the steric effect is small ($SE \approx 1$). For a medium-sized barrier ($E_a > \Delta E_a$), a considerable *SE* can be expected (Scheme 15b); this situation is assigned to most of the cases discussed above. If E_a is rather small, however, the diastereotopic TSs may not be rate-determining anymore in that other factors take over because the central intermediate might not be a true minimum any more (Scheme 15c). In particular, the conformational barriers existing between TS and TS* (for clarity

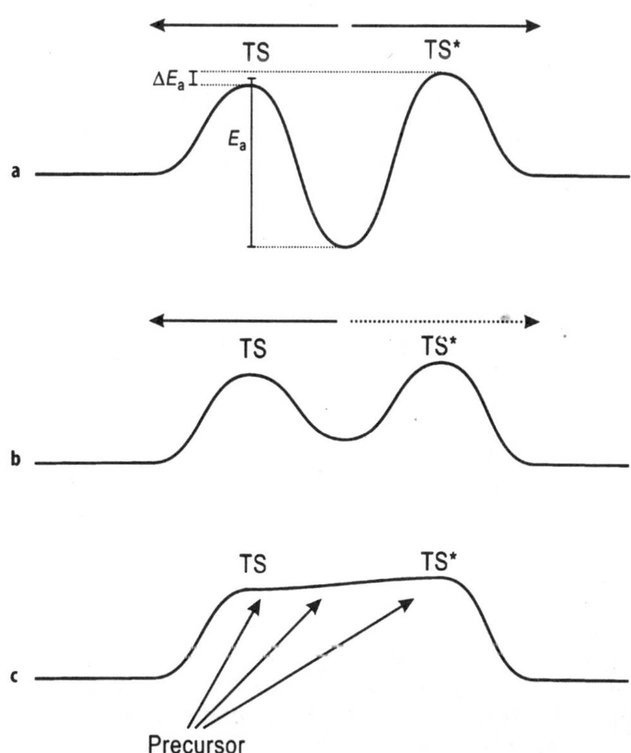

Scheme 15

omitted in the Scheme) may accordingly determine the accessibility of the reactive conformations from the precursor molecule, rather than the TSs themselves. The latter case might appear somewhat artificial. However, we have recently come across an example of this kind in the chemistry of ionized amides where conformational aspects of the neutral precursor determine the course of bond activation processes in the molecular ions [30, 33, 34]. Accordingly, only a small *SE* is observed for the metastable ions of diastereospecifically labeled **9**$^{+\cdot}$ [35]. This finding is consistent with the small *SE*s found in the early study of Audier et al. about the McLafferty-type reactions of diastereospecifically deuterated but-3-en-1-ols [10].

Another more general comment on the interplay of *KIE* and *SE* is indicated. Primary *KIE*s associated with hydrogen migrations depend on the intrinsic structural details of the key step of bond activation where ring sizes and steric constraints of the intermediates can be regarded as minor perturbations. In contrast, the local details of the TSs are more or less identical for transfer of diastereotopic H(D) atoms while conformational aspects of the backbone play a pivotal role. Therefore, *KIE*s and *SE*s provide complementary information on the reactions observed. With respect to the level of sophistication of contemporary ab initio methods, an explicit computational treatment of the *KIE*s and *SE*s in one of the above systems is therefore desirable.

Upon a more critical inspection of the above examples, an experienced mass spectrometrist might ask why most of our results on diastereoselective bond activation are derived from metastable ion (MI) studies in which the internal energy of the incident ion beam is poorly defined if not more or less unknown. In contrast, well-controlled ion/molecule reactions (IMR) or low-energy collision-induced dissociation (CID) would provide more insightful information and also permit a direct comparison with theoretical data. While we have in fact used other methods, e.g., IMR in the case of cyclohexane and NIDD data of alkoxide ions (see above), the preference for MI studies is due to the particularly high precision and reproducibility of these experiments in conjunction with a dedicated energy balance that prevails in metastable ions. Thus, provided that the MI decomposition is reasonably intense (that is a low E_a), error bars in the sub-percent range can be achieved. Small experimental errors are in turn of prime importance for the reliable determination of the relevant parameters, particularly when competing reactions demand explicit kinetic modeling. Likewise, while the actual internal energy of the decomposing ions in unknown, MI ensures that the decomposition takes place in a certain time scale, thereby precisely matching the decisive energy regime for a mass spectrometric distinction of diastereoisomers by means of their fragmentation behavior in that the decomposing ions have just enough energy to dissociate in a microsecond time scale. Other experimental techniques might, of course, allow similar or even superior measurements, but worrisome systematic errors, e.g., isotope corrections and problems in measuring the precise pressures of the reactants in IMR or the collision-gas pressure and the absolute collision energy in low-energy CID, impose additional uncertainties. In contrast, MI studies of properly mass-selected ions bear no variable other than the stereochemistry of the precursors. Therefore, MI experiments are particularly suited for the examination of steric effects.

7 Synthetic Aspects

Next we will address a non-trivial aspect of the isotopologues **7a**–**7c** with a critical remark on the synthetic procedures used in diastereospecific deuteration, which forms an essential part of the whole approach. Diastereospecific labeling is not just limited by the cost of the isotope and the corresponding reagents but also by the types of synthetic methods available. In this respect, stereoselective deuteration is not always that unambiguous. For example, the reduction of alkyl halides or tosylates by complex metal hydrides, e.g., $NaBD_4$ or $LiAlD_4$, generally proceeds as an S_N2 reaction and thus is often assumed to involve inversion of configuration [11]. While this is true in most cases, there exist several counter examples [36]. Hence, it is fair to ask how many seemingly non-selective reactions of deuterated samples may instead be ascribed to a non-selective synthesis. This objection is quite serious because the otherwise powerful NMR methods may fail in differentiating diastereoisomers such as **7a** and **7b** where a flexible alkyl backbone is deuterated. Therefore, most of our studies started with the reduction of *syn*- and *anti*-2,3-epoxybutane by $LiAlD_4$ which can quite safely be assumed to occur as a S_N2 reaction. The resulting racemic [3-D_1]-2-butanol (**21d**) is then tosylated and converted to the desired substrates using well established synthetic procedures as shown in Scheme 16 for the *Z*-butene.

However, the lack of analytical methods to distinguish clearly diastereoisomers such as **7a** and **7b** casts some doubt on the above assumptions. Specifically, it cannot be ensured with ultimate certainty that the reactions have really occurred in the desired manner or whether epimerization has taken place to a certain extent. In the case of **7a**/Fe^+ and **7b**/Fe^+, for example, the experimental data clearly reveal the occurrence of diastereoselective dehydrogenation, but the intrinsic *SE* might even be larger if the samples were not the pure diastereoisomers **7a** and **7b** but only enriched samples. Epimerization at C(3) is in fact facile via keto/enol tautomerism of the carbonyl compound occurring upon work-up. Therefore, the isotopolog **7c** was prepared via an independent synthetic route (N_2D_2 reduction of 2-methyl-but-*E*-2-enoic acid). As the results obtained with

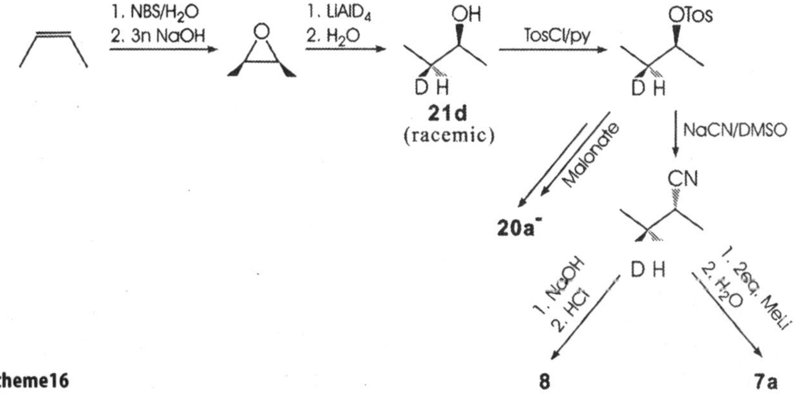

Scheme 16

7c/Fe^+ are consistent with those of **7b**/Fe^+ within reasonable error margins, we simply assume that **7a**–**7c** were obtained as pure diastereoisomers. Quite obviously, there is a need for alternative techniques to differentiate diastereospecifically deuterated molecules.

8 Enantioselective Processes

Most of the above examples deal with diastereospecific reactions involving 1,2-interactions in flexible backbones that can be probed by mass spectrometric means when appropriate labeling schemes are applied. However, what about the existence of 1,*n* interactions or even enantioselective reactions in the gas-phase chemistry of ions?

Whereas diastereospecific 1,2-deuteration is relatively straightforward (see above), stereoselective 1,n-deuteration is all but trivial considering the ambiguities in the reduction of R–X compounds (X=halogen, tosylate, etc.) by complex metal hydrides. In fact, it seems that enantioselective approaches are more suitable for 1,*n*-deuterations than the synthesis of pure, but racemic diastereoisomers. However, we have not realized this option so far. The few relevant data available for flexible molecules imply that 1,n-interactions with $n>2$ attenuate the *SE*s [37]. Nevertheless, it is certainly worthwhile to go beyond 1,2-interactions because 1,n-interactions would uncover structural details of cyclic intermediates provided that significant *SE*s are observed. For example, the comparison of the diastereoselectivity in the Fe^+-mediated remote functionalizations of [4-D_1]-3-methyl-valeramide, **9**, and [4-D_1]-2-methyl-valeramide may provide more detailed insight into the intermediates involved.

As far as genuine enantioselective processes are concerned, one first needs to specify this term because in traditional mass spectrometry one can only distinguish diastereoisomeric reactions. In analogy to the general terminology, we therefore define an enantioselective gas-phase reaction as a process in which a chiral ion reacts with a chiral neutral molecule in a diastereoselective manner. As mass spectrometry is insensitive to the absolute configuration, only one of both partners needs to be varied, e.g., reaction of a chiral ion with the *R*- and *S*-enantiomers of the neutral reagent or vice versa.

Having said this, we have explored several reactions of transition-metal ions with chiral ligands and substrates. For example, an in-depth investigation of the reactions of Fe^+ with mixtures of enantiomeric 2-pentanols (**23**) has been performed [38]. Much like with *n*-alkanols [39], 2-pentanol undergoes remote functionalization by Fe^+ to afford the corresponding pent-4-en-2-ol/Fe^+ complex. Assuming that the stereocenter a C(2) does not epimerize during the reaction, chiral Fe^+-complexes are thus formed. Mass-selection of these product ions and subsequent reactions with mixtures of the enantiomeric 2-pentanols of which one is fully deuterated in the C(5) position leads to the corresponding adducts (Scheme 17). Experimentally, careful investigation of a set of epimers reveals a slight preference for the formation of the homochiral adducts [*R*-**24**/*R*-**23**] and [*S*-**24a**/*S*-**23a**], respectively, with $k_{RR}/k_{RS}=1.05\pm0.03$ and $k_{SS}/k_{SR}=1.04\pm0.02$ [38]. While the preferred formation of homochiral ions is in

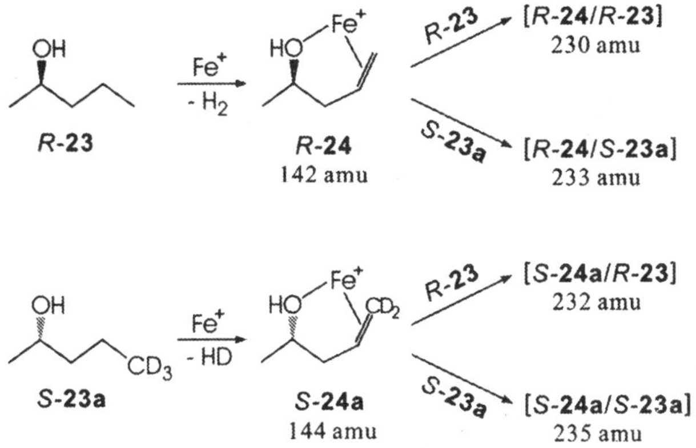

R-23 —(Fe⁺, - H_2)→ R-24 (142 amu) —R-23→ [R-24/R-23] 230 amu; —S-23a→ [R-24/S-23a] 233 amu

S-23a —(Fe⁺, - HD)→ S-24a (144 amu) —R-23→ [S-24a/R-23] 232 amu; —S-23a→ [S-24a/S-23a] 235 amu

Scheme 17

agreement with previous findings for gaseous dimers [40], the magnitude of the observed effects is much too small to draw any decisive mechanistic conclusions. Quite obviously, the stereogenic centers in the adducts are way too far apart from each other in this case to bring about larger stereochemical differentiation. Reactions of several other chiral metal complexes have been examined in a similar manner; yet, we have so far failed to discover a system with pronounced "chiral" effects in the gas-phase chemistry of transition-metal ions [4]. For recent articles on these extremely important as well as related stereochemical topics, see [41–46].

Acknowledgement. Continuous financial support by the Deutsche Forschungsgemeinschaft, the Fonds der Chemischen Industrie, the European Commission, and the Gesellschaft der Freunde der Technischen Universität is gratefully acknowledged. Further, we are indebted to all coworkers quoted in the references, with particular thanks to Dr. G. Hornung and Dipl.-Chem. J. Loos. This contribution is dedicated, with admiration, to Professor Asher Mandelbaum, TECHNION Haifa, an esteemed friend, a stimulating colleague, and one of the world-leading mass spectrometrists.

9
References

1. Splitter JS, Turecek F (eds) (1994) Applications of mass spectrometry to organic stereochemistry. VCH, Weinheim
2. Mandelbaum A (1994) In: Splitter JS, Turecek F (eds) Applications of mass spectrometry to organic stereochemistry. VCH, Weinheim, p 299, and numerous references therein
3. Reetz MT (2002) Angew Chem 114:1391; Angew Chem Int Ed 41:1335
4. Guo J, Wu J, Siuzdak G, Finn MG (1999) Angew Chem 111:1868; Angew Chem Int Ed 38:1754
5. Reetz MT, Becker MH, Kleim H-W, Stöckigt D (1999) Angew Chem 111:1872; Angew Chem Int Ed 38:1758
6. Winkler FJ, Splitter JS (1994) In: Splitter JS, Turecek F (eds) Applications of mass spectrometry to organic stereochemistry. VCH, Weinheim, p 353
7. Fales HM, Wright GJ (1977) J Am Chem Soc 99:2339
8. Selva A, Redenti E, Amari G, Ventura P (1992) Org Mass Spectrom 27:63
9. Splitter JS (1994) In: Splitter JS, Turecek F (eds) Applications of mass spectrometry to organic stereochemistry. VCH, Weinheim, p 83

10. Audier HE, Felkin H, Fetizon M, Vetter W (1965) Bull Soc Chim Fr 3236
11. Morizur JP, Taphanel MH, Mayer PS, Morton TH (2000) J Org Chem 65:381
12. Eller K, Schwarz H (1991) Chem Rev 91:1121
13. Seemeyer K, Schröder D, Kempf M, Lettau O, Müller J, Schwarz H (1995) Organometallics 14:4465
14. Huang Y, Profilet RD, Ng JH, Ranasinghe YA, Rothwell IP, Freiser BS (1994) Anal Chem 66:1050
15. Freiser BS (1994) Acc Chem Res 27:353
16. Schwarz H (1989) Acc Chem Res 22:282
17. Breslow R (1972) Chem Soc Rev 1:553; (1980) Acc Chem Res 13:170
18. Schröder D, Schwarz H (1993) J Am Chem Soc 115:8818
19. Schröder D, Zummack W, Schwarz H (1994) J Am Chem Soc 116:5857
20. Loss J, Schröder D, Zummack W, Schwarz H (2002) Int J Mass Spectrom 217:169
21. Hornung G (1998) Regio- and stereochemistry in the transition-metal mediated remote functionalization in the gas phase. Verlag Dissertation.de, Berlin
22. Hornung G, Schröder D, Schwarz H (1995) J Am Chem Soc 117:8192
23. Hässelbarth A, Prüsse T, Schwarz H (1990) Chem Ber 123:213
24. Hornung G, Schröder D, Schwarz H (1997) J Am Chem Soc 119:2273
25. Schalley CA, Hornung G, Schröder D, Schwarz H (1998) Int J Mass Spectrom Ion Processes 172:181
26. Schalley CA, Hornung G, Schröder D, Schwarz H (1998) Chem Soc Rev 27:91
27. Hornung G, Schalley CA, Dieterle M, Schröder D, Schwarz H (1997) Chem Eur J 3:1866
28. Trage C, Zummack W, Schröder D, Schwarz H (2001) Angew Chem 113:2780; Angew Chem Int Ed 40:2708
29. Thibblin A, Ahlberg P (1989) Chem Soc Rev 18:209
30. Loos J, Schröder D, Zummack W, Schwarz H, Thissen R, Dutuit O (2002) Int J Mass Spectrom 214:105
31. Seemeyer K, Prüsse T, Schwarz H (1993) Helv Chim Acta 76:1632
32. Prüsse T, Fiedler A, Schwarz H (1991) Helv Chim Acta 74:1127
33. Semialjac M, Loos J, Schröder D, Schwarz H (2002) Int J Mass Spectrom 214:129
34. Loos J, Schröder D, Semialjac M, Weiske T, Schwarz H, Höhne G, Thissen R, Dutuit O (2002) Int J Mass Spectrom 214:155
35. Loss J (2000) Diplomarbeit. TU Berlin
36. Smith MB, March J (2001) March's advanced organic chemistry, 5th edn. Wiley, New York, p 524
37. Green MM, Boyle BA, Varaimani M, Mukhopadhyay T, Saunders WH Jr, Bowen P, Allinger NL (1986) J Am Chem Soc 108:2381
38. Schröder D (1992) Dissertation. TU Berlin D83
39. Prüsse T, Schwarz H (1989) Organometallics 8:2856
40. Nikolaev EN, Goginashvili GT, Tal'rose VL, Kostyanovsky RG (1988) Int J Mass Spectrom Ion Processes 86:249
41. Filippi A, Giardini A, Piccirillo S, Speranza M (2000) Int J Mass Spectrom 198:137
42. Lebrilla CB (2001) Acc Chem Res 34:653
43. Schalley CA (2000) Int J Mass Spectrom 194:11
44. Schalley CA (2001) Mass Spectrom Rev 20:253
45. Schalley CA, Weis P (2002) Int J Mass Spectrom 221:9–19
46. Trage C, Schröder D, Schwarz H (2003) Organometallics 22:693

Top Curr Chem (2003) 225: 153–203
DOI 10.1007/b10473

Metalorganic Chemistry in the Gas Phase: Insight into Catalysis

Dietmar A. Plattner

Laboratorium für Organische Chemie, Eidgenössische Technische Hochschule Zürich, ETH-Hönggerberg, HCI G203, Wolfgang-Pauli-Strasse 10, 8093 Zürich, Switzerland
E-mail: plattner@org.chem.ethz.ch

Applications of electrospray mass spectrometry (ESMS) to the study of reactions mediated by transition-metal complexes are reviewed. ESMS has become increasingly popular as an analytical tool in inorganic and organometallic chemistry, in particular with regard to the identification of short-lived intermediates of catalytic cycles. Going one step further, the coupling of electrospray ionization to ion-molecule techniques in the gas phase yields detailed information about single reaction steps of catalytic cycles. This method allows the study of transient intermediates that have previously not been within reach of condensed-phase techniques on both a qualitative and quantitative level.

Keywords. Electrospray ionization, Ion-molecule chemistry, Transition-metal coordination compounds, Organometallic catalysis

1 Introduction

Electrospray mass spectrometry (ESMS) is a technique that allows ions present in solution to be transferred to the gas phase with minimal fragmentation, followed by conventional ion-molecule techniques and eventual mass-spectrometric characterization. Three features of electrospray set it apart from other mass-spectrometric ionization techniques. The first is the truly unique ability to produce extensively multiply charged ions. This attribute allows the creation of highly charged forms of very large molecular-weight compounds which may be analyzed on virtually all types of mass spectrometers. A second distinguishing feature of ESI is that samples under analysis must be introduced in solution. A third unique feature is the extreme softness of the electrospray process which permits the preservation in the gas phase of weakly binding interactions between molecules which existed in solution.

ESMS has been widely used for the measurement of molecular masses of non-volatile and thermally unstable compounds. Characterized by soft ionization, ESMS in the beginning had its most spectacular successes in the area of mass spectrometry of large biomolecules. Strangely, despite the enormous impact of electrospray on the mass spectrometric characterization of biomolecules, applications to inorganic and organometallic chemistry had been lagging somewhat behind. With hindsight, there is no reason why ESMS should have been restricted to large protonated molecules such as proteins and, in principle, any cation or anion that is present in solution and more or less stable in the solution environment can in principle be transferred to the gas phase by electrospray. Thus, all ionic inorganic coordination and organometallic compounds, together with neutral compounds that can be chemically converted to a closely related ion, are potential candidates for ESMS studies. There are a number of practical advantages in ESMS with inorganic/organometallic compounds as compared to biochemical samples. Charges on most inorganic cations are small, and often only singly charged ions are observed in the ES mass spectrum, because multiply charged ions frequently form ion-pair adducts with either the counter ion in the solution or other anions available. Hence, the characteristic isotopic patterns of metal ions are readily resolved even by quadrupole mass analyzers, and unambiguous identification of metal-containing species is thus possible.

Straightforward analytical applications of ESMS to organometallic and transition-metal coordination chemistry have become more or less standard nowadays. Because of the ease and speed of transferring all solution-phase ions to the gas phase, ESMS has become increasingly popular as a mechanistic tool for studying short-lived reactive intermediates in organometallic catalytic cycles. More recently, the potential of ESMS for mechanistic studies of ion-molecule reactions by electrosprayed organometallic ions has started to be exploited. The focus of the present account will be on the latter two applications of ESMS, highlighting the unique advantages of this method as compared to traditional solution-phase characterization techniques.

2
A Brief History of Electrospray

Electrospray mass spectrometry has developed into a well-established method of wide scope and potential over the past 15 years. The softness of electrospray ionization has made this technique an indispensable tool for biochemical and biomedical research. Electrospray ionization has revolutionized the analysis of labile biopolymers, with applications ranging from the analysis of DNA, RNA, oligonucleotides, proteins as well as glycoproteins to carbohydrates, lipids, glycolipids, and lipopolysaccharides, often in combination with state-of-the-art separation techniques like liquid chromatography or capillary electrophoresis [1, 2]. Beyond mere analytical applications, electrospray ionization mass spectrometry (ESMS) has proven to be a powerful tool for collision-induced dissociation (CID) and multiple-stage mass spectrometric (MS^n) analysis, and – beyond the elucidation of primary structures – even for the study of noncovalent macromolecular complexes [3].

The electrospray process itself is by no means a recent development. Electrostatic spraying of liquids, i.e., applying a high potential to liquids pressed through a fine capillary, whereby a highly dispersed aerosol consisting of droplets of relatively uniform size is formed, was already known in the eighteenth century [4]. In 1914, John Zeleny rediscovered the electrospray phenomenon and investigated it from the point of view of an electrical discharge from a liquid point [5–8]. Further experiments with electrified streams of water drops showed that the electrified water thread split up into narrower jets while the large electrified drops split up into smaller ones. A detailed experimental and theoretical study on the dispersibility of liquids at high electric potentials was published by Drozin in 1955 [9]. He was able to demonstrate that the dispersibility of a liquid can be predicted from knowledge of the value of its dielectric constant, thus explaining why nonpolar organic liquids could not be dispersed. The utilization of electrospray led to developments of considerable technical importance such as paint and crop spraying or ink-jet printing. In the late 1960s and the 1970s intensive research took place on colloid thrusters for space-flight applications, i.e., ion engines that produce thrust by electrostatically accelerating metallic ions such as cesium or mercury and could potentially be used as secondary propulsion systems.

Although a well-established technique [4], electrospray was not considered for mass spectrometric applications before the late 1960s. Interested in determining molecular weight distributions of synthetic polymers, Dole and co-workers assembled an apparatus designed to stage the following scenario: a dilute solution of the polymer analyte in a volatile solvent is introduced through a small tube into an electrospray source chamber through which nitrogen as bath gas flows at atmospheric pressure. A potential difference of a few kilovolts between the tube and the chamber walls produces an intense electrostatic field at the tube exit and disperses the emerging solution into a fine spray of charged droplets. As the droplets lose solvent by evaporation their charge density increases until the so-called Rayleigh limit is reached at which Coulomb forces overcome surface tension and the droplet breaks up into smaller droplets ("Coulomb explosion") that repeat that sequence. The rationale behind those experiments was that a succession of such Coulomb explosions would finally lead to droplets so small that each would contain only one molecule of analyte. As the last of the solvent evaporates that residual molecule would retain some of the droplet charge, thereby becoming a free macroion. With polystyrene samples ranging in mass between 50 and 500 kDaltons in a benzene/acetone solvent mixture Dole and co-workers were able to demonstrate the formation of distinct macroions which were concentrated in a supersonic jet and detected in a Faraday cage after small-molecular-weight ions had been repelled by suitable repeller voltages [10]. Even at this early stage, the authors mentioned the plan to build a time-of-flight mass spectrometer with an electrospray source. In subsequent studies, the method was further refined ("molecular beams of macroions") [11, 12] and extended to the protein zein and polyvinyl pyrrolidone [13], but the limitations of the comparably low mass resolution in their setup could not be overcome due to experimental problems with the mass analysis and detection of macroions.

After those first attempts to establish analytical applications of electrospray, it took more than ten years for the first bona fide electrospray mass spectrometer to emerge [14]. Yamashita and Fenn published the first electrospray MS experiment in a 1984 paper which was appropriately part of an issue of the Journal of Physical Chemistry dedicated to John Bennett Fenn [15]. They electrosprayed solvents into a bath gas to form a dispersion of ions that was expanded into vacuum in a small supersonic free jet. A portion of the jet was then passed through a skimmer into a vacuum chamber containing a quadrupole mass filter. With this setup, a variety of protonated solvent clusters as well as solvent-ion clusters (Na^+, Li^+) could be detected. Moreover, they extended the application to molecular beams of negatively charged ions [16]. A very similar, independent development was reported at approximately the same time by Aleksandrov and co-workers [17]. Using an improved design of their electrospray ion source, Fenn and co-workers were the first to couple HPLC to a mass spectrometer via an electrospray interface [18].

3 Mechanistic Aspects of the Electrospray Process

3.1 Electrospray – A Method for the Transfer of Ions from Solution to the Gas Phase

A technique that allows the transfer of ions from solution to the gas phase is of the greatest importance because a major part of the chemical and biochemical processes involve ions in solution. Electrospray is such a technique. It affords ion transfer of a wide variety of ions dissolved in a wide variety of solvents. These ions include singly and multiply charged inorganic ions such as (1) the alkali ions; (2) the alkaline earths; (3) singly and doubly charged transition metal ions and their complexes with mono and polydentate ligands; (4) anions of inorganic and organic acids such as NO_3^-, Cl^-, HSO_4^-, etc.; (5) singly and multiply protonated organic bases such as amines, alkaloids, peptides, and proteins; and (6) singly and multiply deprotonated organic acids or organophosphates such as the nucleic acids. The solvents include practically all polar, protic as well as aprotic solvents.

The transfer of ions from solution to the gas phase is a strongly endothermic and endoergic process, due to the strongly binding interactions between the ion in solution and a number of solvent molecules that form a solvation sphere around the ion. The energy required, e.g., for Na^+ (aq)$\rightarrow$$Na^+$ (g), is larger than the energy required to break a C–C single bond (~85 kcal/mol), and this suggests that the process of freeing an organic ion from the solvent can also lead to fragmentation. For earlier ionization methods in which the ions are transferred from solution to the gas phase, such as fast atom bombardment (FAB) and plasma desorption, abundant energy is supplied in a highly localized fashion over a short time. These methods lead not only to ion desolvation but also to fragmentation or even net ionization, i.e., the formation of ions from neutrals. Compared to these methods, ESMS, in which the desolvation is achieved gradually by thermal energy at relatively low temperatures, is a far softer technique. One could even go so far as to call the commonly used term "electrospray ionization" (ESI) misleading, since the electrospray process normally does not involve net ionization, but desolvation of ions already existing in solution. When defining the softness of the ion-transfer method as the degree to which fragmentation of the ions is avoided, ESMS is the softest technique available.

3.2 Production of Gas-Phase Ions in the Electrospray Process

There are three major steps in the production of gas-phase ions from electrolyte ions in solution by electrospray: (1) production of charged droplets at the ES capillary tip; (2) shrinkage of the charged droplets by solvent evaporation and repeated droplet disintegrations, leading ultimately to very small highly charged droplets capable of producing gas-phase ions; and (3) the actual mechanism by which gas-phase ions are produced from the very small and highly charged droplets.

For the production of charged droplets, a voltage V_c, typically 3–5 kV, is applied to the metal capillary which is ~0.1–0.2 mm o.d. and located 1–3 cm from the counter electrode (Fig. 1). The counter electrode in ESMS may be a plate with an orifice leading to the mass spectrometric sampling system or a sampling capillary, mounted on a plate, which leads to the MS. Because the electrospray capillary tip is very thin, the electric field E_c in the air at the capillary tip is very high ($E_c \approx 10^6$ V m^{-1}).

A typical solution present in the capillary consists of a polar solvent in which electrolytes are soluble. Low electrolyte concentrations, 10^{-5} to 10^{-3} mol l^{-1}, are typically used in ESMS. When turned on, the field E_c will penetrate the solution at the capillary tip and the positive and negative electrolyte ions in the solution will move under the influence of the field until a charge distribution results which counteracts the imposed field and leads to essentially field-free conditions inside the solution. When the capillary is the positive electrode, positive ions will have drifted downfield in the solution, i.e., toward the meniscus of the liquid, and negative ions will have drifted away from the surface. The mutual repulsion between the positive ions at the surface overcomes the surface tension of the liquid and the surface begins to expand, allowing the positive charges and liquid to move downfield. A cone forms, the so-called Taylor cone [19], and if the

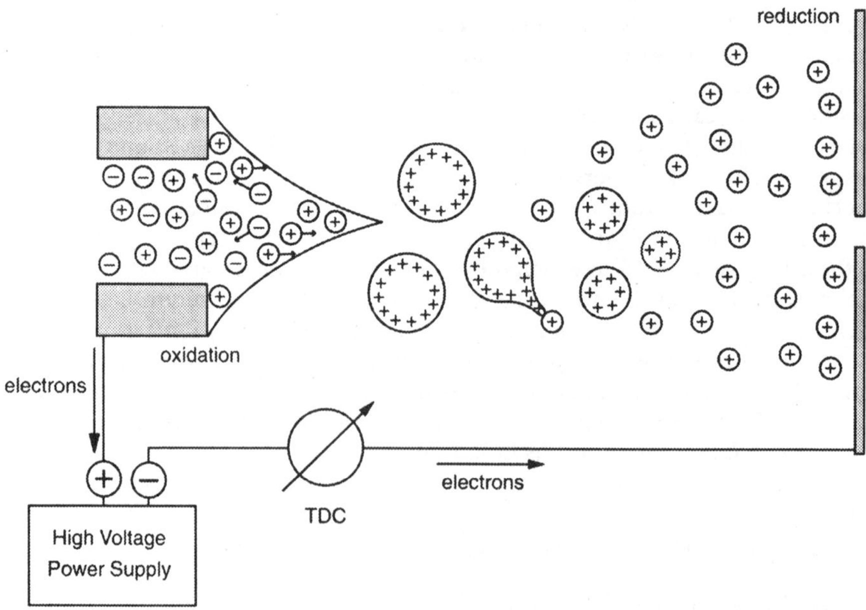

Fig. 1. Schematic of major processes occurring in electrospray (positive ion mode)

applied field is sufficiently high, a fine jet emerges from the cone tip which breaks up into small charged droplets.

The droplets are positively charged owing to the excess of positive electrolyte ions at the surface of the cone and the cone jet. The charged droplets produced by the spray shrink, owing to solvent evaporation while the charge remains constant. The energy required for solvent evaporation is provided by the thermal energy of the ambient gas, i.e., air in most cases. The charge of the droplets is expected to remain constant because the emission of ions from the solution to the gas phase is highly endoergic. The shrinkage of droplets at constant charge has been confirmed by direct observation of the droplets in special experiments [20–22]. The decrease of droplet radius at constant charge leads to an increase of the electrostatic repulsion of the charges at the surface until the droplets reach the so-called Rayleigh stability limit. The Rayleigh equation [23] gives the condition at which the electrostatic repulsion becomes equal to the force due to the surface tension, which holds the droplet together.

It is experimentally observed that the droplets undergo fission when they are close to the Rayleigh limit. The fragmentation is generally referred to as Coulombic fission. Even fission, where the droplet fragments into two or three particles of nearly equal mass and charge, is generally not observed and occurs only with nonpolar solvents whose conductivities are low [24]. Since ESMS is normally practiced with polar solvents with higher conductivities, jet fission ('uneven fission'), where a fine jet of droplets is ejected, can be expected to be the dominant process. Droplets that have relieved the Coulombic stress through jet fission will continue to evaporate solvent until they reach the Rayleigh stability limit and undergo another jet fission.

Two mechanisms have been proposed to account for the formation of gas-phase ions from very small and highly charged droplets. The first mechanism, proposed by Dole et al. [10, 11], depends on the formation of extremely small droplets which should contain only one ion. Solvent evaporation from such a droplet will lead to a gas-phase ion. Mass spectrometric determinations by Dole and co-workers were by and large unsuccessful, but the *charge residue model* (CRM) proposed by them survived. A more detailed consideration of, and support for, the mechanism was later provided by Röllgen et al. [25, 26].

Later on, Iribarne and Thomson proposed a different mechanism for the production of gas-phase ions from charged droplets [27, 28]. Interestingly, the motivation for their studies was far removed from a concern for the needs of mass spectrometry. Instead, it stemmed from their interest in charged droplets as a possible source of ions in the atmosphere. They proposed a model for such ion formation based on the idea that, on charged droplets that were small enough, evaporation could make the surface field sufficiently intense to lift solute ions from the droplet into the ambient gas before the Rayleigh limit is reached. This model is nowadays usually referred to as *ion desorption model* (IDM).

The basic features of the two models can be summarized as follows. In the CRM, the droplets shrink until the resulting instability breaks up the parent droplet into a hatch of offspring droplets, each of which continues to evaporate until it too reaches the Rayleigh limit. This sequence continues until the offspring droplets ultimately become so small that they contain only one analyte ion. In the IDM, the charged droplet commences the same sequence of evaporation and Coulombic fission steps as in the CRM scenario. The fundamental difference between the CRM and IDM is in how an analyte molecule with a charge becomes separated from all its companions in the droplet. In the IDM this separation comes about by desorption of a single analyte ion from a droplet that contains many other solute and solvent molecules. At the present stage, a clear experimental distinction between the two ion-release mechanisms is not possible. Indeed, it has been suggested that both mechanisms may be operative, depending on the chemical nature of the analyte [29].

4 Analytical Applications of Electrospray Mass Spectrometry in Inorganic and Organometallic Chemistry

4.1 Introduction

Because of the ease of transferring large polar organic molecules into gas-phase ions, applications of electrospray MS soon focused on the analysis of large and fragile macromolecules that play vital roles in biology and medicine. The first applications of electrospray to the mass spectrometric characterization of biopolymers, including oligonucleotides and proteins, were pioneered by J.B. Fenn [30]. ESMS has developed at a tremendous pace since the end of the 1980s. A statistical evaluation of the Proceedings of the American Society for Mass Spectrometry Conferences on Mass Spectrometry and Allied Topics (ASMS Proceedings) over the period 1979–1995 revealed the following trends [31]: while new developments such as FAB/SIMS and CI steadily gained in popularity during the 1980s, both techniques slowly leveled off in the early 1990s. Matrix assisted laser desorption/ionization (MALDI) is a relatively recent phenomenon in mass spectrometry and since 1990 it launched into a steep, geometric increase in the number of publications. ESI, after the longest gestation period of all major ionization techniques, has even jumped ahead of electron ionization (EI), traditionally the most dominant ionization technique, and increased its share ever since. ESMS has made a major impact on the bio-related area of pure and applied science and technology. There is no reason to suspect that the popularity of ESMS in these fields will decline in the near future.

Despite the enormous success of ESMS in the mass spectrometric characterization of biomolecules, it took some time for inorganic and organometallic chemistry to follow suit. Electrospray as a method of transferring ions directly from solution to the gas phase has quickly proven to be ideally suited for in organic and organometallic compounds. In general, the ions observed in the gas phase by ESMS for inorganic systems are the same as originally present in solution, unlike molecular ions generated from neutral species in other forms of mass spectrometry, and, therefore, the electrospray technique may be used to explore the solution chemistry directly. Analytical applications of ESMS to transition-metal chemistry can be roughly divided into two groups: the mass spectrometric characterization of known and defined solution-phase species, and the identification of crucial intermediates in transition-metal mediated reactions, whereby structural information can be obtained by CID, often in combination with other methods (electrochemistry, NMR spectroscopy). Such applications were already extensively reviewed by Colton, D'Agostino, and Traeger in 1995 [32]. More recently, applications of ESMS in organometallic chemistry have been summarized with a focus on the ionization technique itself and its utilization in the characterization of transition-metal compounds [33]. The present account will only cover significant recent work in this area and mainly focus on the gas-phase ion chemistry of transition metal complexes transferred to the gas phase by electrospray ionization.

ESMS has become increasingly important as a primary characterization tool in "small-ion" solution chemistry. In order to assess the impact of ESMS on the characterization of transition-metal compounds, we conducted a survey based on the entries in the CAS database (STN CA, July 3, 2002). A search for "electrospray" plus the common acronyms ("ES", "ESI", etc.) yielded 23,219 publications. Extracting all the compounds indexed in these publications and restricting them to those containing transition metals, lanthanides, and actinides, gave a total of 9150 compounds. Taking into account only these compounds restricted the above publications to 584 which were then inspected on the basis of their abstracts. This analysis confirmed that the overwhelming majority of these articles indeed described ES mass spectrometric characterization of transition-metal compounds. Due to the vagaries of indexing in the CAS database [34], this number just represents a lower limit.

4.2
Early Analytical Work

Before ESMS could be applied on a broad scale to explore novel solution-phase species in inorganic and organometallic systems, it was mandatory to prove that the method produces ions in the gas phase that faithfully reflect the species present in the electrosprayed solution. Electrospray mass spectra of tetraalkylammonium ions, among the simplest of inorganic cations, were already reported by Yamashita and Fenn in their first electrospray publications [15, 16]. Rafaelli and Bruins confirmed the relative ease of observing tetraalkylammonium cations by ESMS under a variety of conditions and in different solvents [35].

The beginning of the general and routine use of electrospray ionization as a tool for the characterization of ionic transition-metal coordination compounds is marked by a report on Ru^{II}-bipyridyl and 1,10-phentanthroline complexes by Chait and co-workers [36]. When spraying an acetonitrile solution at low ion-source collision energies, they observed a signal for the intact cation $[Ru^{II}(bpy)_3]^{2+}$ (bpy=bipyridyl) together with a series of lower intensity peaks resulting from the attachment of one to four acetonitrile molecules. At slightly higher energies, when the collisional activation is just sufficient for complete desolvation, the mass spectrum was completely dominated by a single intense peak of $[Ru^{II}(bpy)_3]^{2+}$ ions. Upon further increase of collisional activation, the doubly charged ion dissociated, giving rise to signals for the fragments $[Ru^{II}(bpy)_2]^{2+}$ and $[Ru^{II}(bpy)]^{2+}$. As will be seen from the following paragraphs, it is highly characteristic for ES mass spectra of coordination complexes that fragmentation by collisional activation typically causes loss of a complete ligand rather than fragmentation of a ligand. This feature is particularly helpful for the identification and structural assignment of signals in the mass spectra.

Pseudotetrahedral complexes of Cu^{I}, $[Cu(bpy)_2]BF_4$ and similar derivatives with substituted dipyridyl ligands all gave intact cations in their ES mass spectra [37, 38]. In an extension of this chemistry, bipyridyl groups were introduced into amino acid residues and, after reaction with Cu^{I}, the intact cations of the copper(I) complexes of these species were observed [37]. The analogous $[Cu(phen)_2]^+$ ion and others derived from substituted phenanthrolines have also been observed [39]. For both the $[Cu(bpy)_2]^+$ and $[Cu(phen)_2]^+$ species, collisional activation led to loss of one ligand.

Although negative ion ESMS has not been used as frequently as the positive ion mode, a number of anions have been observed. Early reports include observations of simple anions such as halides, $[CN]^-$, $[SCN]^-$, $[ClO_4]^-$, etc. [32]. The emphasis of that work was directed toward analytical and speciation considerations rather than chemical reaction studies. Going to the more complex anions, the diversity of oligomeric anions in aqueous solutions like isopolyoxovanadates and isopolyoxomolybdates made this class of compounds a particularly rewarding field for ESMS research [40–43].

4.3 Fishing for Intermediates

ESMS opens a straightforward approach to trap and identify short-lived intermediates, because bimolecular processes involving ionic species in solution are greatly attenuated when the ions pass into the gas phase [44]. Though still not as firmly established as traditional solution-phase characterization techniques like NMR, ESMS has become increasingly popular as a tool to identify intermediates in transition-metal catalysis. Due to the increasing number of applications in this field, only a few classic and recent representative examples of such work will be covered here.

ESMS has been used to characterize the intermediate Ni^{II}-complexes formed in the coupling reaction of 2-bromo-6-methylpyridine in the presence of Raney nickel (Scheme 1) [45]. The composition of the intermediate had already been determined previously by elemental analysis, but the ES mass spectra, showing a strong peak for the ion $[Ni_2(dmbp)_2Br_3]^+$, pointed to a dimeric structure. It was concluded that this ion was formed by the loss of Br^- from the dimeric structure **1**. An alternative explanation is that the intermediate has the more common four-coordinate structure **2**, and that the observed peak was due to the ion-paired species $\{[Ni_2(dmbp)_2Br_2]^{2+}+Br^-\}$. The dimeric nature of the intermediate was confirmed by a cross experiment: when mixtures of differently substituted pyridines were reacted, mixed ligand dinickel species were observed in the ES mass spectra.

In a seminal paper, Aliprantis and Canary described the use of ESMS for the characterization of catalytic intermediates in the palladium(0)-catalyzed cross-coupling reaction of alkyl halides and aryl boronic acids known as the Suzuki reaction [46]. The now accepted reaction pathway for the Suzuki coupling is out-

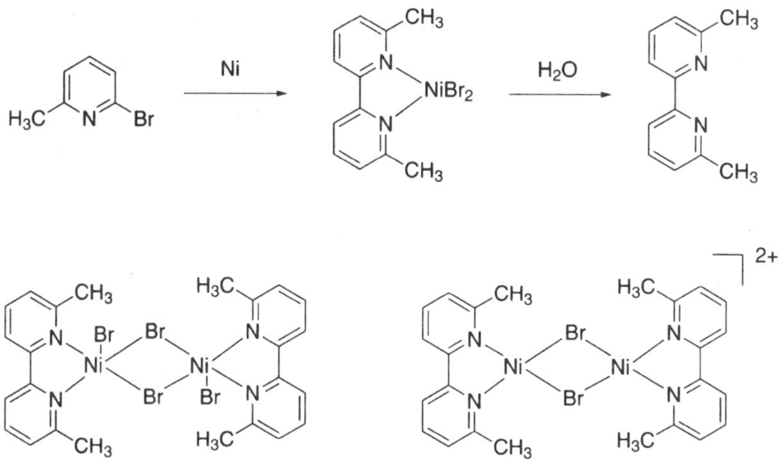

Scheme 1. Raney nickel mediated coupling of 2-bromo-6-methylpyridine

lined in Scheme 2. In the ESMS study, substituted pyridines were used as the aryl halide components so that the intermediates could be observed as the protonated pyridinium salt $[(pyH)Pd(PPh_3)_2Br]^+$. The intermediates **3** and **4** could both be detected for the first time in authentic reaction mixtures. A similar approach was taken by Ripa and Hallberg in order to detect intermediates in the palladium-catalyzed Heck arylation of cyclic enamidines [47]. ESMS was also employed by Brown and co-workers in conjunction with NMR spectroscopy to identify crucial intermediates in the Pd- and Ru-catalyzed Heck arylation of acrylates [48, 49].

Lipshutz and co-workers characterized anionic copper(I) complexes by ESMS [50]. The preparation of organocopper reagents relies on the combination of a copper(I) salt and two equivalents of an organolithium reagent to yield R_2CuLi (Gilman cuprates [51]). Solubilization of cuprate precursors, e.g., CuCN, in ethereal media is usually effected with lithium salts (LiCl, LiBr). Such mixtures of salts are usually termed "CuX·*n*LiX" which describes the stoichiometry but imparts little information about composition or structure. Negative-ion ES mass spectra of THF solutions of CuCN with two equivalents if LiX (X=Cl, Br, I) showed the same types of homo- and mixed halocyanocuprate anions. Particularly noteworthy was the presence of $[CuCN{\cdot}LiX_2]^-$, where Li^+ is part of the anionic complex. The species can be conceived to arise by either adding X^- or removing Li^+ from the neutral parent species XCu(CN)Li and Cu(CN)·2LiX, respectively. The same monoanions were observed by ESMS starting from a copper halide to which had been added one equivalent of LiCN and LiX. Interestingly, no such species $[CuX{\cdot}LiX_2]^-$ was detected from mixtures of a copper halide and two equivalents of lithium halide, attesting to the unique role of cyanide for drawing Li^+ into these anionic aggregates. In positive-ion mode, no species $[Li_2CN]^+$ could be detected.

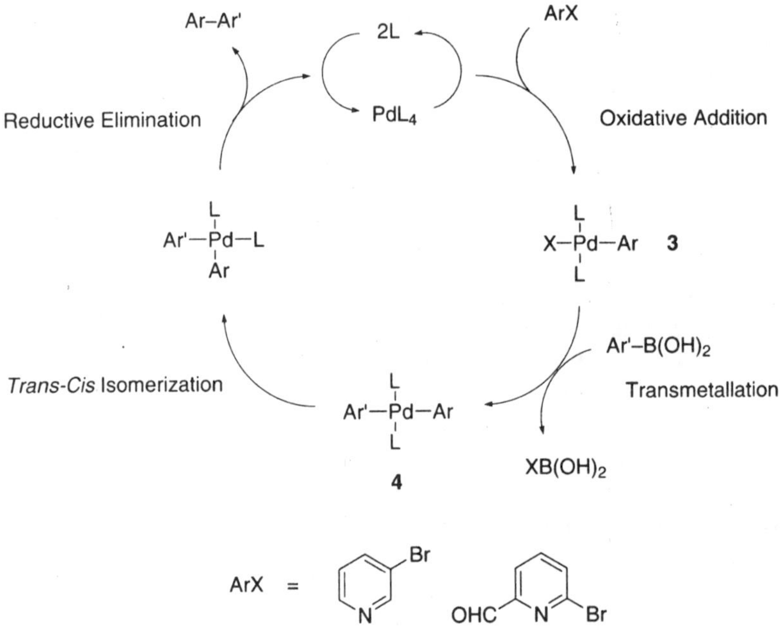

Scheme 2. The catalytic cycle of the Suzuki reaction

Going one step further, Lipshutz and co-workers studied the composition of synthetically valued mixed lithioorganocuprates by ESMS [52]. They prepared cuprates of the type "RR′CuLi" derived from CuI, $CuBr{\cdot}SMe_2$, and CuCN with various organolithium reagents. The negative-ion ES mass spectra revealed a surprising complexity of species: lower mass species from monomers to trimers were prevalent, whereas tetramers and pentamers were present in much smaller amounts. At low cone voltages, the base peak was typically of dimeric constitution. When the cone voltage was increased, a gradual decrease occurred in signal intensities for the aggregates observed at lower cone voltages, while an increase of monomeric species was observed. Cuprates derived from CuCN, however, afforded much cleaner and simpler spectra, uncomplicated by the presence of halide ions.

Electrospray mass spectrometry has also been used as a tool to explore the structural diversity of the $\{Pt_2Se_2\}$ core by a highly effective approach [53–55]: (1) the reaction products of $Pt_2(\mu\text{-}Se)_2(PPh_3)_4$ with a wide range of metal complexes were screened using ESMS; (2) the potentially stable and isolable intermetallic aggregates formed in situ were identified based on ion distribution and simulated isotope patterns; and (3) 'promising' reactions were repeated on the laboratory scale, the target products finally being isolated and characterized. Full characterization of the products isolated from preparative scale synthesis corresponded to the respective species deduced from the ESMS data. These results again emphasized the ability of ESMS and its simplicity to monitor inorganic solution chemistry.

ESMS was employed to identify reactive intermediates in the enantioselective hydrogenation of ethyl pyruvate on Pt-alumina, Pt black, and Pt black+alumina catalysts modified by dihydrocinchonidine in acetic acid [56]. The ESMS spectra of the raw product revealed a large number of species which fell into four groups: (1) dihydrocinchonidine and its hydrogenated derivatives; (2) the adducts of ethyl pyruvate and its oligomers; (3) (*R*)-ethyl lactate, the product of the enantioselective hydrogenation, and its adducts; and (4) oxonium compounds formed from alumina. The latter most likely play a decisive role in the development of the chiral environment of the catalyst surface. As suggested by the authors, these oxonium cations could make the so-called electrostatic catalysis [57], based on electrostatic acceleration, possible.

Kutal and co-workers used ESMS to probe the solution photochemical behavior of $[CpFe(\eta^6\text{-benzene})]PF_6$, a visible-light-sensitive photoinitiator for the polymerization of epoxides [58]. While control experiments showed that in the absence of light $[CpFe(\eta^6\text{-benzene})]^+$ emerged unchanged from the nanospray tip, irradiation of an acetonitrile solution of the complex produced two major series of ionic products: $[CpFe(CH_3CN)_{1\text{-}3}]^+$ and $[Fe(CH_3CN)_{3\text{-}6}]^{2+}$, obviously reflecting the photodissociation of benzene and subsequent thermal loss of the cyclopentadienide anion. Photolysis of $[CpFe(\eta^6\text{-benzene})]PF_6$ in acetonitrile solutions containing cyclohexene oxide (cho) yielded the same products as well as $[Fe(CH_3CN)_3(cho)]^{2+}$ and $[Fe(CH_3CN)_4(cho)]^{2+}$. Irradiating $[CpFe(\eta^6\text{-benzene})]PF_6$ and cyclohexene oxide in a poorly coordinating solvent like 1,2-dichloroethane, however, yielded the series $[(H_2O)Fe(cho)_{4-12}]^{2+}$ and $[(H_2O)CpFe(cho)_{1\text{-}4}]^{2+}$. The coordinated water originated from traces of moisture introduced during sample preparation or electrospraying. Signals from the water-free products also appeared in the mass spectrum, but with diminished intensities. The structures of these species were assigned following the polymer model, i.e., products such as $[(H_2O)Fe(cho)_{12}]^{2+}$ or $[Fe(cho)_8]^{2+}$ contain a growing polymer chain bound directly to the metal center.

Intermediates in the copper(I) mediated asymmetric insertion reaction of diazoacetates into the Si–H bond of silanes were identified by ESMS [59]. Starting from $[Cu(CH_3CN)_4]PF_6$ and the chiral ligand (*R,R*)-bis((2,6-dichlorobenzylidene)diamino) cyclohexane (bdc), NMR and ESMS studies established that only $[Cu(bdc)(CH_3CN)]^+$ was present in solution in high yields. By ESMS, it was found that $[Cu(bdc)(C(CO_2Me)Ph))]^+$ and $[Cu(bdc)(C(CO_2Me)Ph)(CH_3CN)]^+$ were formed upon addition of $PhC(N_2)CO_2Me$.

ESMS, in combination with low-temperature NMR techniques, was used to characterize reactive intermediates in an enantioselective titanium alkoxide mediated sulfoxidation reaction [60, 61]. Catalyst formation in situ by mixing titanium(IV) isopropoxide and chiral trialkanolamine ligands was monitored by ESMS. When a 1:1 stoichiometric ratio was used, monomeric, highly symmetric complexes were obtained. A slight excess of the trialkanolamine ligand produced discrete 2:1, 3:2, and 4:3 oligomers in which the excess trialkanolamine were bridging multiple titanatrane units. In the presence of an oxygen donor like *tert* butyl hydroperoxide, all the precatalyst species were converted to a mononuclear titanium(IV) peroxo complex which serves as the active species in the asymmetric sulfoxidation.

Finally, Hinderling and Chen used ESMS to screen the activity of a small library of eight Brookhart-type palladium(II) complexes in the solution-phase polymerization of ethylene [62]. The crude reaction mixture was quenched with DMSO, diluted, and electrosprayed in order to analyze the growing polymers chains. Upon CID in the gas phase, the polymer chain was fragmented from the catalyst by β-hydride elimination, thus facilitating the identification of the most active catalysts in an otherwise dauntingly complex mass spectrum of a polymer mixture. Since this analysis can be performed simultaneously for a whole catalyst library, ESMS was hereby proven the method of choice for an assay of multiple, competitive and simultaneously occurring catalytic reactions.

5 Organometallic Gas-Phase Chemistry

5.1 Advantages of Electrospray Tandem Mass Spectrometry

Gas-phase studies of metal ions and ionic transition-metal compounds and their reactivity have a long and distinguished history of over 25 years. The advancement of mass spectrometric ion-molecule techniques has made it possible to study single, elementary reaction steps mediated by transition-metal ions on both qualitative and quantitative levels [63]. Unfortunately, due to the limitations of "conventional" ionization techniques, the gas-phase chemistry of transition metal-compounds has for a long time been restricted to the study of compounds that would hardly be considered as authentic catalysts by synthetic organometallic chemists. In other words, most studies had to confine themselves to singly and multiply charged "naked" ions and very simple organometallics of the type $[M(L)_n]^+$, with L some simple ligand, e.g., hydride, O, alkyl, CO, H_2O, etc.

When considering the possibilities of transferring real-world organometallics to the gas phase, one arrives at the conclusion that electrospray ionization (ES) should be the ideal technique due to its softness, i.e., even weakly bound ligands will remain intact. Once in the gas phase, the species of interest can be singled out by mass spectrometric techniques, and its reactivity subsequently studied without perturbation by any of the other compounds present. For further reactivity studies and in order to get structural information, at least two mass selection steps should be available. By transferring reactive intermediates to the gas-phase, they become almost infinitely stable due to the high-vacuum conditions. Their reactivity can then be probed by directed collision with reaction partners. Finally, thermochemical parameters of single reaction steps can be obtained by quantitative collision-induced dissociation (CID) threshold measurements.

5.2 Simple Gas-Phase Reactions

The simplest gas-phase reactions of transition-metal compounds just involve an inert reaction partner as the collision gas, leading to plain bond rupture (fragmentation) or elimination reactions in the analyte ion. Since the fragmentation pattern of a species of known mass-to-charge ratio is practically the only piece of information that allows the assignment of a specific structure, these gas-phase reactions are routinely performed in order to corroborate that the species detected in the mass spectrometer correspond to the solution-phase composition. These form a special subgroup of gas-phase reactions which will be just mentioned en passant in the context of this account. Triggering a chemical reaction upon collisional activation in such cases is just a sort of by-product in the mass spectrometric characterization. One of the earliest observations of a chemical reaction induced by gas-phase collisional activation of an electrosprayed transition-metal compound that has a well-known analogy in solution phase was a reductive elimination from palladium(IV) complexes [64].

Dalgaard and McKenzie studied the CO_2 loss from inorganic carbonato compounds upon collisional activation in the gas phase [65]. This is not simply the reversal of the synthetic reaction of metal hydroxides with CO_2 since the lack of available protons in the gas phase makes the regeneration of the starting hydroxide complexes impossible. This way, novel oxide species rather than the more common hydroxide species were generated; e.g., $[Co(tpa)O]^+$ (tpa=tris(2-pyridylmethyl)methylamine) was found as a major fragment ion upon CID of the simple carbonate $[Co(tpa)CO_3]^+$. Similarly, unusual platinum(II) complexes of the type $[Pt(dien)O]^+$ and $[Pt(dien)N]^+$ (dien=diethylenetriamine) were obtained by collisional activation of electrosprayed $[Pt(dien)NO_2]^+$ and $[Pt(dien)N_3]^+$, via loss of NO and N_2, respectively [66]. Hor and co-workers studied the fragmentation pathways of the μ-oxo bridged dinuclear rhenium complexes $[Re_2(\mu\text{-}OH)_3(CO)_6]^-$ and $[Re_2(\mu\text{-}OMe)_3(CO)_6]^-$ by ESMS [67]. The hydroxy complex underwent dehydration followed by CO loss, whereas for the methoxy complex *β*-hydride elimination and CO loss was observed.

One of the earliest examples of non-analytical applications of electrospray ionization was the coupling of ES to ion-molecule techniques in the gas phase by Posey and co-workers. They electrosprayed methanolic solutions of $Fe(bpy)_3(ClO_4)_2$ to produce molecular beams of complexes of divalent transition metal ions [68–70]. The residual electrospray solvent was removed from the gas-phase ions by collisional activation, and clusters were formed by association of solvent molecules from the purge in an expansion as the ions passed through the first and second stages of differential pumping in the ES source. By this method, clusters of $[Fe(bpy)_3]^{2+}$ with a variety of solvents (acetone, acetonitrile, *N*,*N*-dimethylformamide, alcohols, etc.) were prepared by introducing the solvent of choice to the nitrogen purge [68]. Metal-to-ligand charge transfer in the gas-phase ions and clusters was then probed by laser photofragmentation mass spectrometry. Excitation in clusters triggers evaporation of neutral solvent molecules, e.g., methanol, as the system relaxes. The wavelength-dependent photodissociation yield was measured between 530 and 600 nm for mass-selected $[Fe(bpy)_3 \cdot (CH_3OH)_n]^{2+}$ clusters (n=2–6). The resulting photodissociation action spectra reflected the onset of the metal-to-ligand charge transfer absorption band as a function of cluster size [69]. The solvent dependence of the charge transfer was further studied for terpyridyl iron complexes $[Fe(terpy)_2]^{2+}$ in clusters with one or four molecules of polar, organic solvents (acetone, acetonitrile, dimethylsulfoxide, *N*,*N*-dimethylformamide, etc.) [70].

Reactions of coordinatively unsaturated transition metal complexes with molecular oxygen in the gas phase have been reported by McKenzie and coworkers [71]. Using a triple-quadrupole setup with an atmospheric pressure ionization source operating in electrospray mode they investigated the reactivity of $[M(bipy)_2]^{2+}$ (M=Cr, Mn, Fe, Co, Ni, Cu, Ru, Os) and $[M(bipy)]^+$ (M=Ni, Co, Cu) toward O_2. In the case of the bis(pyridine) ions, the complexes with M=Cr, Ru, and Os yielded the dioxygen adducts $[M(bipy)_2O_2]^{2+}$ upon collision with O_2 in the collision octopol. This behavior correlates well with the propensity of Cr, Ru, and Os for attaining the high oxidation states known from solution and solid-state chemistry. As for the mono(pyridine) complex ions, only the Ni- and Co-containing species gave detectable dioxygen adducts of the composition $[M(bipy)O_2]^+$. The structural assignment of the oxidized products was attempted by MS/MS/MS experiments. $[M(bipy)_2]^{2+}$ and $[M(bipy)]^+$ were generated in the skimmer region and allowed to react with O_2 in the source octopole in a similar experimental setup as described below. The adducts $[M(bipy)_2O_2]^{2+}$ and $[M(bipy)O_2]^+$ were then subjected to CID in the collision octopole (Ar as collision gas). The loss of a mass equivalent to two oxygen atoms was found as the predominant process for $[M(bipy)_2O_2]^{2+}$ (M=Cr, Ru) and $[M(bipy)O_2]^+$ (M=Ni, Co), thus indicating an intact O–O bond in the structure of the dioxygen adducts. In the case of $[Os(bipy)_2O_2]^{2+}$, however, no formation of $[Os(bipy)_2]^{2+}$ was observed. Instead, weak signals assigned to $[Os(bipy)O_2]^+$, $[Os(bipy)O_2]^{2+}$, and the bipyridyl radical cation indicated loss of a bipyridine ligand in preference to O_2, suggesting that the dioxygen adduct was of an "osmyl" type complex cation with both oxygen atoms independently bound to the osmium center.

5.3 Experimental Setup for the Gas-Phase Reaction Studies

An experimental setup that meets the demands posed in Sect. 5.1 is an electrospray ionization tandem mass spectrometer. The apparatus which was used in the experiments described in the following chapters is a modified Finnigan MAT TSQ7000 spectrometer. In a typical experiment, a 10^{-4} to 10^{-5} mol l^{-1} solution of the transition-metal compound in a polar organic solvent (preferentially acetonitrile or dichloromethane) is pressed through a capillary at a flow rate of 7–15 μl min^{-1} and electrosprayed at a potential of 4–5 kV using N_2 as sheath gas. The ions are then passed through a heated capillary (typically at 150–200°C) where they are declustered and the remaining solvent molecules evaporate. The extent of desolvation and collisional activation can further be controlled by a tube lens potential in the electrospray source, which typically ranges from 35 (mild conditions) to 100 V (hard conditions). The first octopole (O1) acts as an ion guide to separate the ions from neutral molecules which are pumped off by a turbo pump located underneath the octopole. O1 is fitted with an open cylindrical sheath around the rods into which, depending on the setup used, a collision gas can be bled for thermalization or reaction at pressures up to 100 mTorr. The ions then enter the actual mass spectrometer, which is at 10^{-6} Torr and 70°C manifold temperature during operation. The configuration is quadrupole/octopole/quadrupole (Q1/O2/Q2), with the two quadrupoles as mass selection stages and the second octopole operating as a collision-induced dissociation (CID) cell. Spectra can be recorded in three different modes. In the normal ESMS mode, only one quadrupole is operated (either Q1 or Q2), and a mass spectrum of the electrosprayed ions is recorded. This mode serves primarily to characterize the ions produced by a given set of conditions. In the daughter-ion mode, Q1 is used to mass-select ions of a single mass-to-charge-ratio from among all of the ions produced in O1, which are then collided or reacted with a target gas in O2, and finally mass-analyzed by Q2. This mode is used to obtain structural information (by analysis of the fragments) or the specific reactivity of a species of a given mass. The third mode of operation, the RFD mode (radio frequency daughter mode), is used to record CID thresholds for quantitative thermochemical measurements (see Sect. 5.8). Ions produced in O1 are mass-selected, with Q1 acting as a high-pass filter. The selected ions are then collided with a target gas in O2, and mass-analyzed in Q2. In the RFD mode, the reactions of ions in a particular mass range are isolated by setting the mass cutoff above and below the desired range and substracting the latter from the former spectra.

5.4 C–H Activation by Cationic Iridium(III) Complexes

5.4.1 *Introduction*

The search for a homogeneous transition metal-catalyst capable of selective insertion into non-activated C–H bonds of alkanes and arenes is generally acknowledged to be one of the most challenging and important tasks in chemistry [72]. In the early 1980s, rhodium and iridium Cp*-complexes with the metal in the oxidation states +1 and +3, respectively, were found to be the most promising candidates to achieve this task [73–75]. The search for a catalyst active under milder conditions led Bergman to the introduction of cationic iridium(III) complexes [76]. These complexes perform C–H activation without prior photochemical or thermal activation, and are therefore particularly interesting as model systems ultimately leading to catalytic cycles. The actual reactive species was presumed to be the 16-electron complex $[Cp^{*}Ir(PMe_3)(CH_3)]^+$, which would be formed from the 18-electron precursor by dissociation of a weakly bound ligand in solution. Two mechanisms have been suggested by Bergman and co-workers [76–78] for the observed σ-bond metathesis reaction by cationic Ir^{III} complexes (Scheme 3): an oxidative addition/reductive elimination sequence of reactions and a concerted σ-bond metathesis reaction. This concerted mechanism comes into consideration for the iridium(III) complex because it might be unfavorable for the strongly electron-deficient 16-electron Ir^{III}-complex to engage in an oxidative addition step resulting in a formally iridium(V) species. On the basis of the increase in reactivity as the ligand binding is weakened, going from triflate to dichloromethane, the actual reactive species was presumed to be **6**, which would be formed from **5** in a pre-equilibrium in the solution-phase studies. For the dichloromethane-ligated complex, activation of C–H bonds was observed well below room temperature. A similar pattern of reactiv-

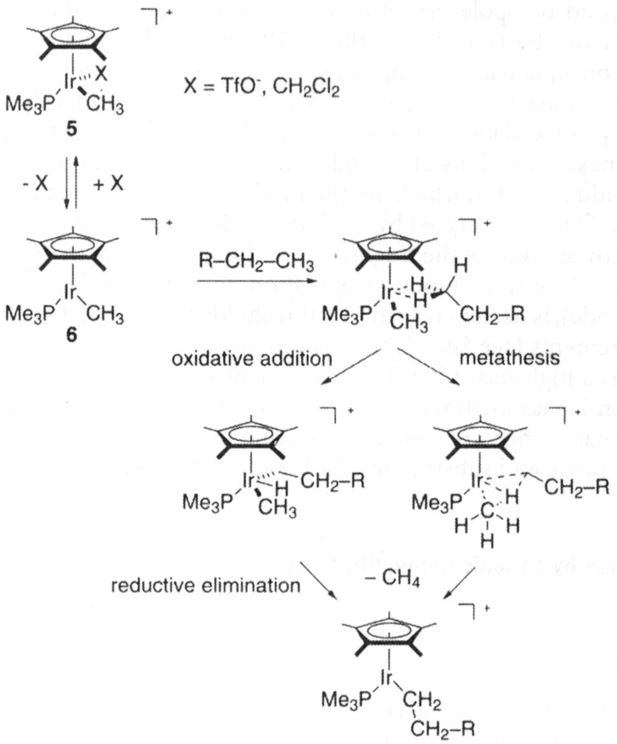

$X = TfO^-, CH_2Cl_2$

Scheme 3. Mechanism for the C–H activation by cationic iridium(III) complexes proposed by Bergman and co-workers

ity, with the same set of mechanistic possibilities, was reported by Bercaw and co-workers for platinum(II) complexes [79].

5.4.2
ESMS Studies

In order to obtain insight into the detailed mechanism of C–H activation ESMS was used for the study of the gas-phase reactivity of both $[Cp^*Ir(PMe_3)(CH_3)]^+$, the proposed reactive species in the Bergman system, and the Cp-analogue $[CpIr(PMe_3)(CH_3)]^+$ [80, 81]. The Bergman system was deemed particularly suitable for an ESMS study since all the species of interest are present as ions in solution without the need for further derivatization. The 16-electron complex cations were generated by electrospray of acetonitrile solutions of $[Cp^*Ir(PMe_3)(CH_3)(CH_3CN)]^+ ClO_4^-$ and the corresponding salt of the Cp analogue. Electrospray ionization produced the molecular ion of organometallic complexes, with even weakly-bound ligands still intact, i.e., electrospray under mild desolvation conditions produces the unfragmented cation. A slight increase in the tube lens potential, i.e., higher collisional activation in the desolvation region of the electrospray source, led to almost complete loss of the acetonitrile ligand, leaving predominantly the bare, unsolvated 16-electron complex cation $[Cp^{(*)}Ir(PMe_3)(CH_3)]^+$. With some small differences (see below), the reactions of the $[Cp^*Ir(PMe_3)(CH_3)]^+$ and $[CpIr(PMe_3)(CH_3)]^+$ complexes **6** and **10** were entirely parallel, with the Cp series of complexes showing qualitatively higher reactivity than their Cp* analogs (Scheme 4).

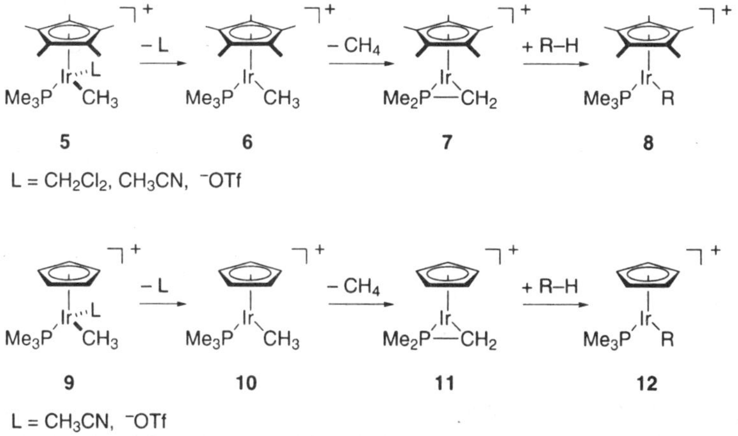

5 —(− L)→ 6 —(− CH_4)→ 7 —(+ R–H)→ 8

L = CH_2Cl_2, CH_3CN, ^-OTf

9 —(− L)→ 10 —(− CH_4)→ 11 —(+ R–H)→ 12

L = CH_3CN, ^-OTf

Scheme 4. Dissociative-associative mechanism for the C–H activation by cationic iridium(III) complexes observed in the gas phase

The gas-phase chemistry of the coordinatively unsaturated complex cations can be summarized as follows. (1) Properties related to the coordination chemistry behavior in solution were probed in the gas phase. The ion assigned as **10**, prepared by collisional removal of CH_3CN from **9** and thermalized, picked up neutral 2-electron ligands such as acetonitrile or CO with high efficiency at very low collision energies. (2) Neither $[Cp^*Ir(PMe_3)(CH_3)]^+$ nor the more electron-deficient $[CpIr(PMe_3)(CH_3)]^+$ showed any appreciable reactivity towards hydrocarbons in the gas phase. This behavior clearly contradicts the pathways outlined in Scheme 3. (3) Higher collisional activation of $[Cp^*Ir(PMe_3)(CH_3)(CH_3CN)]^+$ and $[CpIr(PMe_3)(CH_3)(CH_3CN)]^+$ not only caused complete dissociation of acetonitrile, but also induced a chemical reaction leading to a loss of 16 mass units. The mass loss was interpreted as methane elimination in an intramolecular cyclometalation reaction. (4) In the gas phase, all detectable cyclometalation proceeded exclusively on the methyl groups of the phosphine, and not on the Cp or Cp*. (5) The cyclometalation products **7** and **11** reacted readily with pentane and benzene to yield the products of an overall σ-bond metathesis. Based on deuterium-labeling experiments (Scheme 5), the structure of these species was assigned as $[Cp^{(*)}Ir(\eta^2\text{-}CH_2PMe_2)]^+$, the product of cyclometalation with one of the phosphine methyl groups and concomitant loss of methane. (6) The addition of $[Cp^*Ir(\eta^2\text{-}CH_2PMe_2)]^+$ or $[CpIr(\eta^2\text{-}CH_2PMe_2)]^+$ to benzene is reversible upon further CID. An elegant proof for the structure of the cyclometalation intermediate was provided by the reactivity of the Cp complex

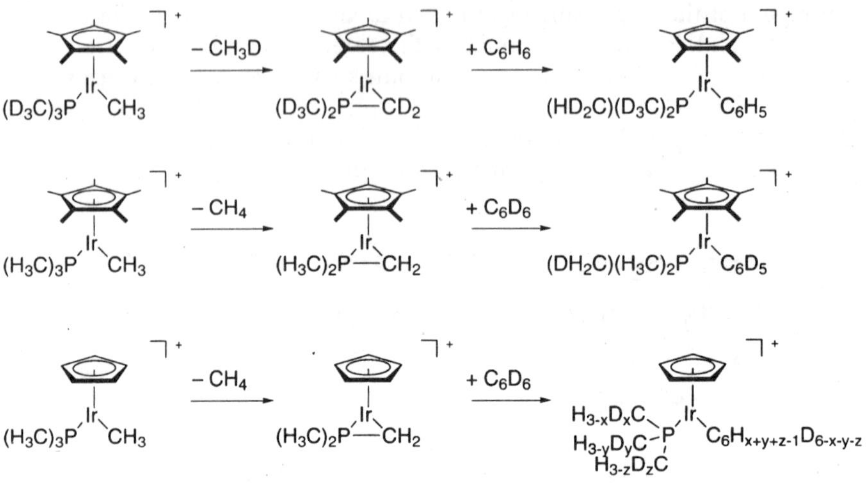

Scheme 5. Deuterium-labeling experiments suggesting the presence of a three-membered ring intermediate

$[CpIr(PMe_3\text{-}d_9)(CH_3)]^+$. In this case, not only one deuterium is incorporated in the ligand, but the six deuterium atoms of the incoming benzene were completely scrambled over the phosphine (Fig. 2). (7) The three-membered ring intermediate did not only react with benzene, but also added to less reactive hydrocarbons such as methane, pentane, or cyclohexane. However, the C–H activation products displayed a different reactivity pattern. The product of the addition to pentane, $[Cp^*Ir(PMe_3)(C_5H_{11})]^+$, underwent β-hydride elimination when collided with a target gas to yield $[Cp^*Ir(PMe_3)(H)]^+$. With alkyl ligands at Ir^{III}, the reverse cyclometalation is obviously not a feasible pathway in competition with the facile elimination of pentene.

The differences in the gas-phase chemistry of the Cp* complexes **5–8** and the Cp analogs **9–12** are confined to two areas. While the Cp-substituted complexes added to pentane, cyclohexane, or benzene, the Cp*-substituted analogs reacted with pentane or benzene only. No cyclohexane adducts were observed in the gas-phase reactions of $[Cp^*Ir(PMe_3)(CH_3)]^+$ or $[Cp^*Ir(\eta^2\text{-}CH_2PMe_2)]^+$. Second, in the CID of the two phenyl-substituted complexes **8** and **12** (R=Ph), there was complete deuterium scrambling in **12**, but only partial scrambling in **8**.

5.4.3
The Gas-Phase Mechanism of C–H Activation by the Iridium(III) Complex Cations

The gas-phase reactions of the cationic Ir^{III} complexes follow a previously unreported mechanism for their observed σ-bond metathesis reactions. Previous discussions had considered a two-step mechanism involving intermolecular oxidative addition of either $[Cp^*Ir(PMe_3)(CH_3)]^+$ or $[CpIr(PMe_3)(CH_3)]^+$ to the C–H bond of an alkane or arene producing an Ir^V intermediate, followed by reductive elimination of methane, or a concerted σ-bond metathesis reaction sim-

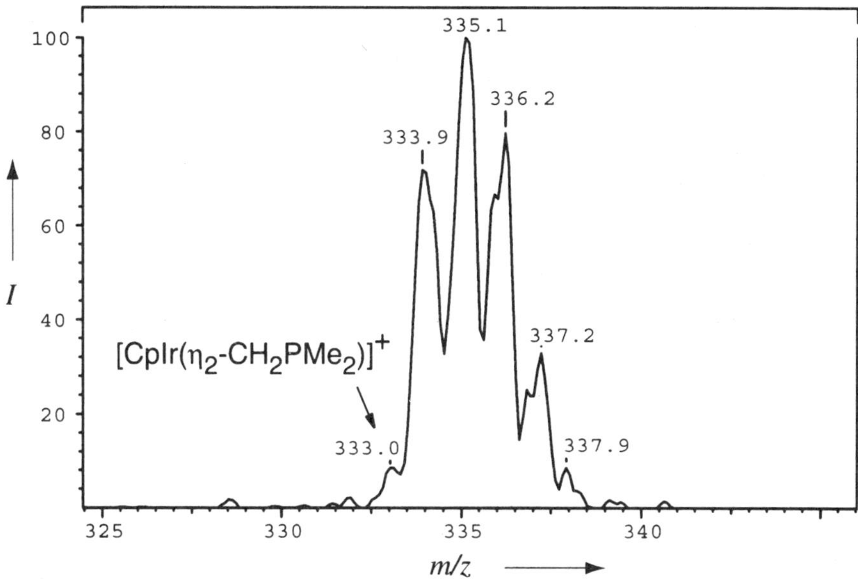

Fig. 2. Expansion of the CID spectrum of $[CpIr(PMe_3)(Ph)]^+$-d_6, showing the formation of various deuterated three-ring species $[CpIr(\eta^2\text{-}CH_2PMe_2)]^+$-$d_x$ (x=0–6)

ilar to that seen in early transition metals (Scheme 3). The electrospray mass spectra show another mechanism involving initial elimination of methane, followed by addition to an alkane or arene, with all detected reactive intermediates retaining a formal +3 oxidation state (Scheme 4). The structures of the intermediate species, $[Cp^*Ir(\eta^2\text{-}CH_2PMe_2)]^+$ and $[CpIr(\eta^2\text{-}CH_2PMe_2)]^+$, were assigned on the basis of experimental observations from ESMS, indicating a cyclic structure with cyclometalation occurring on the phosphine and not the Cp or Cp* ligand. The observed threshold for methane loss from $[CpIr(\eta^2\text{-}CH_2PMe_2)]^+$ of 13.6 kcal/mol (see Sect. 5.8.2) is also consistent with the structural assignment. The facile addition of $[CpIr(\eta^2\text{-}CH_2PMe_2)]^+$ to alkanes and arenes indicates that the reactions of $[CpIr(\eta^2\text{-}CH_2PMe_2)]^+$ occur over a rather low barrier.

The difference in the extent of isotopic exchange between the trimethylphosphine and the phenyl ligand upon CID of **8** vs **12** (R=Ph) comes from a general feature of ion-molecule reactions. In the gas phase, ion-molecule reactions proceed through a charge-dipole or charge-induced dipole complex bound by anywhere from a few to several kilocalories per mole relative to the separated molecules. The non-negligible attractive potential amounts to a potential well in which the reacting partners of a bimolecular ion-molecule reaction are trapped prior to reaction. In the absence of subsequent thermalizing collisions, the charge-dipole or charge-induced dipole interaction can provide the energy needed to surmount an activation barrier for the trapped reactants as long as the transition state is not substantially higher in energy than the separated reactants. By microscopic reversibility, CID of **12** (R=Ph) to produce **11** and benzene should go through the same charge-induced dipole complex. As long as the transition state for the conversion of the complex to the adduct lies below the asymptotic limit for separate molecules, the first step of the dissociation should be reversible, leading to the extensive isotopic scrambling.

Going to the Cp* series, a shallower well is to be expected for the complex relative to the separated molecules. One would justify this change by invoking both the increased steric bulk of the Cp* ligand and the greater degree of charge-delocalization in $[Cp^*Ir(\eta^2\text{-}CH_2PMe_2)]^+$ (relative to $[CpIr(\eta^2\text{-}CH_2PMe_2)]^+$) as factors making the charge-induced dipole interaction less stabilizing. Having lowered the asymptote relative to the two minima and the transition state for reaction, a small activation energy now appears for the bimolecular gas-phase reaction. The reverse reaction, CID of **8** (R=Ph), would now proceed without reversibility in the first step of the dissociation, limiting the isotopic exchange to zero or one label. As a consequence, benzene is highly reactive with the cationic metallaphosphacyclopropanes because, with its large π-system, it is highly polarizable. Pentane reacts because the primary C–H bonds at its termini can approach even a sterically hindered ion. Cyclohexane, with its secondary C–H bonds, cannot approach as closely, so the reaction is worse. Last, methane, although it is small, is the least polarizable of the aliphatic hydrocarbons, and therefore forms only weakly-bound charge-induced dipole complexes, giving the lowest reactivity.

The question may arise how the knowledge of the intrinsic reactivity of cationic iridium(III) complexes displayed in the gas-phase studies can be applied to the design of more efficient catalysts. It did not take too long before an answer to this question was given. Less than one year after publication of the gas-phase results summarized above, Luecke and Bergman reported the solution-phase reactivity of $Cp^*Ir(\eta^2\text{-}CH_2PMe_2)(OTf)$ (TfO^-=trifluoromethanesulfonate), a catalyst which had been designed according to the reactivity scheme found in the ESMS work (Scheme 6) [82]. The rate of disappearance of the precursor complex $Cp^*Ir(\eta^2\text{-}CH_2PMe_2)(Cl)$, from which $Cp^*Ir(\eta^2\text{-}CH_2PMe_2)(OTf)$ was formed by adding AgOTf, in neat benzene was extremely rapid ($t_{1/2}$=5 min at 25°C), much faster than the analogous reaction of $Cp^*Ir(PMe_3)(CH_3)(OTf)$ ($t_{1/2}$=24 h at 25°C). The iridaphosphacyclopropane complex is much superior in

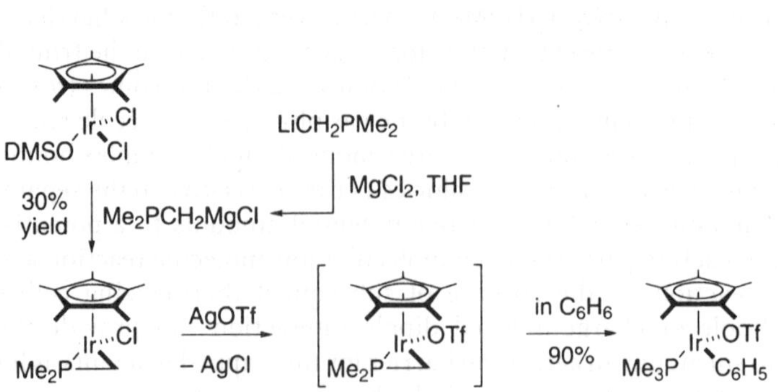

Scheme 6. Reactivity of an iridaphosphacyclopropane catalyst

C–H activation than is the 16-electron complex [Cp*Ir(PMe_3)(CH_3)]OTf. This solution-phase reactivity resembles exactly the order of reactivity as found in the gas-phase experiments.

Both the similarities and the differences between the observed gas-phase and solution-phase chemistry hold information useful to the dissection of complex reaction pathways. In the particular case of C–H activation by cationic iridium(III) complexes, a novel mechanism involving metallaphosphacyclopropane intermediates was found. The balance favoring this mechanism over competing mechanisms not involving cyclometalation depends on the presence or absence of solvent, and may be adjustable by perturbations in structure or surroundings. The multiplicity of competing mechanisms and energetically similar structural types in organometallic chemistry has often been blamed for the poor state of mechanistic understanding in the field, but with the introduction of new, powerful physical techniques, this problem can potentially become an asset. We see that changes in mechanism can be induced by small changes in the molecule or its surroundings, which if properly understood can then be used to fine-tune reactivity rationally.

5.5 Ziegler-Natta-Like Olefin Oligomerization by Alkylzirconocene Cations

5.5.1 *Introduction*

The Ziegler-Natta polymerization of olefins is a technical process of utmost importance for the synthesis of high-molecular-weight polymers [83]. The technical process is based on the activation of a suitable zirconocene precursor, typically Cp_2ZrMe_2 or Cp_2ZrCl_2, by a strong Lewis acid as co-catalyst (e.g., methylaluminoxane MAO, the product of a partial hydrolysis of $AlMe_3$). The co-catalyst has a twofold function in the technical process: (1) it assures partial or complete alkylation if the precursor is a zirconocene dihalide; and (2) due to its strong Lewis acidity, MAO removes one of the methyl groups of dimethylzirconocene, thus forming an ion pair $[Cp_2ZrCH_3]^+$ CH_3–MAO^- where the cation is not free, but stabilized by $Al_2O \rightarrow Zr$ or $AlCH_3 \rightarrow Zr$ contacts.

Mechanistic studies on Ziegler-Natta-like polymerization soon focused on the role of cationic alkylzirconocene species in the catalysis. In the 1980s, Jordan and co-workers were the first to isolate and characterize an alkylzirconocene cation in the form of the salt $[Cp_2Zr(CH_3)(THF)]^+$ $[BPh_4]^-$ (Scheme 7) [84, 85].

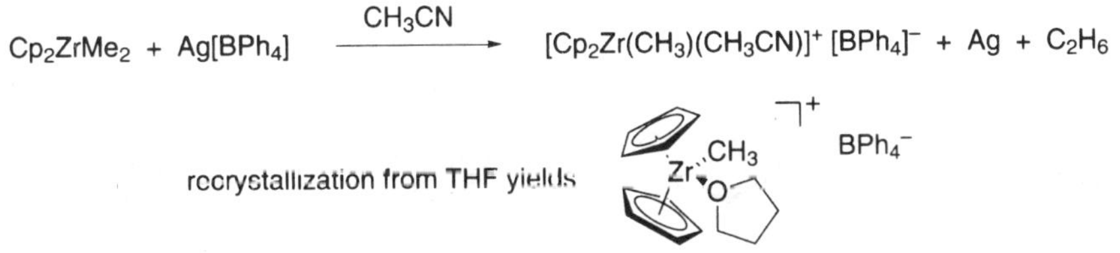

$$Cp_2ZrMe_2 + Ag[BPh_4] \xrightarrow{CH_3CN} [Cp_2Zr(CH_3)(CH_3CN)]^+ [BPh_4]^- + Ag + C_2H_6$$

recrystallization from THF yields $[Cp_2Zr(CH_3)(THF)]^+$ BPh_4^-

Scheme 7. Preparation and isolation of a methylzirconocene salt by Jordan et al. [84, 85]

Due to the predissociation of THF being the rate-limiting step, the ethylene polymerization activity of this methylzirconocene salt was relatively low. In solvents acting as donor ligands (THF, acetonitrile) its polymerization activity was shut down completely. These results supported the original proposal (the so-called Long-Breslow-Newburg mechanism) that a cationic alkylzirconocene is the active species in Ziegler-Natta polymerization. A breakthrough was achieved by the introduction of perfluorinated tetraphenylborate as a counterion (Scheme 8) [86]. Those salts were the first well-defined zirconocene complexes that could polymerize propene and higher olefins without addition of an activator.

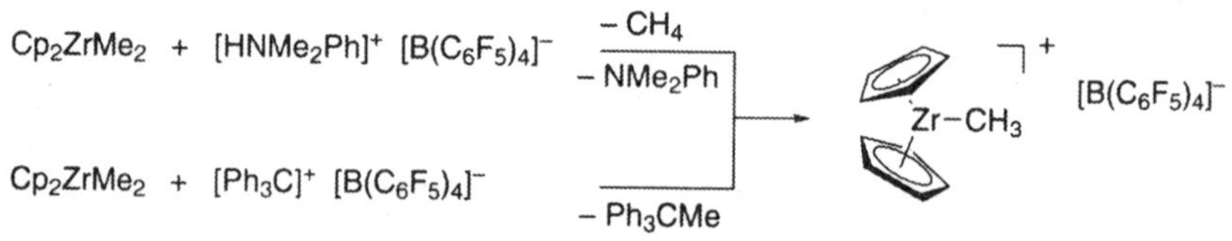

Cp_2ZrMe_2 + $[HNMe_2Ph]^+$ $[B(C_6F_5)_4]^-$ ($-CH_4$, $-NMe_2Ph$)

Cp_2ZrMe_2 + $[Ph_3C]^+$ $[B(C_6F_5)_4]^-$ ($-Ph_3CMe$)

→ $[Cp_2Zr{-}CH_3]^+$ $[B(C_6F_5)_4]^-$

Scheme 8. Synthesis of the perfluorinated tetraphenylborate of methylzirconocene

5.5.2
ESMS Studies

There had been previous mass spectrometric studies of isolated metallocene ions in the gas phase, most notably by Eyler, Richardson, and co-workers [87–89]. They prepared $[Cp_2ZrCH_3]^+$ by electron impact on dimethylzirconocene and studied its reactivity towards olefinic substrates in an ICR spectrometer. For the wide range of α-olefins investigated by Richardson and Eyler, allylic complexes, formed by loss of H_2, were the sole observed products of reaction of $[Cp_2ZrCH_3]^+$ with olefins bearing a β-hydrogen (Scheme 9). In the case of other α-olefins lacking a β-hydrogen, e.g., isobutylene and α-methylstyrene, Richardson and Eyler report exclusive conversion to products coming from loss of CH_4. The H_2-loss and the CH_4-loss products, assumed to have π-allyl structures, are catalytically inactive in further Ziegler-Natta steps. Accordingly, addition of more than one unit of an olefin was not seen. On the other hand, reaction of olefins with laser-vaporized Ti^+ and Al^+ in high-pressure mass spectrometric experiments [90–93] gave multiple additions, frequently accompanied by exten-

$$[Cp_2ZrCH_3]^+ + C_nH_m \rightarrow [Cp_2ZrC_{n+1}H_{m+1}]^+ + H_2$$

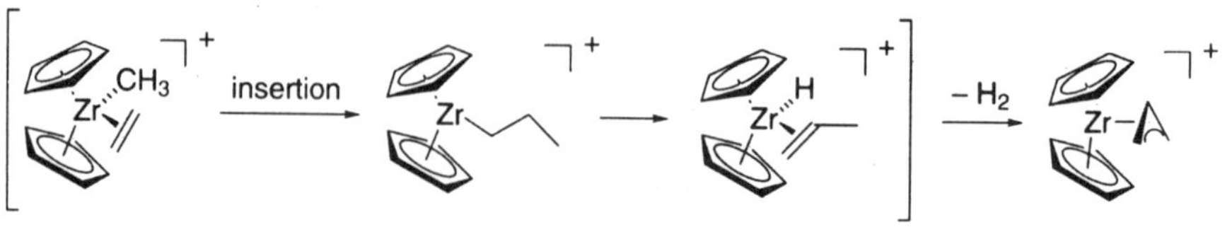

insertion → → $-H_2$ →

Scheme 9. Reaction of $[Cp_2ZrCH_3]^+$ with α-olefins in an ICR spectrometer

sive dehydrogenation. There had been no solution-phase analogues for these atomic cations or their reactions. Richardson and Eyler attributed the difference between their results and the condensed-phase reactivity to incomplete thermalization of the intermediate ions formed by olefin insertion in the very low-pressure environment of an ICR spectrometer cell. Since the reaction chamber in the ESMS setup is operated at significantly higher pressures, the conditions to carry off the reaction energy and thus to observe the Ziegler-Natta-type reactivity were much more favorable.

Stable solutions of the tetrakis(pentafluorophenyl)borate salt of the methylzirconocene cation $[Cp_2ZrCH_3]^+$, suitable for electrospray, were prepared by treatment of a 10^{-3} mol l^{-1} solution of $Cp_2Zr(CH_3)_2$ in acetonitrile with dimethylanilinium tetrakis (pentafluorophenyl)borate [94]. The solution was then diluted with CH_2Cl_2 and electrosprayed. Due to the extremely high sensitivity of methylzirconocene with regard to traces of water, large peaks consistent with oxygen-bridged binuclear complexes $[Cp_2Zr(L)O(L')ZrCp_2]^+$ with the additional ligands L being methyl, acetonitrile, or triflate could not be suppressed completely. Nevertheless, a dominant peak corresponding to $[Cp_2Zr(CH_3)(CH_3CN)]^+$ could be reproducibly observed. More severe desolvation conditions (higher tube lens potential) converted $[Cp_2Zr(CH_3)(CH_3CN)]^+$ to $[Cp_2ZrCH_3]^+$, from which collision-induced dissociation (CID) in the second octopole gave loss of methyl.

Selection of $[Cp_2ZrCH_3]^+$ in the first quadrupole and reaction with 1-butene generated the species shown in Fig. 3. The products observed reflected the reactivity which had already been seen in the ICR experiments: the insertion prod-

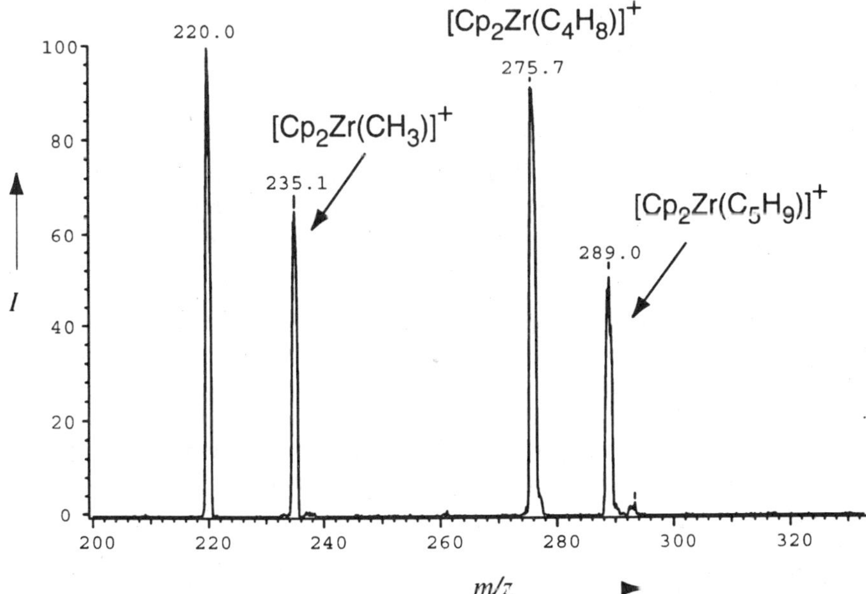

Fig. 3. Daughter-ion spectrum of the products from the gas-phase reaction of mass-selected $[Cp_2Zr(CH_3)]^+$ (m/z=235) and 1-butene

uct of the Ziegler-Natta reaction $[Cp_2Zr(C_5H_{11})]^+$ (*m*/*z*=291) was not detected, but instead the allylic complex $[Cp_2Zr(C_5H_9)]^+$. The remaining prominent peaks in the spectrum were due to methyl loss from $[Cp_2ZrCH_3]^+$ yielding the zirconocene cation $[Cp_2Zr]^+$ (*m*/*z*=220). This highly unsaturated complex readily picks up 1-butene, thus forming the olefin complex $[Cp_2Zr(C_4H_8)]^+$ (*m*/*z*=276).

Use of 1-butene in the first octopole region, i.e., use of 1-butene for reaction and thermalization, and a tube lens potential set to produce $[Cp_2ZrCH_3]^+$ gave new peaks, prominent among which were the masses corresponding to **13** (n=1) and **14**. The adduct **13** (n=1) corresponded to the product of insertion of the olefin into the Zr–C bond of the methylzirconocene cation, while **14** corresponded to the product arising from H_2 loss from **13** (Scheme 10). Characterization by CID of the mass-selected ions in the second octopole gave the expected results. For **13** (n=1), loss of H_2 was the predominant process when the gas pressure in the second octopole was low (below a few millitorr). Mass selection of **13** (n=1) in the first quadrupole and collision with more 1-butene (10 mTorr, near-zero collision energy) in the second octopole produced the daughter-ion mass spectrum shown in Fig. 4. The reaction of mass-selected **13** (n=1) in the second octopole was remarkably clean. As expected from a Ziegler-Natta-type reaction, addition of up to three units of the olefin had occurred.

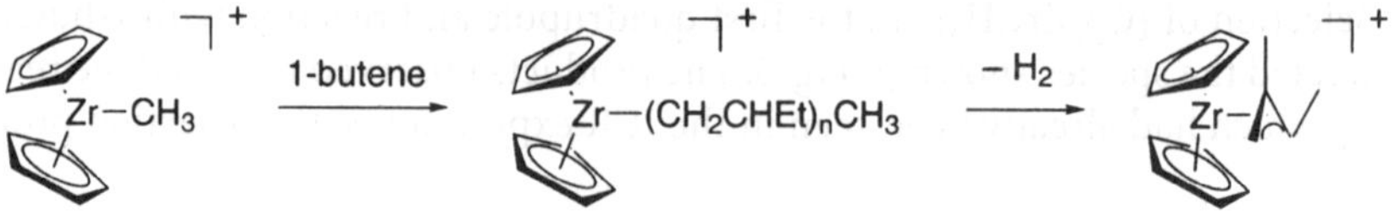

$[Cp_2Zr{-}CH_3]^+$ —1-butene→ $[Cp_2Zr{-}(CH_2CHEt)_nCH_3]^+$ (**13**) —$-H_2$→ **14**

Scheme 10. Ziegler-Natta oligomerization of 1-butene by $[Cp_2ZrCH_3]^+$

Reaction of the methylzirconocene cation with isobutylene instead of 1-butene in the first octopole gave a prominent peak due to the addition product, with no accompanying loss of H_2 and only a small peak corresponding to CH_4 loss. Experiments with ethylene and propylene gave similar results, although H_2 loss from the analogue of **13** (n=1) was more difficult to suppress. For ethylene, propylene, 1-butene, and isobutylene, a novel alkene dehydrogenation reaction could also be observed for methylzirconocene, but not for **13**. These results constituted the first observation of multiple olefin additions to a gas-phase metallocene cation that also performs Ziegler-Natta polymerization in solution. While electrospray mass spectrometry has been used to study catalytic reactions (see Sect. 4.3), reports in the literature have been limited to analysis of solutions which were catalytically active. The gas-phase ions themselves have been never shown to be competent as catalysts in those studies.

5.5.3 The Gas-Phase Reactivity of the Methylzirconocene Cation

By and large, the gas-phase reaction resembles that for the condensed phase. The complete absence of H_2 loss from **13** (n>1) indicates that the increased density

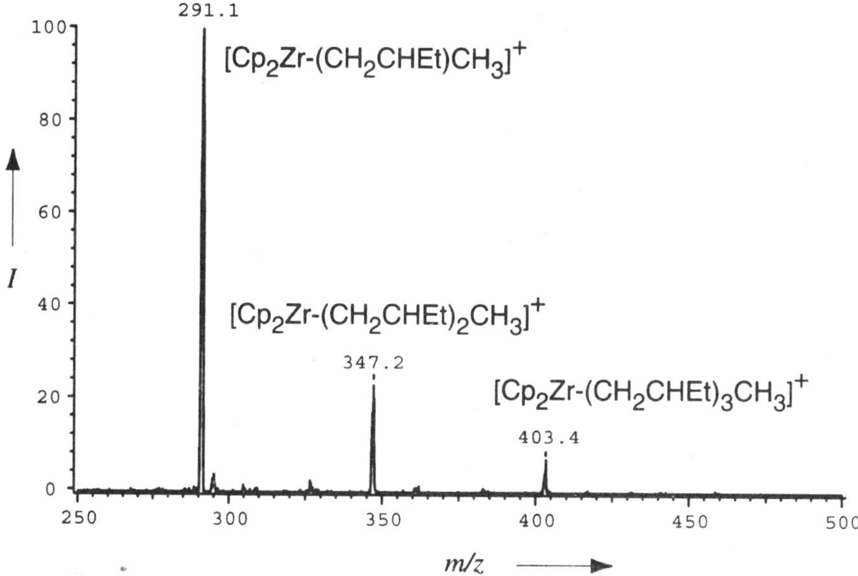

Fig. 4. Daughter-ion spectrum of the products from the gas-phase reaction of $[Cp_2Zr(C_5H_{11})]^+$ (m/z=291) and 1-butene. The peaks at m/z=347 and 403 correspond to the addition of one and two olefin molecules

of states associated with the growing alkyl side chain sufficiently slows the H_2-loss reaction, allowing collisional stabilization of the adduct and/or further addition reactions under conditions where higher pressures and shorter reaction times work against chemically activated reactions. The alkene dehydrogenation reaction, also a chemically activated reaction, is similarly quenched by higher pressure and longer chain lengths.

The most notable difference between the gas-phase results and the analogous reaction in solution is the large rate acceleration. A kinetic study of the Cp_2ZrCl_2/MAO system in solution-phase olefin polymerization found k_p=168–1670 M^{-1} s^{-1} at 70°C with an absolute upper-bound of $k_p \leq 5000$ M^{-1} s^{-1} with 75–100% of the available zirconium available in a catalytically active form [95]. To estimate the second-order rate constant for addition of 1-butene to $[Cp_2ZrR]^+$ in the ESMS setup, ion mobility in the octopole collision cell at "high" pressure, i.e., tens of millitorr, was modeled using a Monte Carlo simulation assuming a Langevin cross section for the collision rate in the presence of the radiofrequency field. The collision cell was found to behave similarly to an ion drift cell, with the incident ions undergoing (depending on conditions) up to 100,000 collisions with gas molecules before exit. The model was validated by measured ion residence times in the collision cell and found to give excellent quantitative agreement. Comparing the measured product yield with the number of collisions gave a reaction probability per collision of about 10^{-3}, which means that the 1-butene addition to $[Cp_2ZrR]^+$ occurs with a second-order rate constant of $k \sim 10^8 – 10^9$ M^{-1} s^{-1} (at 70°C), which is entirely consistent with a composite rate constant of $(3.5\pm0.4)\times10^8$ M^{-1} s^{-1} reported by Richardson et al. [89]. The addition of 1-butene to the unsolvated cations occurs at a rate up to a few orders of magnitude slower than diffusion-controlled, which is still approximately 10^5 times faster than the corresponding solution-phase addition, for which the metallocene cation is part of an ion pair, often with one or more additional coordinating ligands (e.g., solvent). The result is qualitatively reasonable on two grounds: (1) lacking a counterion or any charge donation by even a weakly coordinating solvent molecule, and furthermore lacking any preequilibria to form the active species, the isolated cations should be intrinsically more reactive; and (2) the electrostatic interaction, either ion dipole or ion induced-dipole, which is screened out in solution, effectively lowers the activation energy of a bimolecular ion-molecule reaction by perhaps 10 kcal/mol relative to the same reaction in solution.

5.6
High-Valent Oxomanganese-Salen Complexes

5.6.1
Introduction

The transition metal-catalyzed oxidation of various organic substrates is of utmost importance in synthetic applications as well as in biochemical transformations. High-valent oxomanganese and oxoiron complexes play a crucial role in oxygen transfer to otherwise unreactive organic substrates. The paradigmatic system for oxidative transformations in biological systems is the enzyme cytochrome P-450 [96]. Numerous studies point to oxoiron intermediates being involved in the mechanism of cytochrome P-450. The study of cytochrome P-450 and porphyrin model complexes designed to mimic its reactivity has yielded a number of synthetically useful catalysts for the epoxidation and hydroxylation of organic substrates [97, 98]. In addition to their importance in enzymatic oxygenations, oxometal complexes have often been suggested as the catalytically active species in epoxidations catalyzed by metal-salen and porphyrin complexes. Designed for synthetic purposes, metal-salen complexes (salen=*N,N'*-bis(salicylidene)ethylenediamine), which can be viewed as simplified synthetic models of their porphyrin analogs, were introduced by Kochi and co-workers as versatile epoxidation catalysts in the 1980s [99–103]. A breakthrough was achieved in the field of enantioselective epoxidation through the introduction of chiral manganese-salen catalysts by Jacobsen and co-workers [104], with a closely related but somewhat less effective system by Katsuki and co-workers developed at about the same time [105]. The Jacobsen-Katsuki reaction is now recognized as one of the most useful and widely applicable methods for the epoxidation of non-functionalized olefins [106].

The mechanistic scheme adopted for oxygen transfer to organic substrates by salen complexes is based on the isolation and characterization of the oxochromium(V) species by Kochi and co-workers (Scheme 11) [99, 101]. The (salen)Cr-oxo complex was prepared by oxidation of the chromium(III) precursor with PhIO, isolated, and shown to be competent in epoxidizing alkenes

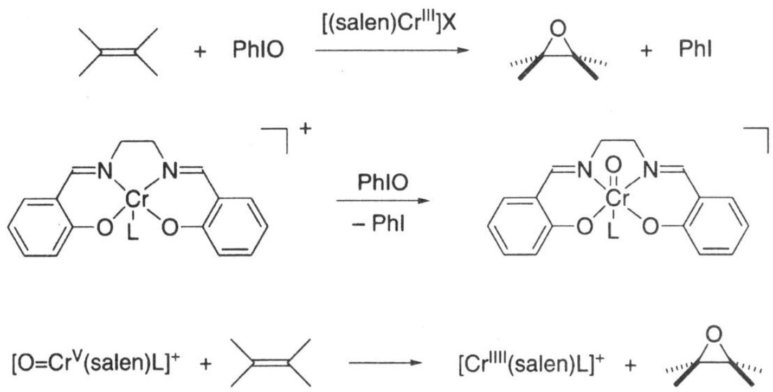

Scheme 11. The mechanism of alkene epoxidation catalyzed by chromium-salen complexes

under stoichiometric and catalytic conditions [100]. This result was in accordance with the properties and reactivity of an analogous oxoporphyrinatochromium(V) complex studied earlier by Groves and Kruper [107]. The metal-oxo species in these reactions are acting as a staging post for oxygen transfer, which is why the mechanism was termed "oxygen-rebound". The chromium complexes were useful for mechanistic studies, but their utility as epoxidation catalysts is limited to electron-rich olefins. Switching to manganese as the metal center, Kochi and co-workers established a much more versatile salen-based oxidation catalyst [102, 103]. However, mechanistic studies on these systems were always hampered by the fact that the catalytically active oxomanganese species appear as fleeting putative intermediates.

In view of the fact that oxygen transfer reactions are so common and important in biological systems as well as in organic synthesis it may come as a surprise that our knowledge of the detailed mechanisms of these reactions leaves so much to be desired. Two factors are mainly responsible for the lack of mechanistic insight. (1) Oxygen transfer to the transition metal by dioxygen or some oxygen atom donor can yield several different species with quite diverse reactivities. In the case of manganese-porphyrin complexes, Groves and Stern were able to isolate several different oxomanganese(IV) complexes, none of which showed the reactivity typical for the intermediate in question [108, 109]. Moreover, insight into the detailed mechanism of oxygen transfer by cytochrome P-450 is enormously complicated by the different reaction steps which can be envisioned after the initial formation of the ferric peroxocomplex: heterolytic O-O cleavage (→(porphyrin)Fe^{V}=O), homolytic O-O cleavage (→(porphyrin) Fe^{IV}=O), or direct nucleophilic attack on enzyme-bound substrate. (2) The transient nature of the catalytically active species. Groves and co-workers measured the conversion of a reactive intermediate in the manganese-porphyrin catalyzed oxidation, to which they assigned an oxomanganese(V) structure. They determined that the reaction followed a first-order rate constant of 5.7 s^{-1} [110].

Both problems can easily be overcome by transfer of the metal oxo complexes to the gas-phase. By mass selection, the species of interest can be singled out and studied separately without interference by additional complexes present in solution. The reactivity of the crucial intermediates can be monitored by directed collision with an appropriate substrate. Their short solution-phase lifetimes do not pose a problem anymore due to high-vacuum conditions.

5.6.2
Gas-Phase Coordination Chemistry of Manganese(III)-Salen Complexes

Many open questions concerning the mechanism of the catalytic oxidation can be addressed in a unique way by using electrospray tandem mass spectrometry. In the beginning, our gas-phase studies were focused on two basic questions: (1) how is the oxygen transfer to the metal center achieved; and (2) what are the species produced upon oxidation under typical reaction conditions. Electrospray of unsubstituted manganese(III) salen complexes in acetonitrile gave mainly the singly charged, five-coordinate cation $[(salen)Mn^{III}(CH_3CN)]^+$ [111]. When the ions were collisionally activated by applying higher tube lens potentials of ~130 V, some of the ligated acetonitrile molecules could be dissociated from the manganese centers resulting in the detection of the "naked" salen complex $[(salen)Mn^{III}]^+$. It is well known from numerous crystallographic studies that manganese(III)-salen complexes can bind one or two ligands (usually the solvent used for recrystallization, i.e., acetone, ethanol, etc.) in the axial positions, thus forming five- or six-coordinate species in the solid state [112]. In the tetra- and penta-coordinate modes of these complexes, the salen ligand is always in a planar conformation. In the octahedral coordination mode, an alternative arrangement has been occasionally observed, where one of the oxygens of the salen ligand moves out of the plane and occupies an apical position. The axial ligands can easily be exchanged in solution, indicating that they are only weakly bound to the manganese center [113].

When electrospraying manganese-salen complexes in acetonitrile no six-coordinate species bearing two acetonitrile ligands were detected [111]. The presence of significant amounts of the five-coordinate species, however, posed the question whether or not acetonitrile is bound strongly enough to survive the spraying process. When a 10^{-3} mol l^{-1} stock solution of $[(salen)Mn]ClO_4$ in acetonitrile was diluted to 10^{-5} mol l^{-1} with CH_2Cl_2 and electrosprayed, only small traces of the five-coordinate complex with CH_3CN as a ligand remained. The acetonitrile could then be completely removed by applying high tube lens potentials. The presence of $[(salen)Mn^{III}(CH_3CN)]^+$ was thus probably due to recombination during the electrospray process when acetonitrile was used as the electrospray solvent. Ligand pick-up experiments in the collision cell confirmed this hypothesis. While the "naked" salen complex readily picked up one molecule of acetonitrile, no traces of the acetonitrile bisadduct were detected. Accordingly, when the five-coordinate complex $[(salen)Mn^{III}(CH_3CN)]^+$ was mass-selected in the first quadrupole, no addition of acetonitrile could be induced.

A similar picture emerged from experiments performed with methanol as the axial ligand. The picture changed somewhat when a better coordinating ligand was used, namely pyridine. Pick-up experiments in the collision cell led to exclusive formation of the five-coordinate complex. However, when a solution of [(salen)Mn]ClO_4 with a large excess of pyridine in CH_3CN was electrosprayed, a small amount of the six-coordinate species [py(salen)Mn^{III}py]$^+$ bearing two axial pyridine ligands was also observed. Electronic factors play a decisive role in the formation of six- vs five-coordinate manganese-salen complexes. One would conceive intuitively that Mn-salen complexes with a more electron-deficient metal center bind axial ligands more strongly. With [(5,5′-dinitrosalen)Mn]$^+$, a species corresponding to the complex with two axial acetonitrile ligands was observed in the ES mass spectrum. Accordingly, pick-up experiments with [(5,5′-dinitrosalen)Mn]$^+$ in the collision cell clearly showed addition of one or two molecules CH_3CN.

5.6.3
Preparation of Oxomanganese(V)-Salen Complexes: The Mechanism of Oxygen Transfer

Oxomanganese(V)-salen and porphyrin complexes have so far remained elusive in condensed-phase studies. Electrospray of in situ mixtures of [(salen)Mn^{III}] ClO_4 and a suitable oxygen transferring agent, however, should reveal the nature of the oxidation products, since all ionic species in solution will be transferred to the gas phase and thus be detectable. The oxidizing species were prepared following the two most common literature procedures: (1) addition of a solution of the complex to a suspension of iodosobenzene in acetonitrile; and (2) oxidation of the Mn^{III} complex with bleach in a two-phase system CH_2Cl_2/H_2O. Following procedure (1), the qualitative mass spectrometric analysis of the electrosprayed ions revealed a plethora of new species formed upon addition of the oxidant (Scheme 12, Fig. 5) [114]. The Mn^{III} complexes [(salen)Mn^{III}(CH_3CN)]$^+$ and [(salen)Mn^{III}]$^+$ were still present in the spectrum, albeit with strongly reduced intensities. The two oxidized species most prominent in the spectrum were the parent oxo complex [O=Mn^V(salen)]$^+$ (m/z=337) and the μ-oxo bridged dinuclear complex with two terminal PhIO ligands [PhIO(salen)Mn–O–Mn(salen)OIPh]$^{2+}$ (m/z=549). Following a similar approach, Lindsay Smith and co-workers were able to identify oxomanganese(V) species in the oxidation of 4-methoxyphenol by H_2O_2 catalyzed by manganese 1,4,7-triazacyclononane complexes [115].

The detection of [PhIO(salen)Mn–O–Mn(salen)OIPh]$^{2+}$ in the ESMS experiments was the first direct observation of the conproportionation of Mn^{III} and Mn^V-oxo species as the mechanism for parking the catalytically active complex in a more persistent form, the mechanism postulated earlier by Kochi et al. [102]. The microscopic reverse process, the disproportionation of the μ-oxo bridged dinuclear complex, would lead to the release of [O=Mn^V(salen)]$^+$. This

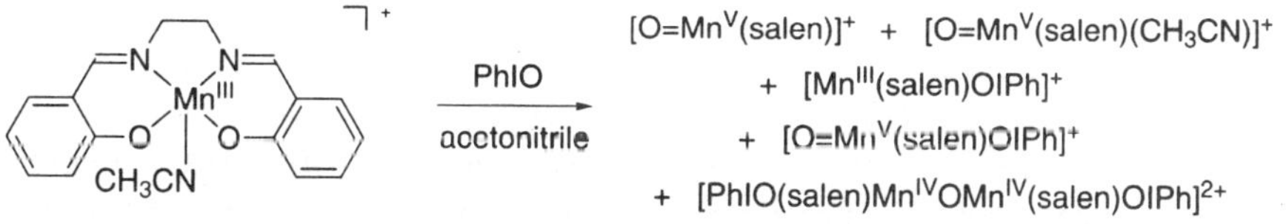

[(salen)Mn^{III}(CH_3CN)]$^+$ —PhIO / acetonitrile→ [O=Mn^V(salen)]$^+$ + [O=Mn^V(salen)(CH_3CN)]$^+$ + [Mn^{III}(salen)OIPh]$^+$ + [O=Mn^V(salen)OIPh]$^+$ + [PhIO(salen)Mn^{IV}OMn^{IV}(salen)OIPh]$^{2+}$

Scheme 12. Species detected by ESMS upon oxidation of [(salen)Mn^{III}]$^+$

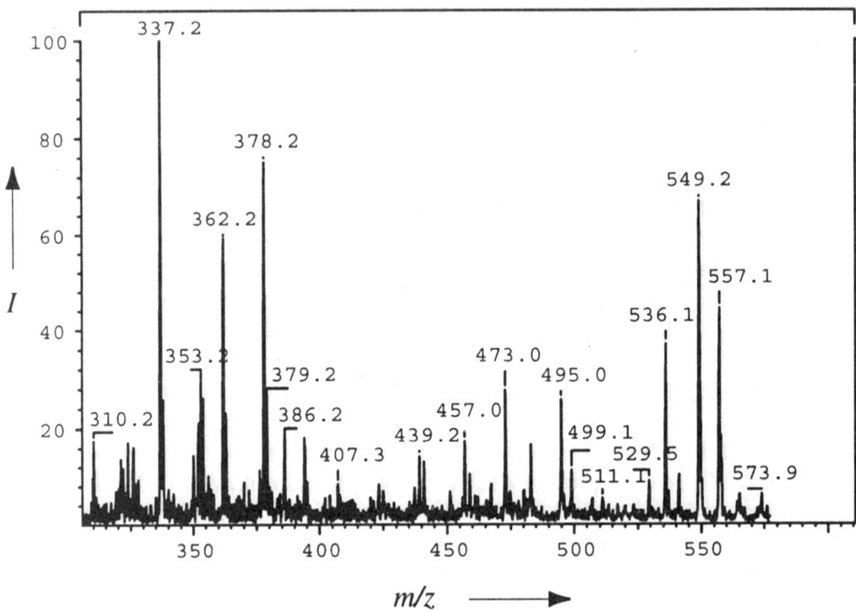

Fig. 5. Electrospray mass spectrum of an authentic epoxidation reagent mixture obtained by adding PhIO to an acetonitrile solution of $[(\text{salen})\text{Mn}(\text{CH}_3\text{CN})]^+$

reaction can be triggered by collision with argon in the gas phase, thus providing a structural proof for the parent dication as well as for the putative reactive oxomanganese(V) cation. The two primary daughter ions formed by collisional activation immediately fragmented further, leading to a quite complex product pattern.

This behavior made it possible to study in detail the ligand properties of iodosobenzene and the mechanism of oxygen transfer to the metal center [116]. By adjusting the tube lens potential in the electrospray source it was possible to generate the complex $[(\text{salen})\text{Mn}^{\text{III}}\cdot\text{PhIO}]^+$ in high enough yields to conduct MS/MS experiments. The daughter-ion spectrum of $[\text{PhIO}(\text{salen})\text{Mn}^{\text{III}}]^+$ gave predominantly $[\text{O=Mn}^{\text{V}}(\text{salen})]^+$ due to the facile fragmentation of the O–I bond. Loss of PhIO leading to $[(\text{salen})\text{Mn}^{\text{III}}]^+$ was only a minor pathway. The following conclusions for the mechanism of oxidation by iodosobenzene could be drawn from this experiment. Due to the polymeric structure (with an O–I–O backbone) in the solid state, the solubility of PhIO in organic solvents is very low, and consequently at any time there are only small amounts available for ligation. The detection of [Mn(salen)·PhIO] proved that iodosobenzene coordinates strongly enough to the metal center to survive electrospray conditions, thus indicating that the major part of dissolved PhIO will be effectively bound to the manganese complex. From the fragmentation experiments it was obvious that O–I bond scission dominates by far over the alternative O–Mn breakdown in the energy range of the CID experiments, thus leading to the preferred formation of $[\text{O=Mn}^{\text{V}}(\text{salen})]^+$ and PhI over $[(\text{salen})\text{Mn}^{\text{III}}]^+$ and PhIO.

5.6.4
Reactivity and Coordination Chemistry of Oxomanganese(V)-Salen Complexes

Does the oxomanganese(V)-salen complex also display oxygen-transfer reactivity in the gas phase? When $[(salen)Mn{=}O]^+$ was collided with an inert gas (Ar, Xe), no fragmentation was observed due to the stable metal-oxo multiple bond [117]. With substrates that are readily oxidized in solution (electron-rich olefins or sulfides), however, the picture changed drastically. As there was no way of observing the oxidized products directly in this MS experiment, the reactivity had to be deduced by correlating the turnover with the appearance of the $[(salen)Mn^{III}]^+$ complex in the daughter spectrum. With 2,3-dihydrofuran, dimethyl sulfide, or phenyl methyl sulfide in the reaction chamber, $[(salen)Mn]^+$ was produced to a considerable amount from $[(salen)Mn{=}O]^+$ [114]. A control experiment with tetrahydrofuran as collision gas showed no such fragmentation at the chosen collision energies.

μ-Oxomanganese(IV) complexes without terminal ligands or with acetonitrile instead of iodosobenzene were conspicuously absent in all the mass spectra recorded. Iodosobenzene is obviously efficient in stabilizing a μ-oxo complex, but the lability of the I–O bond and the problems experienced with different samples of varying properties forced us to look for more reliable alternatives. Amine *N*-oxides have proven to be a viable alternative. The best method to obtain relatively stable μ-oxo complexes was found in the mixing of $[(salen)Mn^{III}]^+$ and an amine *N*-oxide (~1:10) in a slurry of iodosobenzene in acetonitrile [118]. A representative spectrum of the electrosprayed supernatant solution with *p*-CN-$C_6H_4NMe_2O$ as ligand is shown in Fig. 6. The species which appear most prominently in the spectrum are $[p\text{-CN-}C_6H_4NMe_2O(salen)Mn^{III}]^+$ (m/z=483), $[p\text{-CN-}C_6H_4NMe_2O(salen)Mn^{V}{=}O]^+$ (m/z=499), and $[p\text{-CN-}C_6H_4NMe_2O(salen)Mn^{IV}\text{–O–}Mn^{IV}(salen)OMe_2N\text{-}p\text{-CN-}C_6H_4]^{2+}$ (m/z=491). The signal at m/z=325 was assigned to the H^+-bridged *N*-oxide adduct $[p\text{-CN-}C_6H_4\,NMe_2O\text{–H–}ONMe_2\text{-}p\text{-CN-}C_6H_4]^+$. Amine *N*-oxides apparently are much better ligands than iodobenzene or acetonitrile, which were both displaced, and they are very effective in stabilizing the μ-oxo bridged manganese(IV) complexes.

Tertiary amine *N*-oxides were quite useful as stabilizing capping ligands for the μ-oxo bridged dinuclear complexes, but the ambiguities in the fragmentation of ligated *N*-oxides made the search for an alternative neutral ligand desirable. Phosphine oxides turned out to be the ligands of choice because of the oxophilicity of Mn^{III} on the one hand and the strong, fragmentation-stable P=O bond on the other. Since triethylphosphine oxide does not act as an oxidant, a good method to prepare relatively stable μ-oxo complexes was found in the mixing of $[(salen)Mn^{III}]^+$ and the phosphine oxide in a slurry of iodosobenzene in acetonitrile. The species thus generated in substantial yields were $[Et_3PO(salen)Mn^{III}]^+$, $[Et_3PO(salen)Mn^{V}{=}O]^+$, and $[Et_3PO(salen)Mn^{IV}\text{–O–}Mn^{IV}(salen)Et_3PO]^{2+}$. Phosphine oxides apparently are much better ligands than either iodosobenzene or acetonitrile, which are both largely displaced, and they are very effective in stabilizing the μ-oxo bridged manganese(IV) dimers.

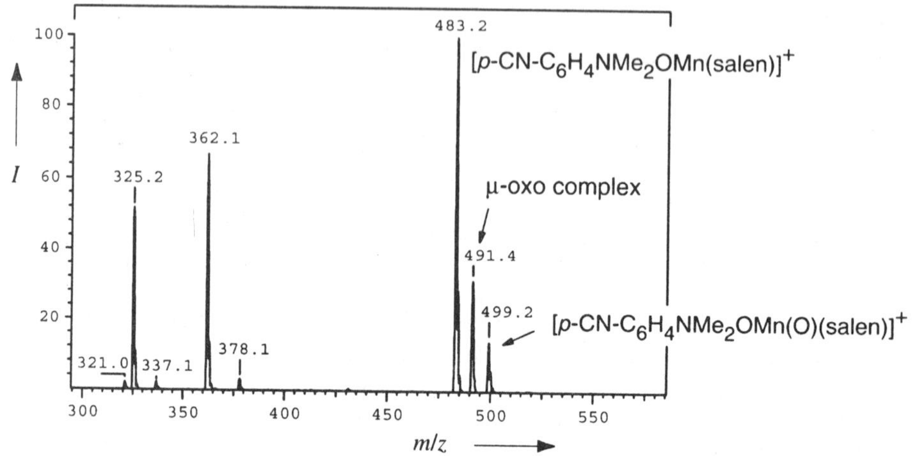

Fig. 6. Electrospray mass spectrum of an acetonitrile solution of $[(salen)Mn]ClO_4$, *p*-CN-*N*,*N*-dimethylaniline *N*-oxide, and iodosobenzene, showing the formation of *N*-oxide ligated manganese(III) and oxomanganese species

5.6.5
Electronic Tuning of the Salen Ligand

Electronic tuning of the salen ligand, i.e., substitution on the aryl portion of the ligand, can drastically alter the catalyst properties. With a source for oxomanganese(V) salen complexes available, the effects of salen substitution on the catalyst reactivity can in principle be measured directly in the gas phase. The obvious way to probe the influence of electronic tuning, namely CID (collision-induced dissociation) threshold measurements (see Sect. 5.8) of differently substituted $[(salen)Mn^{III}{\cdot}PhIO]$ adducts, was precluded by the sheer size of the molecules. In the case of the cyclometallation reaction $[CpIr(PMe_3)(CH_3)]^+ \rightarrow [CpIr(\eta^2\text{-}CH_2PMe_2)]^+ + CH_4$, we had been able to demonstrate that reliable thermochemical data can be obtained by CID threshold measurements within the ESMS setup, but it also became clear from this study that the number of degrees of freedom and the quality of the frequencies obtained from quantum chemical calculations needed for the RRKM correction pose a substantial problem when solution-phase species with a full ligand sphere are studied [81].

The presence of the µ-oxomanganese(IV) complexes opens an alternative route to assess substituent effects on the reactivity of the Mn=O species. When a dinuclear µ-oxo complex with different salen ligands on each side is fragmented, the ratio of the resulting two oxo-manganese(V) complexes (and, accordingly, the corresponding $(salen)Mn^{III}$ fragments) will reflect their kinetic stability (Scheme 13) [118]. On the assumption that the reverse barrier for both reaction channels is roughly equal, entropic factors cancel, and that the energy distribution of the reactant ions can be approximated by a Boltzmann distribution, the observed difference in kinetic stability will reflect the difference in thermodynamic stability as well. Thus, it becomes possible to establish an ordering of

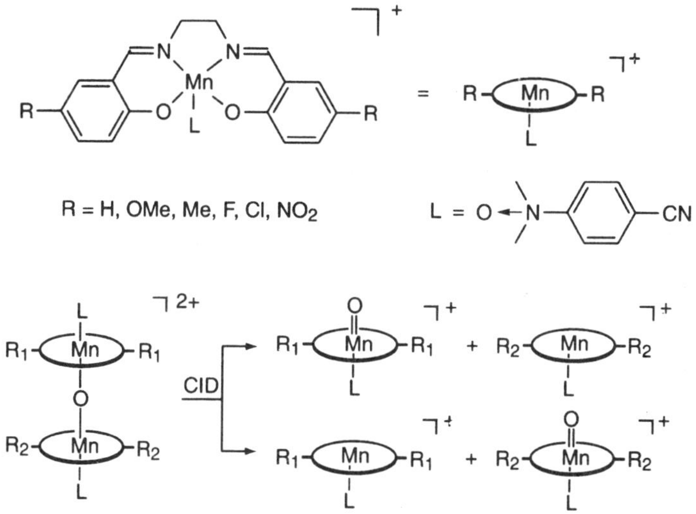

Scheme 13. Possible fragmentation products upon CID of μ-oxomanganese(IV)-salen complexes with salen ligands differently substituted in the 5 and 5′ positions

the relative stabilities of oxomanganese(V) salen complexes depending on the electronic influence of the salen substituents.

It has long been known that substituents in the 5 and 5′ positions of the salen ligand strongly perturb the redox properties of metal salen complexes. Therefore, the following salen derivatives were studied: 5,5′-dinitro; 5,5′-difluoro; 5,5′-dichloro; 5,5′-dimethyl; 5,5′-dimethoxy; and the unsubstituted salen itself. The results of the fragmentation of the "mixed" μ-oxo complexes can be summarized as follows. The destabilizing effects on the oxomanganese(V) complex were in the following order: NO_2>Cl≅H>F>CH_3>OMe. The net effect of the 5,5′-dichloro substituted ligand as compared to the unsubstituted salen was zero within the limits of experimental error. Going to the electron-donating substituents, the 5,5′-dimethyl-oxomanganese(V) complex was twice as stable as the unsubstituted one. The maximum stability was provided by the 5,5′-dimethoxy substitution yielding an oxomanganese(V) species 15 times more stable than the unsubstituted salen complex (Fig. 7). A small stabilizing effect was found with the 5,5′-difluorosubstituted salen. Conversely, substitution with electron-withdrawing groups decreased the stability of the oxygen-transferring species markedly. The dinuclear μ-oxo complex with the 5,5′-dinitro-substituted salen could only be detected right after mixing with amine *N*-oxide/PhIO and disappeared within seconds from the spectrum. Accordingly, the formation of mixed dinuclear complexes with 5,5′-dinitrosalen was very difficult to detect, and they were simply too short-lived in order to conduct subsequent fragmentation experiments.

The product yields seen in the fragmentation experiments reflect the trend in stability of the oxomanganese(V) ions that one would conceive intuitively con-

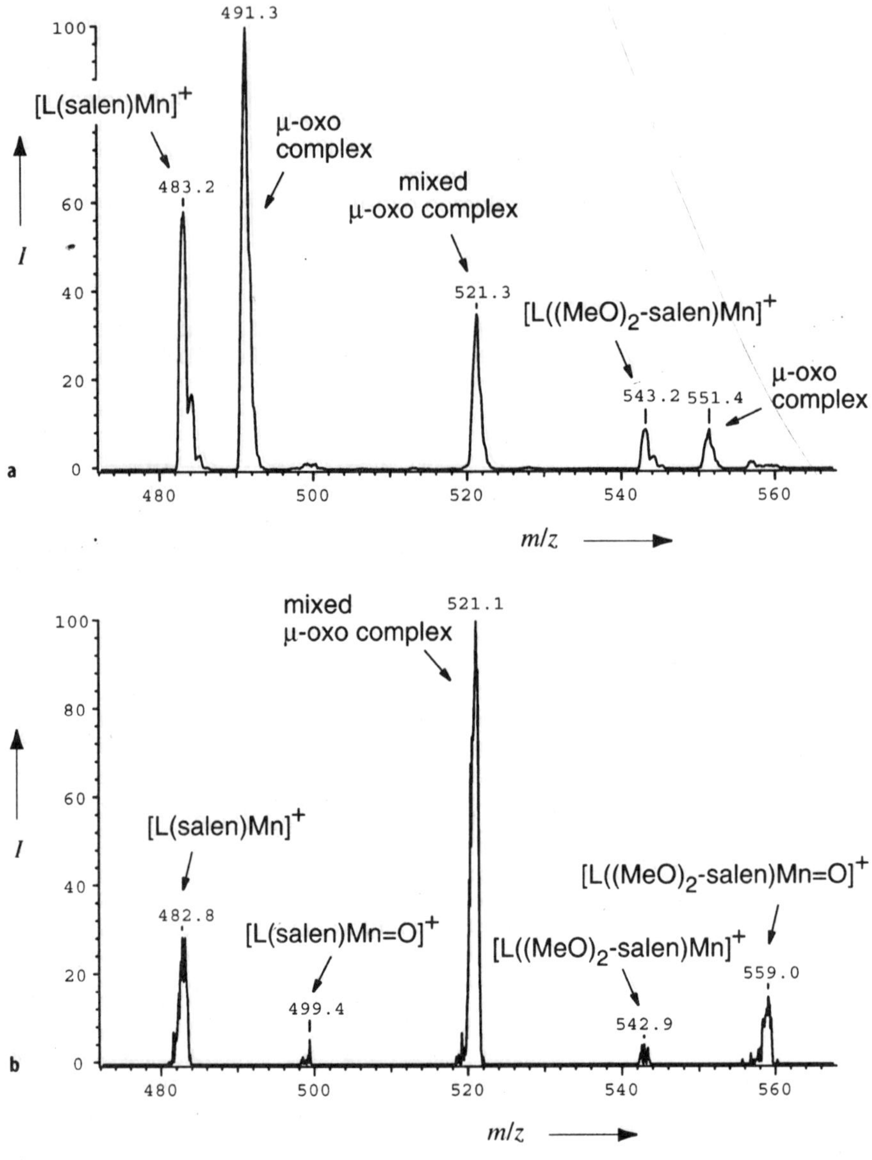

Fig. 7. **a** Electrospray mass spectrum of an acetonitrile solution of $[((MeO)_2\text{-salen})Mn]ClO_4$, $[(salen)\text{-}Mn]ClO_4$, *p*-CN-*N*,*N*-dimethylaniline *N*-oxide, and iodosobenzene, showing the formation of homogeneous and mixed μ-oxo complexes with terminal *N*-oxide ligands. **b** Daughter-ion spectrum (0.7 mTorr Ar, collision energy 34 eV) of the mixed dinuclear complex $[L((MeO)_2\text{-salen})Mn\text{–}O\text{–}Mn(salen)L]^{2+}$ (*m*/*z*=521, L=*p*-CN-*N*,*N*-dimethylaniline *N*-oxide), showing both the fragmentation to $[L((MeO)_2\text{-salen})Mn]^+$ (*m*/*z*=543) and to the corresponding oxo complex (*m*/*z*=559)

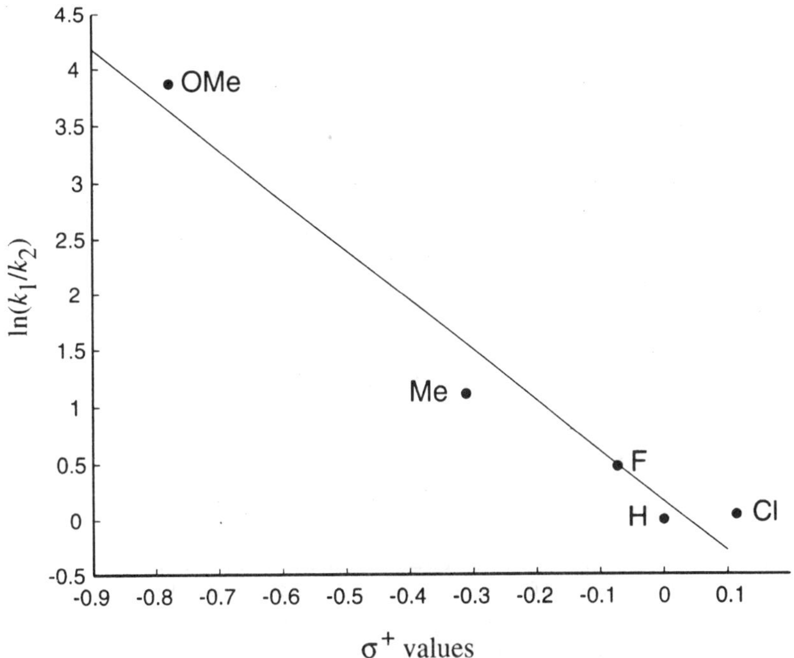

Fig. 8. Correlation of the branching ratios of "mixed" μ-oxomanganese(IV)-salen complexes and σ^+ values of the 5,5′ substituents

sidering the electronic properties of the substituted salen ligands. The Mn=O moiety in these high-valent complexes is stabilized by electron-donating and destabilized by electron-withdrawing substituents. This result can easily be rationalized by the electron deficiency imposed on the manganese center upon oxidation to Mn^V=O. The differences in stability derived from gas-phase experiments are quite pronounced. The ordering of the substituent effects suggests an underlying linear-free-energy relationship. When plotting $\ln(k_1/k_2)$ vs Hammett parameters found in the literature, the best correlation was obtained with the σ^+ values given by Brown and Okamoto (Fig. 8) [119]. The good correlation with the σ^+ values in the gas-phase fragmentation experiments can be rationalized by the analogy between a developing carbocation and the increase in oxidation state of the manganese center in the course of the fragmentation.

The mechanistic picture obtained from the ESMS studies fully explain the solution-phase observations. It was early recognized by Kochi et al. that (salen)Mn complexes with electron-donating substituents such as the 5,5′-dimethoxy derivative effect only poor yields of epoxide, whereas the catalyst with 5,5′-dinitro substituents gave the best product yields [102]. While the reactivity differences seen with differently substituted achiral salens led to the conclusion that the more electron-deficient ligand will give the more effective catalyst, the interplay between epoxidation efficiency and selectivity is much more subtle and less predictable for asymmetric epoxidation. In 1991, Jacobsen reported the dramatic effects of electronic tuning, i.e., using different substituents in the para position to the ligating oxygens (5,5′-disubstitution), on the enantioselectivity of the (salen)Mn catalyzed epoxidation of *cis*-disubstituted olefins [120]. In a later account, Jacobsen and co-workers systematically investigated the correlation of enantioselectivity and the electronic character of the Jacobsen catalyst by varying the substituents in the 5,5′ positions (NO_2, Cl, H, Me, OMe) [121]. In all cases, electron-donating groups on the catalyst were found to give higher enantioselectivities in epoxidation, whereas electron-withdrawing substituents led to decreased enantioselectivity. The possibility that the substituents might induce conformational distortions in the oxygen-transferring complexes or provoke changes in the Mn–O bond lengths and thus in the substrate-ligand interactions were dismissed as highly improbable by these authors. Instead, the electronic effect on the enantioselectivity was attributed to changes in the reactivity of the $Mn^{V}{=}O$ moiety.

The ESMS gas-phase experiments provided the first direct probe for the intrinsic properties of the reactive intermediates in the Jacobsen-Katsuki epoxidation. The intermolecular experiment in which the two metal centers in the "mixed" μ-oxo complex compete for the bridging oxygen atom give a direct measure of the oxygen-transfer reactivities of the oxomanganese(V) species, with the benefit that no olefinic substrate has to be present. Electron-donating substituents stabilize the oxomanganese(V) moiety, which should attenuate its reactivity and give a relatively milder oxidant. Electron-withdrawing substituents, on the other hand, destabilize the oxomanganese(V) moiety, making it a more reactive oxidant. Accordingly, the milder oxidant which delivers the oxygen to the substrate in a more product-like transition state will achieve a higher degree of stereochemical communication between substrate and catalyst, whereas the more reactive oxidant's differentiation of the diastereomeric transition structures will be much poorer. The analogous trends in the reactivities seen in solution phase and in the gas-phase stability measurements indicate that the mechanism of oxygen transfer to the substrate in solution is primarily governed by the intrinsic reactivity of the oxomanganese(V) complexes.

5.6.6
Effects of Axial Ligands

The reactivity of salen catalysts in epoxidation reactions is not only tuned by substitution of the salen, but also by adding donor ligands to the reaction mixture [100, 103]. The basic question that had been posed was whether or not donor ligands directly modulate the reactivity of the oxygen-transferring species by ligation [122]. The methodology employed for the investigation of electronic influences on the stability of the oxomanganese(V) complexes can readily be extended to study the effects of additional axial ligands. For this purpose, a μ-oxomanganese(IV) complex with two different terminal ligands has to be prepared, which upon fragmentation will give rise to two differently ligated $Mn^{V}{=}O$ species (Scheme 14) [118]. This experiment provided a direct probe for possible stabilizing/destabilizing effects of donor ligands on the Mn=O moiety. The following ligands were studied: *p*-CN-*N*,*N*-dimethylaniline *N*-oxide, *p*-Me-

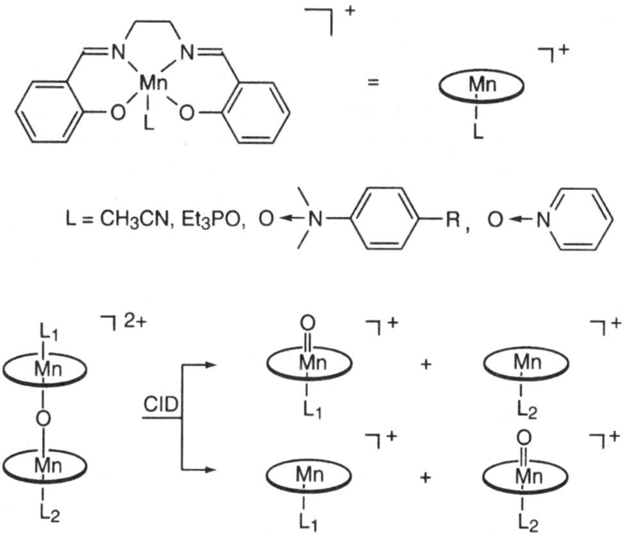

Scheme 14. Possible fragmentation products upon CID of μ-oxomanganese(IV)-salen complexes bearing two different terminal ligands

N,N-dimethylaniline *N*-oxide, *p*-Br-*N,N*-dimethylaniline *N*-oxide, *N,N*-dimethylaniline *N*-oxide, acetonitrile, pyridine *N*-oxide, and triethylphosphine oxide.

Pyridine *N*-oxide was the first axial ligand studied. Kochi had already reported the use of pyridine *N*-oxide in manganese- and chromium-salen catalyzed epoxidations, which resulted in a significant enhancement of epoxide yields [100, 102]. Fragmentation of the asymmetrical dinuclear μ-oxo complex [pyO(salen)Mn–O–Mn(salen)]$^{2+}$ in the gas phase led to [(salen)Mn=O]$^{+}$ and [pyO(salen)Mn^{III}]$^{+}$ as fragmentation products, with no traces of the alternative fragmentation products detectable (Scheme 15). Fragmentation of the bridging μ-oxo bond occurred exclusively on the side where the terminal ligand was bound.

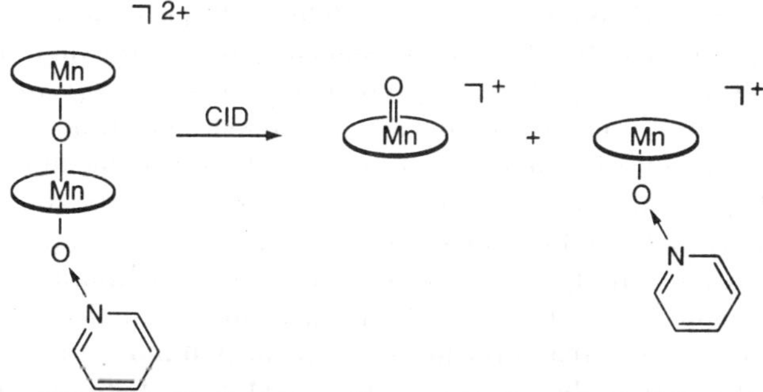

Scheme 15. Fragmentation of the asymmetrical μ-oxo complex [(salen)Mn–O–Mn(salen)Opy]$^{2+}$ upon CID

The relative destabilizing effects of different promoters of the catalytic oxidation on the Mn-oxo moiety can be seen from the fragmentation of the μ-oxomanganese(IV) complex with *p*-CN-*N*,*N*-dimethylaniline *N*-oxide and pyridine *N*-oxide, respectively, as terminal ligands. Fragmentation of the Mn–O bond occurred predominantly on the side where the aniline *N*-oxide was bound, i.e., pyridine *N*-oxide is much less destabilizing and will thus be the less effective promoter for oxidation catalysis. Triethylphosphine oxide, also known to be a good promoter in metal salen catalyzed epoxidations [100], had a similar destabilizing effect as *p*-CN-*N*,*N*-dimethylaniline *N*-oxide.

The following mechanistic picture emerged from the ESMS data. Since μ-oxo complexes without terminal ligands have never been detected in the in situ mixtures, it is obvious that additional axial ligands strongly promote formation of dinuclear complexes. This can simply be rationalized by viewing the formation of the μ-oxomanganese(IV) species as a partial oxygen transfer from $[L(salen)Mn{=}O]^+$ to $[L(salen)Mn]^+$. As evidenced by the stability measurements, axial ligands enhance the oxygen-transfer capability and thus the formation of dinuclear μ-oxo complexes. The conproportionation can be viewed as an escape route for the extremely reactive oxomanganese(V) species to "hide" in a more persistent form. After disproportionation of the dinuclear complex, i.e., after the release of $[L(salen)Mn{=}O]^+$, the coordinating ligand will stay ligated at the metal center and will now promote the reactivity of the catalyst in the epoxidation (or in the back reaction to the μ-oxo complex).

5.7
ESMS Studies of Olefin Metathesis by Ruthenium Carbene Complexes

Olefin metathesis has become one of the most important large-scale technical processes for the manufacture of olefins in the petrochemical industry [123]. When cyclic olefins are used as substrates, high-molecular polymers, which are formed by the so-called ring-opening metathesis (ROM), have found applications as elastomers and plastics. Gas-phase studies on the mechanism of olefin metathesis had been confined to simple metal carbenes, for example $[Mn{=}CH_2]^+$, $[Fe{=}CH_2]^+$, and $[Co{=}CH_2]^+$ [124–127]. Most of the metatheses have been observed with deuterated ethylene.

In order to parallel solution-phase reactivity and ion-molecule reactions in the gas phase, the reactivity of a typical homogeneous catalyst, described earlier by Grubbs and co-workers [128], was studied by ESMS [129]. Electrospray of the dichloride salt of **15** and increasing the collisional activation potential first yielded predominantly the monocation **16**, but with raising the tube lens potential even higher the intensity of **16** decreased due to loss of the second phosphine ligand, loss of trimethylamine, and loss of HCl. The observed fragmentation pattern was consistent with the assumed structure of the ruthenium complex.

When 1-butene was used as reaction partner and thermalization gas in the first octopole, a new signal with the mass corresponding to the propylidene complex **17** appeared in the spectrum, indicating that an olefin metathesis reaction had occurred. This reactivity was confirmed by mass-selecting either **15** or **16** in the first quadrupole and reaction with 1-butene in the collision cell. The

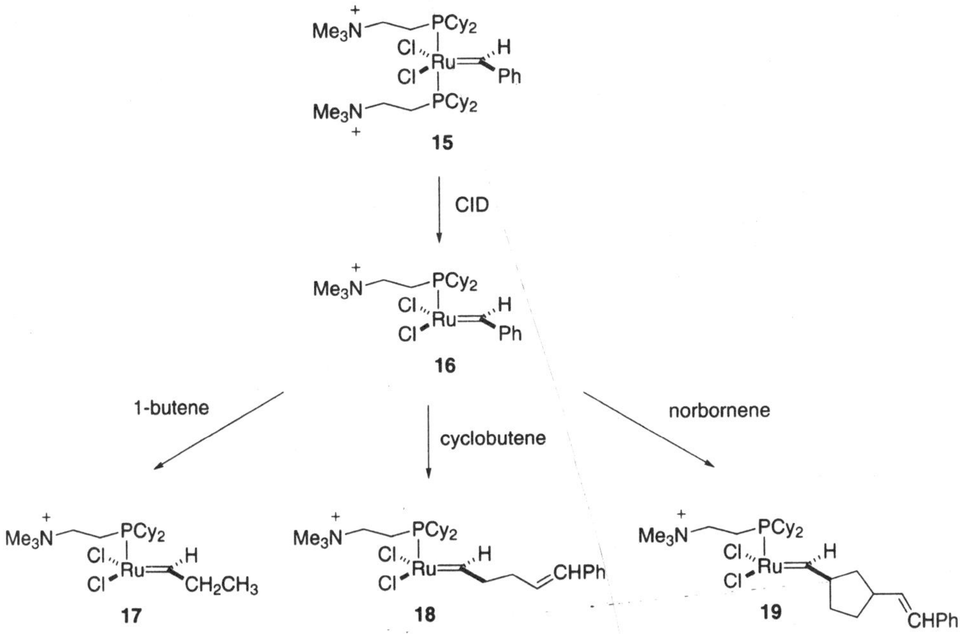

Scheme 16. Reactions of ruthenium carbene **16** in the mass spectrometer

only metathesis product observed in both cases was **17**. With cyclobutene or norbornene in the first octopole, the complexes **18** and **19** were formed predominantly as a result of ROM (Scheme 16). When the reaction was performed in the collision cell, addition of up to three cycloolefin units was detected.

Many features that appear in solution-phase metathesis reactions also appeared in the gas-phase reactions. (1) Dissociation of one phosphine ligand is a prerequisite for metathesis reactions, as evidenced by the complete absence of products containing two phosphine ligands. (2) The monocation **16** showed no metathesis reactivity with carbonyl compounds. (3) In the reaction with 1-butene, only the propylidene complex was formed, while the alternative methylene complex could not be detected. This kinetic preference became evident because of the nonequilibrating conditions in the gas-phase reaction. (4) In ROM, the addition of the first cycloolefin unit was much faster than the subsequent addition steps, which resembles the observation in solution-phase studies that the rates of initiation are much larger than the rates of propagation. The one exception to the similarity of the gas-phase and solution-phase reactions was the absolute rate. As in the case of the Ziegler-Natta-like polymerization by unsolvated $[Cp_2ZrR]^+$ described in Sect. 5.5, the olefin metathesis and ROM reactions were accelerated by up to a factor of 10^4 relative to the solution phase. As with the zirconocene cations, the rate acceleration can be attributed to the complete absence of preequilibria as well as ion-pairing effects, but mainly to the energy gained by the ion dipole or ion induced-dipole interactions common to all ion-molecule reactions.

The perfect parallelism of gas- and solution-phase reactivities shows once again the validity of the ESMS approach as a tool for mechanistic research. Extending the previous work, substituted ruthenium benzylidene monophosphine complexes were prepared by reaction of the electrosprayed parent ruthenium benzylidene with a variety of substituted styrenes [130]. The substituent effects found in the gas-phase reactions can be summarized as follows. For the reaction with 1-butene, a modest acceleration of the acyclic metathesis reaction by electron-withdrawing groups on the aryl group of the substituted ruthenium benzylidene was found. For the reaction with norbornene, on the other hand, essentially no electronic influence on the rate of the ring-opening metathesis reaction was observed. In the solution phase, investigation of the metathesis reaction of ruthenium benzylidenes substituted on the aryl group of the carbene found no clearly interpretable electronic substituent effects [131, 132]. One has to bear in mind that in solution there is a ligand exchange preceding metathesis in which one phosphine ligand must first dissociate, and a further ligand exchange at the end of the reaction after metathesis has occurred. The first ligand exchange enters into the overall solution-phase rate as a multiplicative factor, and is adversely affected by electron withdrawal on the benzylidene. Accordingly, the two effects run in opposite directions, thus reconciling the gas phase results with the solution-phase studies.

An experimental probe of reversibility in ROMP was accomplished by using bifunctional substrates which offer the product of a ring-opening metathesis reaction the chance to undergo a nondegenerate ring-closing metathesis to a isotopically or otherwise substituted complex distinguishable from the original complex by mass [130]. The ROMP substrate 4-[(*Z*)-2-pentenyl-3,4,4,5,5,5-d_6]-cyclopentene, e.g., upon reaction with the ruthenium propylidene complex **20** led to a significant conversion of **20** to **20**-d_6, indicating that the ring-opening metathesis of cyclopentene had occurred in a reversible fashion (Scheme 17).

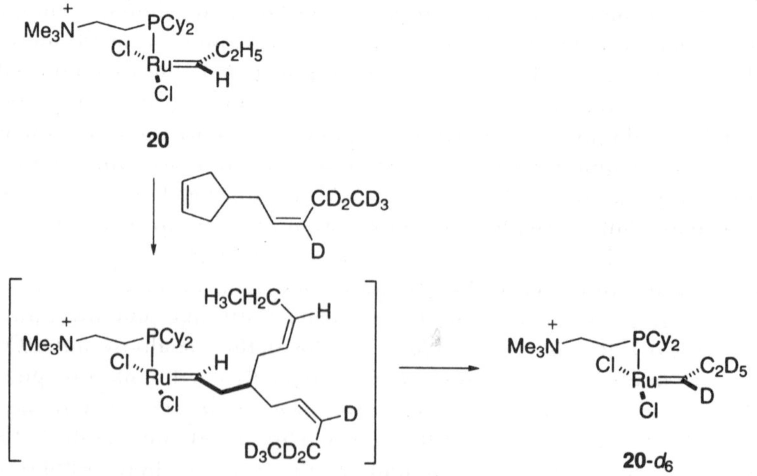

Scheme 17. Reversibility of ROM in the gas-phase metathesis of **20**

Ring-opening metathesis of the substrate 5-vinylnorborn-2-ene, on the other hand, could not be reversed under those experimental conditions. As had been suggested much earlier [133–135], strain release in the substrate is not the principal determinant in the several orders-of-magnitude difference in apparent ROMP reactivity of cyclohexene vs norbornene. Instead, the easy intramolecular π-complexation and concomitant ring-closing metathesis are the controlling factors.

5.8 Quantitative CID Threshold Determinations

5.8.1 *General Remarks*

We have already pointed out that one of the advantages of carrying out organometallic reactions in the gas phase is the possibility to obtain thermochemical data for single reaction steps. This is in strong contrast to measurements in solution, e.g., calorimetry, where reaction enthalpies of a single transformation can in principle be measured, but will always be affected by side reactions, solvent effects (if only caused by reorganization of the solvent sphere), etc. The actual problem with transition metal-chemistry in condensed phases lies in its complexity. One can never be sure about the nature of the actual reactive species, which may even be present at less than 1%. The multiplicity of competing mechanisms and energetically similar structural types in organometallic chemistry has often been blamed for the poor state of mechanistic understanding in this field. One of the main reasons for the lack of quantitative structure-reactivity correlations in transition metal-chemistry which would go beyond the qualitative achievements in this field (e.g., the isolobal analogy) is the paucity of quantitative thermochemical data for metal-ligand bond energies and elemental reaction steps for real-world organometallics. While the binding of simple ligands, e.g., hydride, alkyl, CO, etc. in homoleptic complexes has been approached previously, reliable data for even marginally more complicated complexes is rare [136]. With the powerful physical techniques at hand, we are now in a position to address this task.

A very useful method by which ion-ligand dissociation energies or ion fragmentation energies can be determined depends on threshold measurements for collision-induced dissociation. The CID threshold method utilizing tandem mass spectrometry has experienced significant growth and is currently used for quantitative measurements by several groups [137]. This methodology has been pioneered and continuously refined over the past 15 years by Armentrout [138]. In principal, reduction of the experimentally observed threshold to meaningful chemical information requires two determinations: the structure of the initial ions needs to be determined, and the binding energy needs to be extracted from the experimental threshold. The measured intensities of the reactant and product ions are converted to a total cross section; the energy thresholds E_0 are then obtained by fitting the product ion energy profiles with the CRUNCH program developed by Armentrout [137]. The fitting procedure is based on the equation

$$\sigma = \sigma_0 \Sigma_i g_i (E + E_i + E_{rot} - E_0)^n / E$$

where σ is the cross section, which is proportional to the observed product ion intensity at single collision conditions, E is the kinetic energy in the center-of-mass-frame, E_i is the internal vibrational energy of the precursor ion whose relative abundance at a given temperature is g_i, where $\Sigma_i g_i=1$, and E_0 is the threshold energy, corresponding to the energy required for the dissociation reaction at 0 K. The fitting procedure treats n and E_0 as variable parameters which are determined through the best fit. The vibrational energies E_i are calculated from the frequencies of the normal vibrations of the precursor ion. The frequencies are generally obtained from quantum chemical calculations. An important feature underlying the above equation is the assumption that, due to the relatively long residence time of the excited ions in the collision cell dissociation at the threshold involves the process where the internal energy of the products is close to 0 K, i.e., essentially all the internal energy of the precursor ion is used up in the dissociation at the observed threshold.

5.8.2 *Model Studie*

As a control to check that established CID thresholds could be reproduced with the ESMS setup, the thresholds for the reaction $[(H_2O)_n{\cdot}(H_3O)]^+ \rightarrow [(H_2O)_{n-1}{\cdot}(H_3O)]^+ + H_2O$, which had been determined earlier by Armentrout et al. [139] and Kebarle et al. [140], were measured. Generation of protonated water clusters is particularly simple by electrospray of a 10^{-4} mol l^{-1} aqueous HCl solution (Fig. 9a). Thresholds were measured at four different, decreasing pressures and extrapolated to zero pressure in order to eliminate the contribution of multiple collisions to the foot of the threshold curve. The threshold for n=1, measured in daughter-ion mode, was extracted from our data with Armentrout's CRUNCH program, yielding a value of 1.29 eV (see Fig. 9b) [81], in excellent agreement with the published data.

In a second experiment, the threshold for the reaction $[Mn(CO)_6]^+ \rightarrow [Mn(CO)_5]^+ + CO$ in RFD mode was determined. A meaningful reaction enthalpy can be extracted from the threshold only when the initial kinetic energy distribution of the ions is narrow. With heavier ions, a narrow kinetic energy distribution is achievable when the first quadrupole is operated in RFD mode. The distribution in RFD mode was <0.6 eV fwhm in the laboratory frame, corresponding to <0.1 eV in the center-of-mass frame and, furthermore, nearly Gaussian (argon or xenon as collision gas). Fitting the curve to Armentrout's threshold function yielded a value of 1.48 eV [141], also in good agreement with literature data [138].

When going to the reaction $[CpIr(PMe_3)(CH_3)]^+ \rightarrow [CpIr(\eta^2\text{-}CH_2PMe_2)]^+ + CH_4$, another factor comes into play which has to be taken into account. So far, the fits of the experimental profiles were based on the assumption that the energized reactant ions, whose total energy is equal or larger than E_0, will decom-

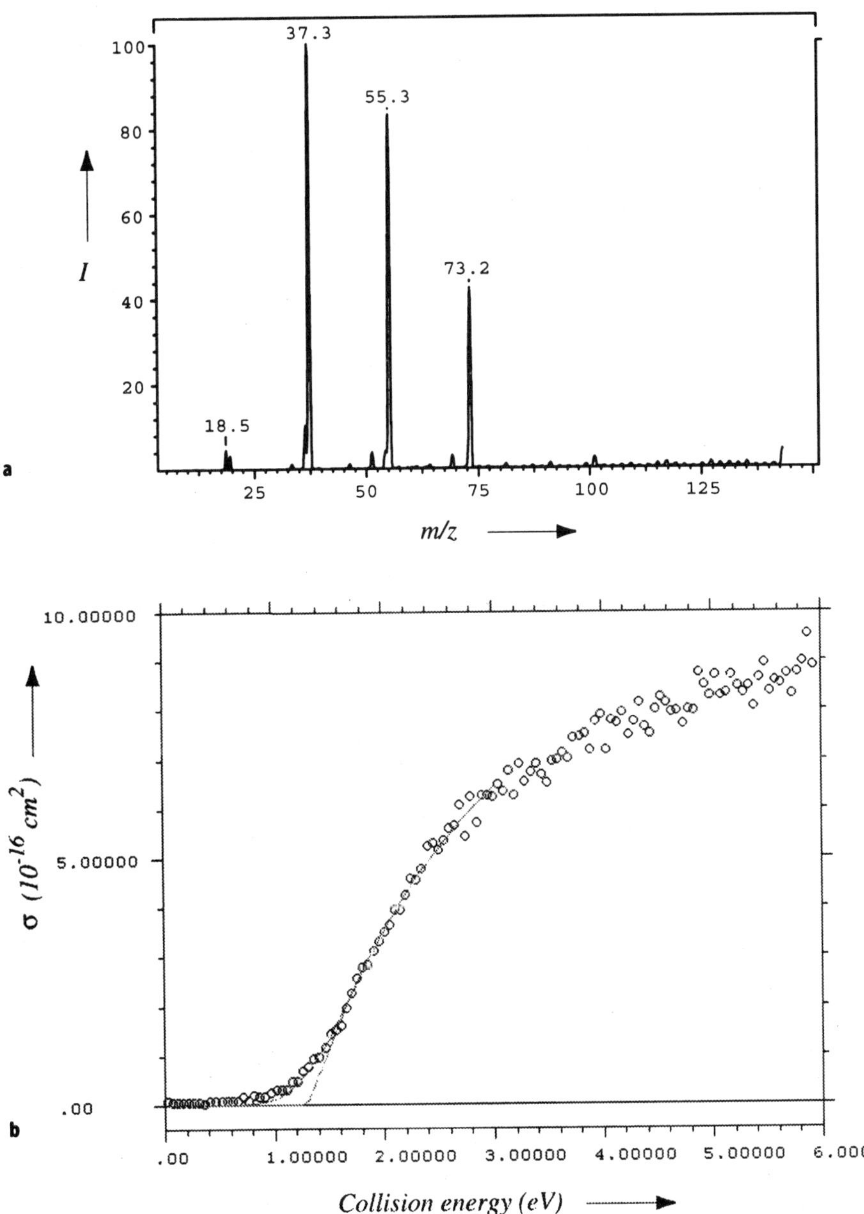

Fig. 9. a Electrospray mass spectrum of a 10^{-4} mol l^{-1} solution of HCl in H_2O under mild declustering conditions, showing the presence of different clusters $[(H_3O)(H_2O)_n]^+$, the clusters with n=1, 2, and 3 being most prominent. **b** CID threshold curve (collision gas Ar), showing the dependence of the dissociation cross section against collision energy, and fit to Armentrout's threshold function

pose within the residence time t of the ion in O2 before entering the mass analyzer Q2. For the experimental setup used, complete dissociation within the residence time can be expected only for small polyatomic ions. Ions with many internal degrees of freedom will attain the required dissociation rate only at internal energies which are somewhat higher than E_0. This "kinetic shift" will lead to E_0 values that are too high. This effect becomes larger as the number of normal modes increases, making it non-negligible for the complexes described here. The kinetic shift can be corrected by including in the fitting procedure only the fraction of the precursor ions which react during the residence time. This fraction can be evaluated with the RRKM formalism [142]. The quantitative CID threshold for the reaction $[CpIr(PMe_3)(CH_3)]^+ \rightarrow [CpIr(\eta^2\text{-}CH_2PMe_2)]^+ + CH_4$ is shown in Fig. 10. The data was deconvoluted using Armentrout's CRUNCH program with the internal energy of the ions set by the 70 °C manifold temperature to which they were thermalized.

Values for transition-state frequencies were assumed following the arguments by Armentrout et al. [143, 144] and Squires et al. [145]. For an endothermic dissociation of gas-phase ions, a loose transition state can be assumed. Accordingly, the product frequencies can be used where available. For the five normal modes that correspond, at the asymptotic limit of dissociation, to relative rotations and translations of the departing fragments, further estimates had to be made. Numerical estimates for these five frequencies were taken from computations of analogous oxidative addition, reductive elimination, and σ-bond metathesis transition states. The deconvoluted threshold, with the methyl

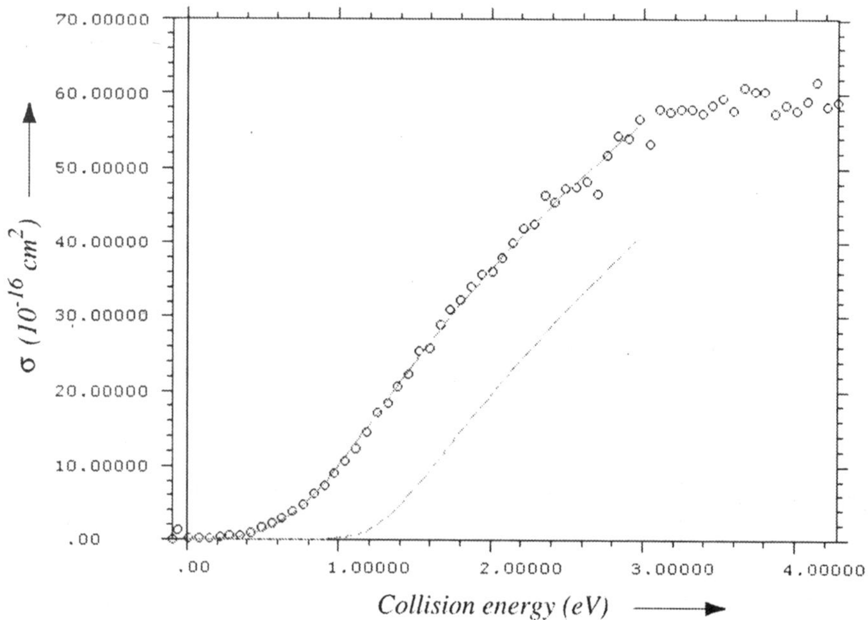

Fig. 10. CID threshold curve (collision gas Ar) for the reaction $[CpIr(PMe_3)(CH_3)]^+ \rightarrow [CpIr(\eta^2\text{-}CH_2PMe_2)]^+ + CH_4$

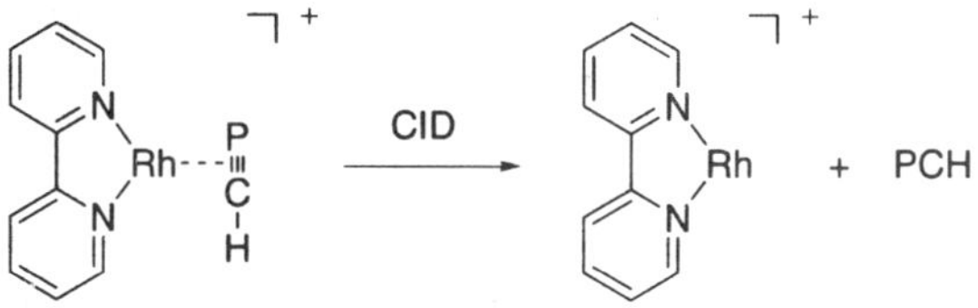

Scheme 18. Fragmentation of $[(bipy)Rh(P{\equiv}CH)]^+$ upon CID

groups, the phosphine as a whole, and the Cp ligand treated as internal rotors, yields a value of E_0=0.59 eV (13.6 kcal/mol) [81].

In a further application of the ES tandem-MS methodology for obtaining quantitative thermochemical data, the ligand binding energy in $[(bpy)Rh(P{\equiv}CH)]^+$ was determined (Scheme 18) [146]. The phosphaethyne complex was prepared in the gas phase by sequential loss of methane from the precursor ion $[(bpy)Rh(PMe_3)_2(H)_2]^+$ by collisional activation. Thermalization was achieved by ~10 mTorr N_2 in the first octopole. For the quantitative threshold measurement, the first quadrupole was operated in RFD mode as a high-pass filter, thereby rendering a narrow kinetic energy distribution of 0.4–0.5 eV fwhm (lab frame, xenon as collision gas). For the RRKM correction, calculations were performed at the B3LYP/LANL2DZ level. The side-on complex is favored over the end-on coordination by 25 kcal/mol at this level of theory. The five remaining transition-state frequencies for the kinetic-shift correction were obtained by taking the appropriate normal modes in the starting ion and reducing the frequencies by a factor of 2. Using these values, the threshold was fitted to yield E_0=2.02±0.15 eV.

As has now been demonstrated convincingly [81, 146, 147], the CID threshold methodology can be applied to molecules with relatively many degrees of freedom, yielding experimental thermochemical data of high quality. It is well known that enthalpies for transition-metal reactions and metal-ligand binding are notoriously difficult to calculate. In contrast to "simple" organic reactions, a systematic approach to the improvement of calculated reaction energies is not feasible for transition-metal compounds with full ligand spheres. Because of the restrictions of post-HF calculations on the system size, semi-empirical and hybrid ab initio/semi-empirical methods like DFT are clearly the methods of choice in this field. Unfortunately, there are severe restrictions on the possibility of a thorough parametrization of semi-empirical methods due to the lack of reliable thermochemical data for transition-metal binding. Thermochemical parameters derived from collision-induced dissociation threshold measurements could open the door to a pool of gas-phase data which can directly be used for such parametrizations. Of course, since the deconvolution of the CID threshold measurement itself requires some kind of quantum chemical calculations for geometries and frequencies, one might conclude that such a procedure represents circular logic. However, previous experience suggests that geometries and frequencies are usually well-modeled by DFT methods, yet the accuracy of these methods for the prediction of energy differences in transition metal complexes can be quite poor. The general success of DFT methods in predicting equilibrium geometries indicated that the local features of a potential surface in the immediate vicinity of a minimum are well treated. This means that a deconvolution scheme based on these frequencies should be sound. The relative energies of two separate minima, however, may not be equally well described. In this light, the use of the more reliable aspects of a computation to deconvolute experimental data that pertain to the less reliable aspects of that same computation is legitimate. Measurements of this kind should serve as benchmarks for the evaluation of the level of theory needed to obtain accurate thermochemistry for transition-metal complexes.

Acknowledgement. The author is very much indebted to Dr. Engelbert Zass, Informationszentrum Chemie Biologie, ETH Zürich, for the extensive literature searches the results of which have been used in the preparation of this article, and to Prof. Peter Armentrout, University of Utah, for his kind support with the use of his CRUNCH program.

6
References

1. Cole RB (ed) (1997) Electrospray ionization mass spectrometry. Wiley, New York
2. Snyder AP (ed) (1995) Biochemical and biotechnological applications of electrospray ionization mass spectrometry. Acs Symp Ser, vol 619
3. Przybylski M, Glocker MO (1996) Angew Chem Int Ed Engl 35:806
4. Bailey AG (1988) Electrostatic spraying of liquids. Research Studies Press, Taunton, UK
5. Zeleny J (1914) Phys Rev 3:69
6. Zeleny J (1917) Phys Rev 10:1
7. Zeleny J (1920) Phys Rev 16:102
8. Zeleny J (1915) Proc Cambridge Phil Soc 18:71
9. Drozin VG (1955) J Colloid Sci 10:158
10. Dole M, Hines RL, Mack LL, Mobley RC, Ferguson LD, Alice MB (1968) Macromolecules 1:96
11. Dole M, Mack LL, Hines RL, Mobley RC, Ferguson LD, Alice MB (1968) J Chem Phys 49:2240
12. Mack LL, Kralik P, Rheude A, Dole M (1970) J Chem Phys 52:4977
13. Clegg GA, Dole M (1971) Biopolymers 10:821
14. For a very personal account on the route to electrospray mass spectrometry see: Fenn JB (1997) Foreword. In: Cole RB (1997) (ed) Electrospray ionization mass spectrometry. Wiley, New York, p ix
15. Yamashita M, Fenn JB (1984) J Phys Chem 88:4451
16. Yamashita M, Fenn JB (1984) J Phys Chem 88:4671
17. Aleksandrov ML, Gall LN, Krasnov VN, Nikolaev VI, Pavlenko VA, Shkurov VA (1984) Dokl Akad Nauk SSSR 277:379
18. Whitehouse CM, Dreyer RN, Yamashita M, Fenn JB (1985) Anal Chem 57:675
19. Taylor GI (1964) Proc R Soc London A280:383
20. Taflin DC, Ward TL, Davis EJ (1989) Langmuir 5:376
21. Davis EJ (1987) ISA Trans 26:1
22. Gomez A, Tang K (1994) Phys Fluids 6:404
23. Lord Rayleigh (1882) Philos Mag 14:184
24. Fernández de la Mora J, Loscertales IG (1994) J Fluid Mech 260:155
25. Schmelzeisen-Redeker G, Röllgen FW, Wirtz H, Vögtle F (1985) Org Mass Spectrom 20:752
26. Schmelzeisen-Redeker G, Bütfering L, Röllgen FW (1989) Int J Mass Spectrom Ion Processes 90:139

27. Iribarne JV, Thomson BA (1976) J Chem Phys 64:2287
28. Thomson BA, Iribarne JV (1979) J Chem Phys 71:4451
29. Fenn JB, Rosell J, Nohmi T, Shen S, Banks FJ Jr (1995) Electrospray ion formation: desorption versus desertion. In: Snyder AP (1995) (ed) Biochemical and biotechnological applications of electrospray ionization mass spectrometry. Acs Symp Ser, vol 619, p 60
30. Fenn JB, Mann M, Meng CK, Wong SF, Whitehouse CM (1989) Science 246:64
31. Snyder AP (1995) Electrospray: a popular ionization technique for mass spectrometry. In: Snyder AP (1995) (ed) Biochemical and biotechnological applications of electrospray ionization mass spectrometry. Acs Symp Ser, vol 619, p 1
32. Colton R, D'Agostino A, Traeger JC (1995) Mass Spectrom Rev 14:79
33. Henderson W, Nicholson BK, McCaffrey LJ (1998) Polyhedron 17:4291
34. Zass E, Plattner DA, Beck AK, Neuburger M (2002) Helv Chim Acta 85:4012
35. Raffaelli A, Bruins AP (1991) Rapid Commun Mass Spectrom 5:269
36. Katta V, Chowdhury SK, Chait BT (1990) J Am Chem Soc 112:5348
37. Wilson SR, Yasmin A, Wu Y (1992) J Org Chem 57:6941
38. Gatlin CL, Turecek F, Vaisar T (1994) Anal Chem 66:3950
39. Colton R, James BD, Potter ID, Traeger JC (1993) Inorg Chem 32:2626
40. Tuoi JLQ, Muller E (1994) Rapid Commun Mass Spectrom 8:692
41. Deery MJ, Howarth OW, Jenings KR (1997) J Chem Soc Dalton Trans 4783
42. Walanda DK, Burns RC, Lawrance GA, von Nagy-Felsobuki EI (2000) Inorg Chim Acta 305:118
43. Walanda DK, Burns RC, Lawrance GA, von Nagy-Felsobuki EI (1999) J Chem Soc Dalton Trans 311
44. Arakawa R, Mimura S, Matsubayashi G, Matsuo T (1996) Inorg Chem 35:5725
45. Wilson SR, Wu YH (1993) Organometallics 12:1478
46. Aliprantis AO, Canary JW (1994) J Am Chem Soc 116:6985
47. Ripa L, Hallberg A (1996) J Org Chem 61:7147
48. Brown JM, Hii KK (1996) Angew Chem Int Ed Engl 35:657
49. Farrington EJ, Brown JM, Barnard CFJ, Rowsell E (2002) Angew Chem Int Ed 41:169
50. Lipshutz BH, Stevens KL, James B, Pavlovich JG, Snyder JP (1996) J Am Chem Soc 118:6796
51. Gilman H, Jones RG, Woods LA (1952) J Org Chem 17:1630
52. Lipshutz BH, Keith J, Buzard DJ (1999) Organometallics 18:1571
53. Yeo JSL, Vittal JJ, Henderson W, Hor TSA (2002) J Chem Soc Dalton Trans 328
54. Fong SWA, Yap WT, Vittal JJ, Hor TSA, Henderson W, Oliver AG, Rickard CEF (2001) J Chem Soc Dalton Trans 1986
55. Fong SWA, Vittal JJ, Henderson W, Hor TSA, Oliver AG, Rickard CEF (2001) Chem Commun 421
56. Bartók M, Balázsik K, Szöllölsi G, Bartók T (2002) J Catal 205:168
57. Náray-Szabó G (1998) Electrostatic catalysis. In: Schleyer PR (ed) Encyclopedia of computational chemistry, vol 2. Wiley, New York, p 905
58. Ding W, Johnson KA, Amster J, Kutal C (2001) Inorg Chem 40:6865
59. Dakin LA, Ong PC, Panek JS, Staples RJ, Stavropoulos P (2000) Organometallics 19:2896
60. Bonchio M, Licini G, Modena G, Moro S, Bortolini O, Traldi P, Nugent WA (1997) Chem Commun 869
61. Bonchio M, Licini G, Modena G, Bortolini O, Moro S, Nugent WA (1999) J Am Chem Soc 121:6258
62. Hinderling C, Chen P (1999) Angew Chem Int Ed 38:2253
63. Freiser BS (ed) (1996) Organometallic ion chemistry. Kluwer, Dordrecht
64. Canty AJ, Traill PR, Colton R, Thomas IM (1993) Inorg Chim Acta 210:91
65. Dalgaard P, McKenzie CJ (1999) J Mass Spectrom 34:1033
66. Styles ML, O'Hair RAJ, McFadyen WD, Tannous L, Holmes RJ, Gable RW (2000) J Chem Soc Dalton Trans 93
67. Jiang C, Hor TSA, Yan YK, Henderson W, McCaffrey LJ (2000) J Chem Soc Dalton Trans 3197

68. Spence TG, Burns TD, Posey LA (1997) J Phys Chem A 101:139
69. Spence TG, Burns TD, Guckenberger GB, Posey LA (1997) J Phys Chem A 101:1081
70. Spence TG, Trotter BT, Burns TD, Posey LA (1998) J Phys Chem A 102:6101
71. Molina-Svendsen H, Bojesen G, McKenzie CJ (1998) Inorg Chem 37:1981
72. Arndtsen BA, Bergman RG, Mobley TA, Petersen TH (1995) Acc Chem Res 28:154
73. Janowicz AH, Bergman RG (1982) J Am Chem Soc 104:352
74. Jones WD, Feher FJ (1984) J Am Chem Soc 106:1650
75. Hoyano JK, McMaster AD, Graham WAG (1983) J Am Chem Soc 105:7190
76. Burger P, Bergman RG (1993) J Am Chem Soc 115:10,462
77. Arndtsen BA, Bergman RG (1995) Science 270:1970
78. Veltheer JE, Burger P, Bergman RG (1995) J Am Chem Soc 117:12,478
79. Holtcamp MW, Labinger JA, Bercaw JE (1997) J Am Chem Soc 119:848
80. Hinderling C, Plattner DA, Chen P (1997) Angew Chem Int Ed Engl 36:243
81. Hinderling C, Feichtinger D, Plattner DA, Chen P (1997) J Am Chem Soc 119:10,793
82. Luecke HF, Bergman RG (1997) J Am Chem Soc 119:11,538
83. Fink G, Mülhaupt R, Brintzinger HH (eds) (1995) Ziegler catalysts. Springer, Berlin Heidelberg New York
84. Jordan RF, Dasher WE, Echols SF (1986) J Am Chem Soc 108:1718
85. Jordan RF, Bajgur CS, Willett R, Scott B (1986) J Am Chem Soc 108:7410
86. Brintzinger HH, Fischer D, Mülhaupt R, Rieger B, Waymouth R (1995) Angew Chem Int Ed Engl 34:1143
87. Christ CS, Eyler JR, Richardson DE (1988) J Am Chem Soc 110:4038
88. Christ CS, Eyler JR, Richardson DE (1990) J Am Chem Soc 112:596
89. Richardson DE, Alameddin NG, Ryan MF, Hayes T, Eyler JR, Siedle AR (1996) J Am Chem Soc 118:11,244
90. Guo BC, Castleman AW Jr (1992) J Am Chem Soc 114:6152
91. Daly GM, El-Shall MS (1994) J Phys Chem 98:696
92. Daly GM, El-Shall MS (1995) J Phys Chem 99:5283
93. Bjarnason A, Ridge DP (1996) J Phys Chem 100:15,118
94. Feichtinger D, Plattner DA, Chen P (1998) J Am Chem Soc 120:7125
95. Chien JCW, Wang BP (1990) J Polym Sci Part A 28:15
96. Ortiz de Montellano PR (ed) (1995) Cytochrome P-450, structure, mechanism and biochemistry. Plenum Press, New York
97. Sheldon RA (ed) (1994) Metalloporphyrins in catalytic oxidations. Dekker, New York
98. Montanari F, Casella L (eds) (1994) Metalloporphyrin catalyzed oxidations. Kluwer, Dordrecht
99. Siddall TL, Miyaura N, Huffman JC, Kochi JK (1983) J Chem Soc Chem Commun 1185
100. Samsel EG, Srinivasan K, Kochi JK (1985) J Am Chem Soc 107:7606
101. Srinivasan K, Kochi JK (1985) Inorg Chem 24:4671
102. Srinivasan K, Michaud P, Kochi JK (1986) J Am Chem Soc 108:2309
103. Srinivasan K, Perrier S, Kochi JK (1986) J Mol Catal 36:297
104. Zhang W, Loebach JL, Wilson SR, Jacobsen EN (1990) J Am Chem Soc 112:2801
105. Irie R, Noda K, Ito Y, Matsumoto N, Katsuki T (1990) Tetrahedron Lett 31:7345
106. Dalton CT, Ryan KM, Wall VM, Bousquet C, Gilheany DG (1998) Top Catal 5:75
107. Groves JT, Kruper WJ (1979) J Am Chem Soc 101:7613
108. Groves JT, Stern MK (1987) J Am Chem Soc 109:3812
109. Groves JT, Stern MK (1988) J Am Chem Soc 110:8628
110. Groves JT, Lee J, Marla SS (1997) J Am Chem Soc 119:6269
111. Plattner DA, Feichtinger D, El-Bahraoui J, Wiest O (2000) Int J Mass Spectrom 195/196:351
112. Calligaris M, Randaccio L (1987) In: Wilkinson G, Gillard RD, McCleverty JA (eds) Comprehensive coordination chemistry, vol 2. Pergamon, Oxford, p 715
113. Oki AR, Hodgson DJ (1990) Inorg Chim Acta 170:65
114. Feichtinger D, Plattner DA (1997) Angew Chem Int Ed Engl 36:1718
115. Gilbert BC, Kamp NWJ, Lindsay Smith JR, Oakes J (1998) J Chem Soc Perkin Trans 2 1841

116. Feichtinger D, Plattner DA (2000) J Chem Soc Perkin Trans 2 1023
117. Nugent WA, Mayer JM (1988) Metal-ligand multiple bonds. Wiley, New York
118. Feichtinger D, Plattner DA (2001) Chem Eur J 7:591
119. Brown HC, Okamoto Y (1958) J Am Chem Soc 80:4979
120. Jacobsen EN, Zhang W, Güler ML (1991) J Am Chem Soc 113:6703
121. Palucki M, Finney NS, Pospisil PJ, Güler ML, Ishida T, Jacobsen EN (1998) J Am Chem Soc 120:948
122. Finney NS, Pospisil PJ, Chang S, Palucki M, Konsler RG, Hansen KB, Jacobsen EN (1997) Angew Chem Int Ed Engl 36:1720
123. Ivin KJ, Mol JC (1997) Metathesis and metathesis polymerization. Academic Press, New York
124. Stevens AE, Beauchamp JL (1979) J Am Chem Soc 101:6449
125. Jacobsen DB, Freiser BS (1985) J Am Chem Soc 107:67
126. Jacobsen DB, Freiser BS (1985) J Am Chem Soc 107:2605
127. Stöckigt D, Schwarz H (1992) Chem Ber 125:2817
128. Mohr B, Lynn DM, Grubbs RH (1996) Organometallics 15:4317
129. Hinderling C, Adlhart C, Chen P (1998) Angew Chem Int Ed 37:2685
130. Adlhart C, Hinderling C, Baumann H, Chen P (2000) J Am Chem Soc 122:8204
131. Schwab P, Grubbs RH, Ziller JW (1996) J Am Chem Soc 118:100
132. Ulman M, Grubbs RH (1998) Organometallics 17:2484
133. Patton PA, McCarthy TJ (1984) Macromolecules 17:2939
134. Patton PA, McCarthy TJ (1985) Polym Prep 26:66
135. Patton PA, McCarthy TJ (1987) Macromolecules 20:778
136. Martinho Simões JA, Beauchamp JL (1990) Chem Rev 90:629
137. Armentrout PB (1992) Thermochemical measurements by guided ion beam mass spectrometry. In: Adams N, Babcock LM (eds) Advances in gas phase ion chemistry, vol 1. JAI Press, Greenwich, CT, p 83
138. Armentrout PB (1995) Acc Chem Res 28:430
139. Dalleska NF, Honma K, Armentrout PB (1993) J Am Chem Soc 115:12,125
140. Anderson SG, Blades AT, Klassen J, Kebarle P (1995) Int J Mass Spectrom Ion Proc 141:217
141. Hinderling C, Plattner DA (unpublished)
142. Holbrook KA, Pilling MJ, Robertson SH (1996) Unimolecular reactions. Wiley, Chichester
143. Schultz RH, Crellin KC, Armentrout PB (1991) J Am Chem Soc 113:8590
144. Khan FA, Clemmer DE, Schultz RH, Armentrout PB (1993) J Phys Chem 97:7978
145. Sunderlin LS, Wang D, Squires RR (1993) J Am Chem Soc 115:12,060
146. Kim YM, Chen P (2000) Int J Mass Spectrom 202:1
147. Klassen JS, Anderson SG, Blades AT, Kebarle P (1996) J Phys Chem 100:14,218

III
Mass Spectrometric Methodology

Top Curr Chem (2003) 225: 207–232
DOI 10.1007/b10470

Gas-Phase Conformations: The Ion Mobility/Ion Chromatography Method

Thomas Wyttenbach[1] · Michael T. Bowers[2]

Department of Chemistry and Biochemistry, University of California at Santa Barbara, Santa Barbara, California 93106, USA
[1] *E-mail: wyttenbach@chem.ucsb.edu*
[2] *E-mail: bowers@chem.ucsb.edu*

Ion mobility spectrometry coupled to mass spectrometry provides a powerful tool to explore the three-dimensional shape of polyatomic ions. Applications include the investigation of cluster ion geometries and conformations of flexible molecules such as biopolymers and synthetic polymers. The ion structure is obtained by measuring collision cross sections in a high pressure drift tube filled with helium and comparing it to model structures obtained by various theoretical methods such as molecular modeling and electronic structure calculations. The temperature of the drift tube is generally adjustable (typically from 80 to 800 K) providing a unique opportunity to address topics such as the thermal motion of floppy molecules, the unfolding process of folded structures, the kinetics of structural interconversion, and the kinetics of dissociation processes. In addition, the ion mobility instrumentation can be used to obtain thermochemical data of ligand addition reactions, giving important additional information about the polyatomic ions under investigation. The theoretical background and the concepts of these ion mobility based experiments and the instrumentation employed are briefly reviewed in this chapter. Furthermore, some detailed examples and a very brief summary of selected applications found in the literature are given.

Keywords. Molecular geometry, Polymer, Protein, Peptide, Cluster

List of Abbreviations and Symbols

A	Arrhenius pre-exponential factor
Ac	Acetyl
ala	Alanine
arg	Arginine
argOMe	Arginine methyl ester
ATD	Arrival time distribution
BPTI	Bovine pancrease trypsin inhibitor
e	Electric charge
E	Electric field
E_a	Arrhenius activation energy
ESI	Electrospray ionization
FT-ICR	Fourier transform ion cyclotron resonance
gly	Glycine
IMS	Ion mobility spectrometry/spectrometer
K	Ion mobility
k	Rate constant
k_B	Boltzman constant
K_{eq}	Equilibrium constant
LHRH	Luteinizing hormone releasing hormone
lys	Lysine
m/z	Mass to charge ratio
MALDI	Matrix assisted laser desorption ionization
MD	Molecular dynamics
MM	Molecular mechanics
MS	Mass spectrometry/spectrometer
N	Particle number density
p	Pressure
PET	Poly(ethylene terephthalate)
POSS	Polyhedral silsesquioxone
pro	Proline
ser	Serine
SIFDT	Selected ion flow drift tube
T	Temperature

t_D	Drift time
TOF	Time of flight
u	Atomic mass unit
val	Valine
v_D	Drift velocity
V_D	Drift voltage
ΔH°	Standard enthalpy change
ΔS°	Standard entropy change
Δt_D	Drift time spread
μ	Reduced mass
σ	Cross section

1 Introduction

This chapter is a brief summary of recent advances in the field of ion mobility–mass spectrometry (IMS-MS) used to obtain structural information of ionic species such as the conformations of biological molecules and the three-dimensional arrangement of atoms in cluster ions. Such IMS-MS applications typically include extensive theoretical work for generating reasonable model structures of the molecular systems studied experimentally, as it is really the synergics of experiment and theory that provides detailed information on molecular structure using the ion mobility technique (or ion chromatography technique as it is sometimes called in this context to set it apart from conventional IMS applications). We will review the concepts and instrumentation used to obtain molecule structure information by IMS-MS and demonstrate the methods on a number of examples. Finally, we will attempt to summarize the major conclusions obtained from the body of work, that emerged in the past five years. IMS applications simply to separate ions of different mobility, generally analytical in nature, are not included in this work. Further, we are not attempting to be totally exhaustive in the review and have chosen detailed examples primarily from our own work, much of which is unpublished at this time.

The area of ion mobility research to obtain structural information started to emerge in the early 1990s and was pioneered by the groups of Bowers [1–5] and Jarrold [6, 7]. Most of the early work was reviewed by Clemmer and Jarrold [8] in 1997 and we focus here on the work that has appeared since then. One major point of interest in the field is clearly the structure of biological molecules in the presence and absence of solvation molecules. Molecules as small as glycine and as large as cytochrome *c*, a heme protein composed of more than 100 amino acids, were studied to address issues such as zwitterion formation in peptides, proteins, and oligonucleotides; protein folding; peptide helix formation; the dynamics of interconversion between conformations; and the effects of hydration. Other significant results were obtained for synthetic polymers, for nonmetal, semiconductor, and metal clusters, and for salt and ion-molecule clusters. In Sect. 5 we will give references to these recent applications together with a short overview of the major conclusions.

2
Concepts

2.1
Ion Mobility and Cross Section

Ions drifting through a buffer gas under the influence of a weak uniform electric field E quickly reach an equilibrium between forward acceleration due to the electric field and retarding effect due to collisions with the buffer gas resulting in a constant drift velocity v_D. The drift field is weak when the steady flow of ions along the electric field is much slower than the random motion leading to diffusion. The low field mobility K is the proportionality constant between v_D and E [9]:

$$v_D = KE\,. \tag{1}$$

Hence, measuring the drift time t_D for a given drift length and given E yields an experimental value for K. For a given pressure (or particle density N) and temperature T of the buffer gas the ion mobility K is given by the collision cross section σ by

$$K = \frac{3e}{16N}\left(\frac{2\pi}{\mu k_B T}\right)^{1/2}\frac{1}{\sigma} \tag{2}$$

where e is the charge of the ion, μ the reduced mass of ion and buffer gas, and k_B the Boltzman constant [9]. Thus, ions with compact structures have a small cross section and a large ion mobility, whereas the opposite is true for large extended structures. Hence, in ion mobility spectrometry (IMS) ions are separated (therefore the term "ion chromatography") by size (collision cross section) in contrast to mass spectrometry, where they are dispersed by mass (mass to charge ratio). The collision cross section is not only determined by the geometry of the ion but also by the interaction between ion and buffer gas. This is particularly true for small ions, large and polarizable buffer gases, and at low temperatures. Therefore, helium is the buffer gas of choice for obtaining structural information by ion mobility techniques and it is the only buffer gas used in all the work reviewed here.

While it is straightforward to measure an experimental collision cross section via drift time and ion mobility, it is far more difficult to obtain a reliable value for a given model structure for comparison with experiment. A number of models have been proposed to calculate theoretical cross sections. A simple approach is to calculate an orientation averaged projection cross section of the model geometry using hard spheres with a specified radius for each atom in the system. In sophisticated methods the radii are adjusted as a function of temperature and ion size (number of atoms) on the basis of ion–buffer gas interaction potentials, which are generally of a (12,6,4) form [10, 11]. For small ions with less than 200 atoms and using He as buffer gas the projection approximation agrees with the ion mobility experiment rather well [10–12]. However, it obviously fails to describe the scattering process that is actually taking place in the experiment, and it is momentum transfer that determines the collision cross section. Hence, models based on calculating collision integrals using trajectory calculations [13] are potentially more accurate. Again, for smaller ions with several hundred atoms the ion-buffer gas interaction has to be treated accurately, usually with a (12, 6, 4) potential, in order to get good agreement with experiment and to get a reasonable temperature dependence. For larger ions (e.g., proteins) the interaction potential is much less important and a hard sphere scattering model yields very good results [14]. Other models including scattering from an iso-electron density surface are presently also being used [15].

2.2 Model Structures

Generating model structures of the ionic molecular species under investigation is essential for interpreting ion mobility data. Computational methods are generally used to obtain candidate model geometries, although crystal structures obtained by X-ray crystallography and solution structures obtained by nuclear magnetic resonance techniques are sometimes useful as well. Molecular mechanics/molecular dynamics (MM/MD) is often the method of choice for calculations because the size of the systems does not allow use of higher level theories. MM is also used for smaller systems to thoroughly search conformational space. However, MM methods are limited to systems where empirical force field parameters are available, i.e., systems composed of elements typically found in organic molecules. For all other types of systems ab initio and density functional calculations are an alternative. However, because of limitations to computer power geometry optimization using such electronic structure calculations are only possible for smaller systems, typically in the range of tens of atoms. Hence, the only possibility for large systems with organically atypical atoms is to parameterize the atypical atom(s) for inclusion in standard force fields. An example of this procedure is parameterization of the silicon atom for use in the AMBER [16] force field allowing MM/MD calculations on polyhedral silsesquioxone (POSS) polymers [17, 18].

2.3 Kinetic and Thermochemical Data

The ion drift tube setup used to measure ion mobilities can also be used to obtain kinetic and thermochemical data often providing important information about the ion under investigation in addition to the collision cross section. If a chemical reaction like an isomerization (change in cross section) or a dissociation (change in cross section or mass to charge ratio) occurs on the time scale of the experiment a reaction rate constant can be measured precisely, since the reaction time (i.e., the drift time) and the temperature are both well defined and readily measured. In addition, since the drift tube temperature can be changed, an Arrhenius type of analysis yields activation energies and pre-exponential factors for the reaction. More details and examples for such experiments are given in Sect. 4.2.

If the time scale of a chemical reaction is much shorter than the experimental time scale a chemical equilibrium will be established. However, an isomerization equilibrium cannot be studied on the basis of cross section (even if different isomers have different cross sections), because the fraction of time an ion travels as one isomer or the other is the same for each ion and therefore every ion exhibits the same cross section, an average of all interconverting isomers.

However, an equilibrium between products and reactants that have different mass to charge ratios can be examined by means of mass spectrometry at the exit of the drift tube. In such an equilibrium, care has to be taken that products and reactants, including any non-ionic species, are present with defined concentrations. The type of equilibrium typically studied using this technique are reactions where ligands are added to ions which were injected into the drift tube. In this case the neutral ligand is present in the drift cell with a defined pressure. Such reactions include ligation of metal ions and solvation of ions by solvent molecules. Studying an equilibrium as a function of temperature yields ΔH° and ΔS° values for the reaction in question. Details and an example of this type of experiment are given in Sect. 4.3.

3 Instrumentation

3.1 Basic Setup

The basic setup allowing for most flexibility to carry out ion mobility experiments is schematically depicted in Fig. 1. Ions are generated in an ion source, mass selected in a first mass filter MS1, and injected into the drift tube. If the ion source produces a continuous ion beam, the beam has to be gated in front of the drift cell, the gate triggering the clock for measuring the drift time of the ion pulse. Sophisticated equipment like a quadrupole ion trap [19–21] or an ion funnel [22] can be used as a gate and ion storage device to convert a continuous ion beam into a pulsed one without significant ion loss.

Ions exiting the drift tube are mass analyzed in mass spectrometer MS2, an important feature if reactions are occurring in the drift cell. Ions are generally detected after MS2 by ion counting techniques. The mass spectrometers MS1 and MS2 are typically quadrupole mass filters, and either one or the other can be run in RF-only mode for better signal but without mass selection, if desired.

The setup in Fig. 1 is conceptually not new. It has been used for decades by a number of groups [23–26] after pioneering work of Hasted and coworkers in 1966 [27]. Also the selected ion flow drift tube (SIFDT) technique [28–33] emerging in the 1980s by combining the features of a drift tube with a flowing afterglow apparatus [34] is equivalent to the setup in Fig. 1. However, all of the

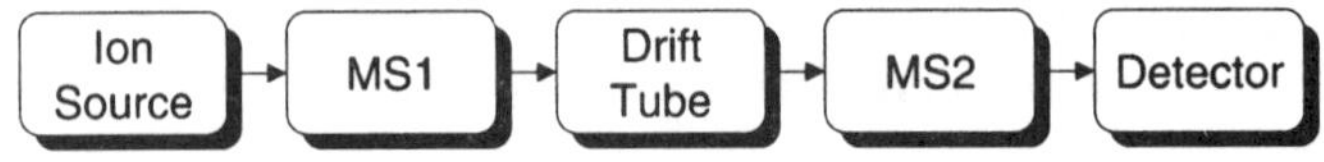

Fig. 1. Schematic outline of the basic ion mobility-mass spectrometry setup using mass selection before (MS1) and after (MS2) the drift tube

instrumentation employed in these earlier studies was used to examine the chemistry and the interaction of various small, often atomic ions with various buffer gases, and a large part of this research was geared towards understanding atmospheric ion chemistry processes.

Injecting ions into a drift tube becomes increasingly more difficult with increasing drift tube pressure. However, high drift tube pressure is desirable for high ion mobility resolution. For these reasons omitting MS1 is a common simplification of the basic setup in Fig. 1 and is typically found in analytical IMS-MS [35] and other high pressure drift tube applications (see Sect. 3.3) [36].

As mentioned above, the mass spectrometers MS1 and MS2 are often quadrupole mass filters, but other types of mass spectrometers have been coupled to IMS as well, including magnetic sectors [37], FT-ICR [38], quadrupole ion trap [21] and time of flight (TOF) mass spectrometers [39–45]. TOF-MS has the advantage that there are no theoretical limits to the mass to charge ratio (*m/z*) covered, whereas quadrupoles typically operate up to a maximum *m/z* of several thousand mass units. Hence, TOF-MS delivering pulses of mass dispersed ions, one of which can be selected by an ion gate, works very naturally as a front end (MS1) to the drift cell to study ions with very high *m/z* values. In this setup a pulsed ion source, such as MALDI, is an ideal choice [45]. Another important feature of TOF-MS is that ions of different *m/z* are non-destructively dispersed in time and not in space like in quadrupole filters, where non-selected ions are lost due to unstable trajectories. Therefore, coupling a TOF-MS to the exit of the drift tube is particularly interesting (MS2) and yields a setup that allows one to obtain two-dimensional mobility-mass information theoretically within seconds (see Sect. 3.4) [39–44].

3.2
Temperature Controlled Drift Cell

Most drift cells used in the applications outlined in this article can be temperature controlled to some degree. In this section we describe the basic design of Kemper and Bowers [22, 37, 46] as an example of a drift cell that is very suitable for applications where high ion transmission of several percent and adjustable temperatures within a large range from typically 80 K (liquid nitrogen) to 800 K are desirable. The upper temperature limit is given by the materials used and by the mechanical design providing a uniform temperature distribution and allowing for expansion of the material at high temperatures.

The temperature controlled drift cell shown in Fig. 2 is fabricated from a copper block with the ring electrodes (providing the drift field) in the interior of the cell. Heating is provided by electrical heaters inserted from the outside into bores in the copper cell. Cooling is achieved by running a pre-cooled gas (typically nitrogen cooled in liquid nitrogen) through a set of bores in the cell block that makes up a cooling cycle. The resistor chain connecting the ring electrodes can be either run in the interior (preferred) or exterior (using feedthroughs) of the cell. The ring electrodes are spaced by ceramic spacers on ceramic rods and by springs to allow for temperature affected contraction and expansion.

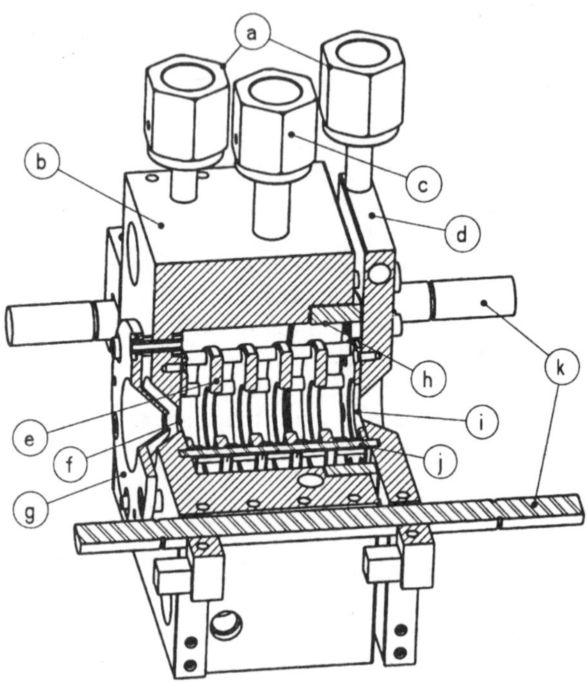

Fig. 2. Perspective cross sectional view of the temperature controlled drift cell after Kemper and Bowers' design [22]: (a) cooling line, (b) cell body, (c) buffer gas inlet, (d) cell end cap, (e) drift guard ring, (f) ion entrance hole, (g) ion focusing lens, (h) ceramic ring, (i) ion exit hole, (j) ceramic rod holding guard rings, (k) ceramic rods holding cell assembly

The ion entrance plate (~0.1 mm thick) with a small hole (typically 0.5 mm diameter) has the same electric potential as the copper cell, whereas the exit plate (same dimensions) is mounted to a cell end cap that is electrically isolated from the cell body by a ceramic ring. The end cap has its own cooling and heating systems.

The entire cell assembly is supported by two large ceramic rods running left and right of the cell parallel to and on the same height as the ion beam. The cell assembly is self-centered between the two rods by a leaf-spring setup that also allows for expansion and contraction of the assembly.

This cell is typically 4–5 cm long and operates at pressures of 3–5 torr and drift voltages of 10–100 V across the cell.

3.3
High Resolution Drift Tube

The resolution $t_D/\Delta t_D$ of a drift tube is proportional to the square root of the drift voltage V_D across the tube [9]:

$$\frac{t_D}{\Delta t_D} = \left(\frac{\pi e V_D}{4 k_B T}\right)^{1/2} \tag{3}$$

where t_D is the average drift time, Δt_D the drift time spread due to diffusion, e the ion charge, k_B the Boltzman constant, and T the temperature. Hence, to increase resolution it is desirable to increase V_D, while still fulfilling the weak field condition (see Sect. 2.1). This is achieved in two ways: by increasing the drift length and by increasing the pressure. In the design used in Jarrold's laboratory and briefly described below, a drift voltage V_D of up to 15,000 V can be used by employing a pressure of 500 torr and a tube that is 63 cm long [36]. Since ions cannot be injected from vacuum into a drift tube maintained at such high pressures, mass selection prior to the high resolution drift tube is not possible. Ions are formed immediately in front of the tube at high pressure and pulled into the drift region through an "ion gate". The ion gate prevents neutral species from entering the drift tube by a counter flow of helium and carries ions through by an electric field.

The high resolution drift tube is fabricated from a six-inch diameter steel tube divided into three sections separated by ceramic breaks. The field in the interior is maintained by copper beryllium drift guard rings, connected to each other by a resistor chain. In Jarrold's apparatus [36] the ion exit hole has a diameter of 0.13 mm. The drift tube temperature is controlled by jackets around each section and by a recirculator with non-conducting temperature regulated fluorocarbon fluids.

Similar drift tubes with similar dimensions are used as medium resolution devices employing a V_D of several hundred to 1000 V and pressures of several torrs [39]. In this lower pressure, medium resolution setup ions are generally injected into the drift tube from vacuum, often with mass selection prior to the drift tube.

3.4
Two-Dimensional Mobility-Mass Data

In this section we would like to point out one of the many possible ion mobility-mass spectrometer combinations, that appear to make up a particularly intriguing setup, the drift tube–TOF MS combination [39–44]. In this setup shape separated ions exiting the drift tube are further dispersed in time by their mass to charge ratio. The timing works out such that ions are first shape separated on a several millisecond time scale and subsequently mass-to-charge separated on the fly on a 10-μs time scale [39]. Thus pulses of ions are injected into the drift tube at a repetition rate of the order of 10 Hz and ions exiting the drift tube are injected into the TOF-MS at a repetition rate of 10^4 Hz. Timing is one of the reasons why the ion mobility-TOF-MS combination works so well together, but it is also the fact that both techniques work with ion pulses and that TOF is not a scanning technique, where all ions are lost except for those with the m/z value selected. However, coupling ion mobility to TOF-MS is technically challenging as nearly ten orders of magnitude of pressure difference between drift tube and TOF-MS have to be overcome. In addition, optimizing the duty cycle of pulsing ions into the TOF-MS is critical for getting good data in a short period of time [47], otherwise long signal accumulation times counteract the extraordinary time advantage gained by TOF-MS.

4 Methods and Examples

4.1 Cross Sections: Experiment and Theory

As mentioned in Sect. 2.1 the mobility measurement of a polyatomic ion provides a value for its orientation averaged cross section and hence information about its shape. Extended, elongated structures of a given molecule have larger rotationally averaged cross sections than compact more spherical structures. For instance, the projection cross section of a right-handed α-helix of the peptide (gly-ala)$_7$Cs$^+$, an extended structure (Fig. 3b), is expected to be 243±10 Å^2, whereas that of a globular, near spherical structure of the same molecule is of the order of 213±5 Å^2 (Fig. 3a) [48]. Most of the "error" bars indicated with these values are due to geometry changes as a function of time during thermal motion at 300 K. The models in Fig. 3 have been obtained by molecular modeling calculations using the AMBER force field [16]. It is of interest that globular structures are only stable if the cesiated (gly-ala)$_7$ peptide is assumed to be a zwitterion. As soon as the proton is transferred from the N-terminus (NH$_3^+$–) to the C-terminus (–COO–) the peptide prefers to fold into a 95% α-helical conformation, where Cs$^+$ caps the helix on the C-terminus side. If Cs$^+$ is replaced by smaller alkali ions more globular charge solvation structures become competitive compared to the helix. An analysis of the Ramachandran plot for a typical (gly-ala)$_7$Li$^+$ charge solvation structure indicates that only 23% is α-helical, just like the protonated form (gly-ala)$_7$H$^+$ (26% helical, σ=213 Å^2). The sodiated, potas-

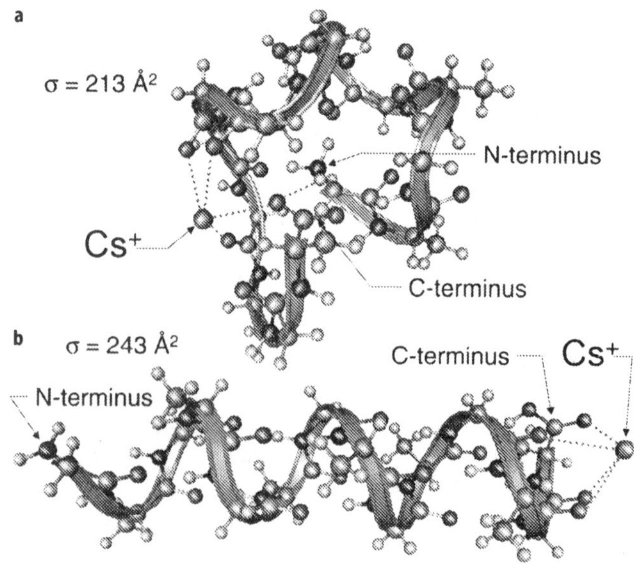

Fig. 3a, b. Structures of (gly-ala)$_7$Cs$^+$ obtained by molecular mechanics calculations: **a** globular zwitterion structure; **b** α-helical non-zwitterion structure

siated, and rubidiated peptides yield conformations which are 59%, 77%, and 90% helical, respectively. The zwitterionic structures for all alkali ions are compact and <20% helical.

Thus, this type of system is ideally set up to be explored by ion mobility. Experimental cross sections for all of the $(gly\text{-}ala)_7X^+$ systems, $X^+=H^+$, Li^+ through Cs^+, are in the range of 211–216 Å^2 at 300 K [48] thus ruling out extensive helix formation for any of the systems. For $(gly\text{-}ala)_7H^+$ the experiment (216±5 Å^2) is in agreement with theory which predicts a non-helical compact charge solvation structure (213 Å^2). For the alkali ion cationized systems both globular and helical conformations are theoretically plausible. However, the experiment indicates that globular structures win out for all alkali ions. For the small alkali ions both compact charge solvation and zwitterion structures are potential candidates. For Cs^+ globular charge solvation structures can be ruled out on theoretical grounds, since they are expected to be less stable than an α-helix by more than 10 kcal mol^{-1}. Hence, experiment leads to the conclusion that the structures of the $(gly\text{-}ala)_7X^+$ are all similar for any choice of X^+ ($X^+=H^+$, alkali ion). However, the combination of experiment and theory makes structural assignments possible and leads in this case to the conclusion that all the structures are non-helical and that the cesiated peptide assumes a compact zwitterion structure.

4.2 Kinetics

If an ion injected into the drift tube is undergoing a reaction from one species with cross section A to another one with cross section B≠A during the drift time, the cross section measured for that ion is between A and B. An ensemble of reacting ions gives a cross section distribution ranging from A (late or no reaction) to B (early reaction). Using kinetic theory the expected cross section distribution for such a system can be calculated for given A, B, and reaction rate constant [49]. Therefore, the only unknown parameter, the rate constant, can be obtained by fitting the theoretical distribution to experiment.

It should be emphasized that the reaction is occurring under thermal conditions at a temperature defined by the helium buffer gas. Measuring rate constants k as a function of temperature yields activation energies E_a and pre-exponential factors A for the reaction, assuming Arrhenius-type behavior:

$$k = Ae^{-E_a/k_B T} \tag{4}$$

Below we will discuss three examples of reactions that have been studied using ion mobility instrumentation. The first example is a dissociation with loss of a neutral fragment (decrease of m/z), the second example is a dissociation where both fragments have identical m/z, and the third example is an isomerization.

4.2.1 *Serine Clusters*

Recent ESI studies of serine showed extensive formation of singly or multiply protonated clusters $(ser)_nH_z^{z+}$ (Fig. 4a) [50–54]. In many cases "magic" numbers

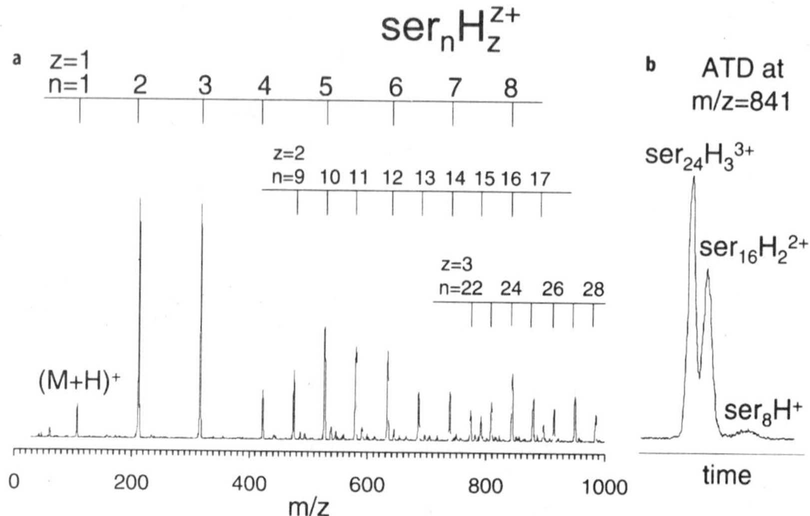

Fig. 4. **a** Electrospray mass spectrum of a 100 mmol l^{-1} solution of L-serine; extensive clustering of serine units is observed for charge states z=1, 2, and 3. **b** Ion mobility spectrum of ions with m/z=841 showing the three features assigned as protonated octamer, doubly protonated 16-mer, and triply protonated 24-mer

for *n* have been reported in the literature. The presence [50–53] or absence [54] (Fig. 4) of "magic" numbers appears to be strongly dependent on source conditions. The mass spectrum shown in Fig. 4a shows a smooth distribution of singly, doubly, and triply protonated clusters. The ion mobility spectrum (or ion arrival time distribution, ATD) recorded at constant m/z=841 (Fig. 4b) indicates the presence of three features, that can readily be assigned as $(ser)_8H^+$, $(ser)_{16}H_2^{2+}$, and $(ser)_{24}H_3^{3+}$ [54].

If serine clusters are injected into an ion mobility cell held at elevated temperature the clusters dissociate. Figure 5a shows mass spectra obtained with source conditions that produce only the small clusters $(ser)_nH^+$, *n*=1, 2, 3, which are injected into the drift cell maintained at the temperature indicated [54]. It can be seen that the trimer dissociates just above room temperature and disappears from the mass spectrum recorded after the ions have left the drift cell. The ion mobility spectrum recorded at an m/z of 211 (serine dimer) at 334 K (Fig. 5b) shows that there is a slow component at longer drift times stemming from the larger trimer that dissociated during the drift time into the dimer. At 464 K there is a slow component for m/z=106 (Fig. 5c) indicating that the dimer is dissociating into the monomer at this temperature. Analyzing the ion mobility and mass spectra as a function of temperature yields a trimer dissociation threshold of ~4 kcal mol^{-1}. The dimer is much more stable and dissociates only at temperatures near 450 K and above. The dimer binding energy is determined to be ~20 kcal mol^{-1}.

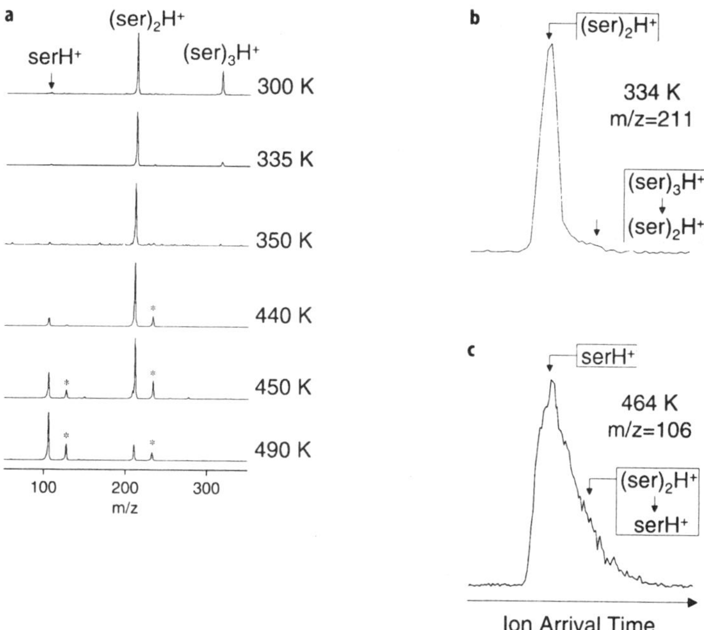

Fig. 5. **a** Mass spectra of L-serine obtained after ions spent ~200 μs in the drift cell at the temperature indicated. With increasing temperature first the trimer $(ser)_3H^+$ and then the dimer $(ser)_2H^+$ disappear from the spectrum; Relative intensities of peaks corresponding to sodiated species (marked with *) are not reproducible. **b** 334 K ion arrival time distribution (ATD) of dimer ions exiting the cell (m/z=211). **c** 464 K ATD of serine monomer (m/z=106)

4.2.2
Bradykinin Dimer

It has been found by ESI mass spectrometry techniques that singly protonated small peptides tend to form aggregates of the type $(nM+nH)^{n+}$ [22, 55–59]. Dissociation of such aggregates into smaller units can be studied by ion mobility techniques. Figure 6 shows ion mobility data for m/z=1061 of bradykinin recorded at different temperatures [22]. The two large features present at 441 K are the monomer $(M+H)^+$ and the dimer $(2M+2H)^{2+}$ with identical m/z=1061 ratio. The dimer feature disappears from the spectrum at a temperature of >500 K. The data at 463 K shows a fill-in between the monomer and dimer feature indicating that dissociation is occurring during the drift time. A fraction of the dimer ions, starting out with the larger dimer mobility, converted into two monomer units drifting for the remaining time with a smaller monomer mobility. The $(M+H)^+$ mobility is smaller than the $(2M+2H)^{2+}$ mobility because the ion mobility is proportional to the charge. Therefore singly charged ions drift half as quickly as doubly charged ions of same size. However, the dimer does not have the same size as the monomer, but the cross section is apparently less than twice that of the monomer ($2^{2/3}$ is expected for a sphere).

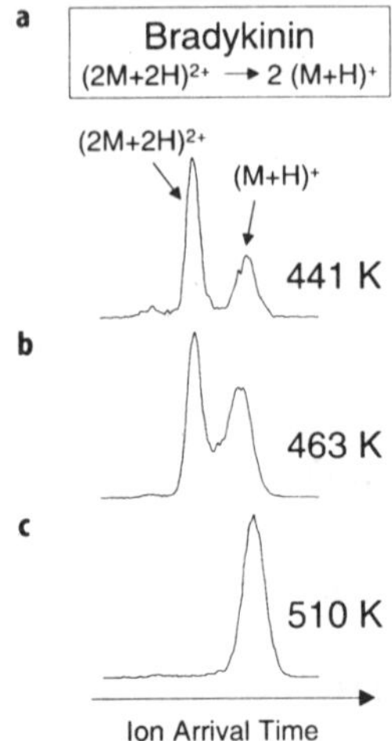

Fig. 6a–c. Ion mobility spectra of bradykinin ions at m/z=1061: **a** the two features seen at 441 K are the singly protonated monomer $(M+H)^+$ and the doubly protonated dimer $(2M+2H)^+$; **b** at a cell temperature of 463 K the dimer dissociates into two monomer units accounting for the fill-in between the two peaks; **c** at 510 K all of the dimer ions have disappeared upon exiting the cell

Fitting theoretical ion arrival time distributions to the experimental data where the rate constant for dissociation is the only adjustable parameter yields the Arrhenius plot shown in Fig. 7. The resulting dimer binding energy is ~30 kcal mol^{-1} and the pre-exponential factor is $\sim 10^{17}\ s^{-1}$.

4.2.3
PET Trimer

PET, poly(ethylene terephthalate), the primary constituent of a commonly used type of plastic, is composed of the rigid planar terephthalic acid units ($HOCO\text{-}C_6H_4\text{-}COOH$), which are esterified with the flexible ethylene glycol links ($HOCH_2\text{-}CH_2OH$):

$$HO\!\left[CH_2CH_2\!-\!O\!-\!\overset{O}{\overset{\|}{C}}\!-\!C_6H_4\!-\!\overset{O}{\overset{\|}{C}}\!-\!O\right]_n\!CH_2CH_2\text{-}OH$$

In a typical MALDI mass spectrum of PET samples a series of equally spaced peaks can readily be assigned as $[HO\text{-}(C_{10}H_8O_4)_n\text{-}C_2H_4OH]Na^+$ [60, 61]. Tuning the mass spectrometer to one oligomer size, n=3, and injecting these ions into the ion mobility cell at 300 K yields one narrow peak in the ion mobility spectrum (Fig. 8, left panel). The corresponding cross section is 175 Å^2. This value is right in between the values obtained theoretically for two plausible families of low energy conformations (164 and 182 Å^2) calculated by molecular mechanics

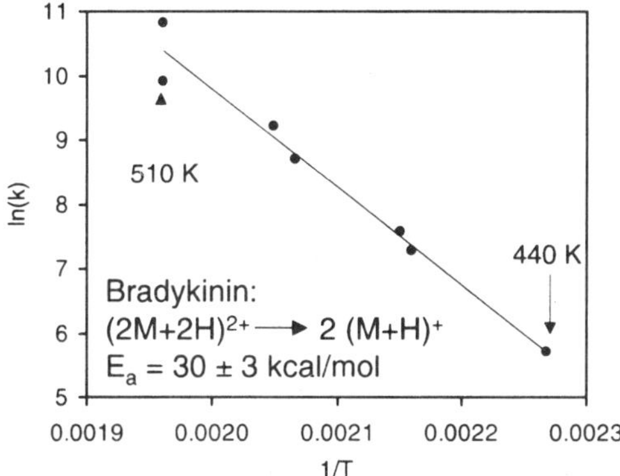

Fig. 7. Arrhenius plot for the dissociation of bradykinin dimer ions $(2M+2H)^{2+}$. Rate constant k in s^{-1}, temperature T in K

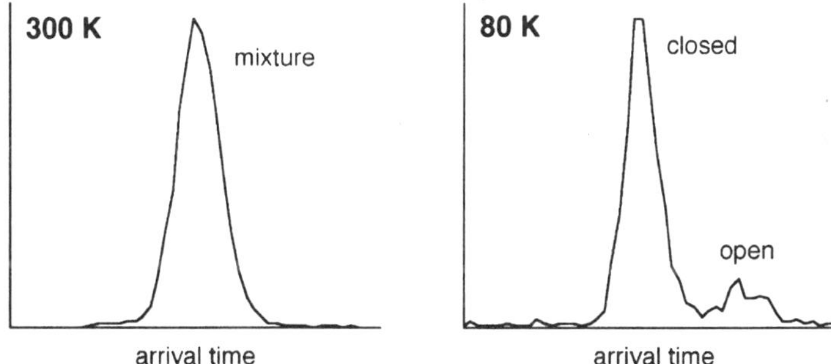

Fig. 8. Ion arrival time distributions of the sodiated PET trimer. The closed and open geometries (see Fig. 9) interconvert rapidly at 300 K (*left*), but are frozen out at 80 K (*right*)

techniques [60]. Typical representatives of the two families are shown in Fig. 9. In the closed structure (164 Å^2) the sodium ion is bound to oxygen atoms of both terminal $-COOC_2H_4OH$ groups, whereas this is not the case in the open (182 Å^2) form. In both families two of the three rigid π-systems stack up, whereas the third one is either bent in towards the sodium ion (closed structure) or extends out away from the sodium ion (open structure).

The 80 K ion mobility spectrum, shown in the right panel of Fig. 8, exhibits two distinct features, which are readily assigned to the two families of closed and open structures [60]. This means that interconversion between the two families of structures is frozen out at 80 K, but is rapid at 300 K. Data taken between 110 and 190 K show that the two peaks present at low temperature melt into one at higher temperatures. Analysis analogous to the previous examples yields an activation barrier for the open-to-closed conversion of 1.6 kcal mol^{-1}. The open-to-closed reaction turns on more rapidly than the reverse reaction, and the closed form is found to be ~0.5 kcal mol^{-1} more stable than the open form [60].

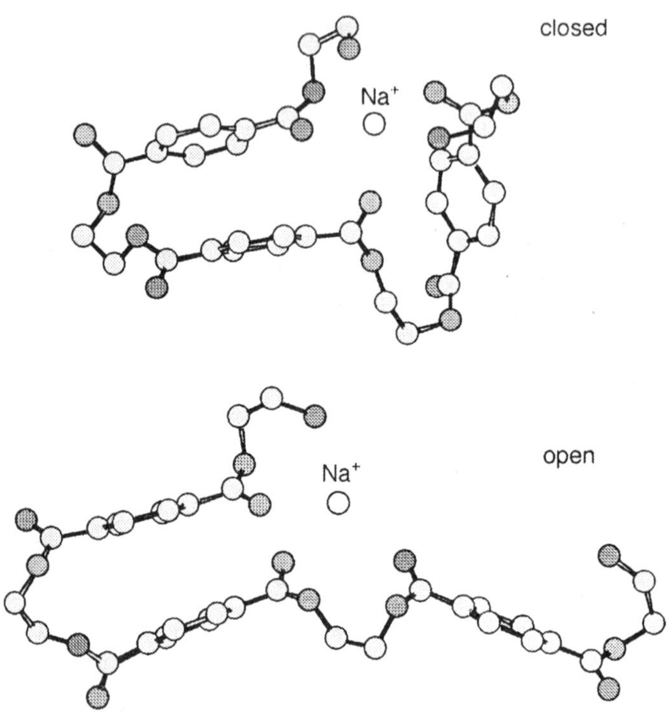

Fig. 9. Closed and open structure for the sodiated PET trimer obtained by molecular modeling

4.3
Equilibrium: Ligand Binding Energy

Ion mobility equipment can be used to obtain precise thermochemical data. The concepts briefly outlined in Sect. 2.3 are here applied to the example of hydration of protonated arginine methyl ester (argOMe)H^+. Molecular mechanics calculations indicate that the water molecule only binds to the guanidinium group and not (simultaneously) to any of the other functional groups [62]. Therefore, (argOMe)H^+ is a good model compound to measure the intrinsic binding energy of water bound to a guanidinium group in an arginine residue of a peptide or protein.

In the experiment the (argOMe)H^+ ions are formed by ESI and injected into an ion mobility cell, which is filled with a defined pressure of water of 0.1–2 torr [22, 62]. The ions undergo collisions with water molecules and an equilibrium of the form

$$(\text{argOMe})H^+ + H_2O \rightleftharpoons (\text{argOMe})H^+ \cdot H_2O \tag{5}$$

is established. Product and reactant ions are analyzed by recording a mass spectrum after ions have left the drift cell. The neutral reactant H_2O enters the equation as the water pressure $p(H_2O)$:

$$K_{eq} = \frac{[(\text{argOMe})\text{H}^+ \cdot (\text{H}_2\text{O})_n]}{[(\text{argOMe})\text{H}^+ \cdot (\text{H}_2\text{O})_{n-1}]\, p(\text{H}_2\text{O})} \tag{6}$$

The presence of an equilibrium is confirmed by verifying that mass spectra recorded at different drift voltages (resulting in different drift times) are identical. Measuring K_{eq} as a function of temperature and generating a van't Hoff plot (Fig. 10) results in values of $\Delta H°=-9$ kcal mol^{-1} and $\Delta S°=-17$ cal mol^{-1} K^{-1} for the reaction (5). These values are relatively small compared to values measured for alkyl amines and other amino acids [63, 64] (Table 1) indicating that water binds less strongly to a guanidinium group than to an ammonium group. This can be rationalized by the fact that the charge is more delocalized in the larger guanidinium group.

The water binding energy of protonated arginine is measured [62] to be very similar to that of (argOMe)H$^+$ (Table 1) suggesting that argH$^+$ is not a zwitterion. If it were, the H_2O molecule would most probably make as strong a bond to the protonated amine as it does with the simple alkyl amines and with pro and val. Instead, a weak bond is observed indicating the water binds to a protonated arginine side chain. The issue is somewhat ambiguous, however, because if there is a zwitterion the deprotonated carboxylate group could potentially reduce the

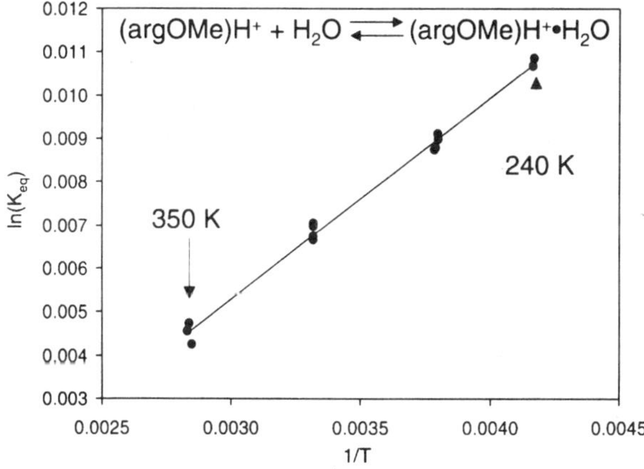

Fig. 10. Van't Hoff plot for the addition of one water molecule to protonated arginine methyl ester

Table 1. Standard enthalpy and entropy of hydration for the protonated species indicated

	$-\Delta H°$ (kcal mol^{-1})	$-\Delta S°$ (cal mol^{-1} K^{-1})
Alkyl amines [64]	~16	~23
pro [63]	18.9	36.8
val [63]	19.3	36.3
arg [62]	9.0	15.4
argOMe [62]	9.2	17.2

amount of positive charge on the amine and therefore reduce the water binding energy of that binding site. Calculations need to be done to draw unambiguous conclusions.

Whatever the detailed interpretation of these results may be in this particular case, the point to make here is that the ion mobility setup provides the opportunity to obtain true thermochemical data, as experiments are carried out under thermal equilibrium conditions.

5 Overview of Applications

5.1 Conformations of Flexible Molecules

For sufficiently small flexible molecules conformational space is fairly limited and only a small number of conformations are energetically reasonable, which makes interpretation of ion mobility data in these cases generally straightforward. If multiple peaks are observed in the ion arrival time distribution (ATD), structural assignment of the peaks is usually unambiguous with help of model structures. If two peaks are present at a given temperature, it can be concluded that the energy barrier between the corresponding two conformations is high compared to the thermal energy available to the molecule. If the two peaks melt into one at higher temperatures, temperature dependent studies can be used to determine the barrier height. Examples for such small flexible molecules with a small number of resolved conformations are oligomers of relatively rigid units. Examples in the literature of this type of systems are the di- and trinucleotides [65] and the trimer of PET (see also Sect. 4.2.3) [60, 61].

A common structural motif of flexible small and medium sized systems, such as synthetic polymers and peptides, is that the charge carrying unit (alkali ion or protonated group) is buried in the interior of the molecule and is well self-solvated by electron rich functional groups [12, 66–71]. As systems get larger multiple shells of self-solvation are observed yielding globular molecular shapes [72]. Functional groups in the second solvation shell are not as tightly bound to the charged center and undergo a fair amount of thermal motion. This is clearly observed as an increase in cross section when temperatures are raised by several 100 K above room temperature [70]. In certain cases structural motifs other than charge solvation can become predominant once the first solvation shell is filled. For instance, in cationized poly(styrene) oligomers two benzene rings fully solvate the metal cation [73]. The remaining benzene rings stack up on top of each other, giving the molecule a distinct "secondary" structure other than spherical. Another very nice example, where maximizing charge solvation is not the only geometry determining factor, is α-helix formation in peptides (10–20 amino acids). In this case hydrogen bonding, alignment of the helix dipole with the charge, and sterical effects are the driving forces for secondary structure formation [74, 75].

Exploring α-helix formation as a function of peptide primary structure enabled Jarrold and coworkers to establish intrinsic helix propensity data for a number of amino acids [76]. Using these data peptides could be designed where the globular charge solvation structure is comparable in stability to the α-helix [77–79]. Temperature dependent studies on these systems allowed examination of the helix–globule transition and determination of the corresponding barrier height. These studies also showed that there is an intermediate in the transition, possibly a helix, which is less tightly twisted than the α-helix [78].

For larger peptides (~30 amino acids) secondary as well as tertiary structure can be observed. For $ala_nH_3^{3+}$ (n=25–35) two helical sections connected by a loop appear to form an antiparallel helical bundle [80].

Another structural motif, the zwitterion, is potentially present in molecules that contain both acidic and basic functional groups, a situation present in many biomolecules [81–90]. However, zwitterions are intrinsically not stable (for common acids and bases) and require a large amount of stabilization by (self-) solvation and Coulomb interaction with other charges. Zwitterions and non-zwitterions of flexible molecules sometimes assume sufficiently different geometries that the ion mobility technique can be used to distinguish between the two forms. The best examples reported to date are the sodiated oligoglycines that were found not to be zwitterions [91].

Ion mobility data have been obtained for peptide aggregates $(nM+zH)^{z+}$, that are formed in electrospray ionization sources [22, 55–59]. Multimers of larger helical peptides show interesting tertiary structure [58, 59]. In the Ac–lysH^+–ala_{19} dimer the charge of one peptide stabilizes the helix of the other in a head to toe fashion, yielding a helical bundle. However, in the Ac–ala_{19}–lysH^+ dimer favorable interaction of the charge of one unit with the helix of the other unit leads to a nearly collinear or vee-shaped arrangement of the two peptides [58]. A similar situation is found in the Ac–$(gly\text{–}ala)_7$–lysH^+ dimer and trimer. In the trimer the three charges are in the center and the three helices extend outwards away from each other forming an interesting pinwheel structure [59].

The largest molecules studied by the ion mobility technique to date are proteins with masses of up to ~15,000 u [92–102]. The most thoroughly studied protein is cytochrome *c* where a number of folded and unfolded structures have been observed [92–95]. Ion mobility data suggests that the most tightly folded structures, observed for the low charge states (≤+5), are very similar to the native structure and that the unfolded structures (charge states ~10–15) agree with structures where the tertiary structure is lost, but most of the secondary structure is in place. The highest charge states (~+20) are very stretched out and essentially no secondary structure is present. Unfolding of the protein has been followed for the charge states +5 and +6 by raising the temperature from 300 to near 600 K. A series of transitions is observed in this temperature range from near fully folded at 300 K (cross section of ~1100 $Å^2$) to substantially unfolded (~2000 $Å^2$) above 500 K. Other proteins that show multiple conformations include ubiquitin [99–101], lysozyme [98], and apomyoglobin [96, 97]. BPTI, on the other hand, is a protein tied together by three covalent disulfide bridges and only one compact structure is observed in ion mobility experiments [95].

In our lab we have looked at calmodulin, a 146 residue protein involved in calcium regulation. This protein also shows multiple conformations at low charge states (Fig. 11) and has the interesting property that essentially identical cross

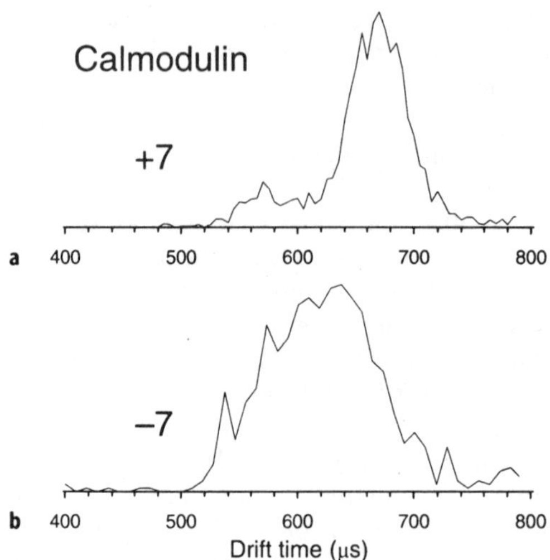

Fig. 11a,b. Ion arrival time distributions of charge states +7 and –7 of the calcium-free protein calmodulin (T=300 K, p=4.9 torr, V_D=90 V, drift length 4.5 cm). For +7 two distinctly different families of structures are present, one with larger cross sections (main peak) and one with smaller cross sections (shoulder to the left). The broad peak for –7 indicates an even distribution of structures with smaller and larger cross sections

sections are observed for equivalent charge states of negative and positive ions [102]. Modeling needs to be done to understand these results, but that is not a simple task for a protein this size.

The study of hydrated biomolecules has only just begun and very few papers have been published on that topic [103–107]. Hydration enthalpies and entropies for the peptides LHRH and bradykinin [108] and for the protein BPTI [106, 107] have been measured and the amount of water addition to folded vs unfolded states of cytochrome *c* has been studied [103–105]. At this point it is too early to draw general conclusions but it is already apparent that only a few water molecules can have significant structural consequences [107, 108].

Adding deuterated water to the ion mobility cell induces H/D-exchange of labile hydrogen atoms in the ion, similar to H/D-exchange observed in FT-ICR mass spectrometers [109–116]. However, this type of ion mobility application has not been widely used. The method has been demonstrated for cytochrome *c* [94, 117], but it is not clear how useful the technique will generally be to obtain structural information of ions.

5.2 Geometries of Clusters

The first applications of ion mobility methods to obtain structural information of polyatomic ions were on cluster ions, that is carbon [1–5] and silicon [6, 7] cluster ions. In carbon clusters atoms are bound to each other by covalent bonds, leading to structures like chains for small clusters (≤10 atoms), rings, polycyclic planar and cup shaped structures, and cage structures, the fullerenes for larger clusters (>30 atoms). Carbon clusters were still an area of active research during the last five years and new and more complex structures have been found, predominantly due to increased ion mobility resolution [118–123]. New structures include possible intermediates for fullerene formation, as well as multimers of C_{60} fullerene. For instance, for fullerene dimers multiple geometries of fused monomer cage units have been found including two cages connected by a chain [121–123]. Recent work on carbon based cage structures also include studies on silicon, niobium, and scandium containing clusters SiC_n^+, $Nb_xC_{60}^+$ (x=1–5), and $Sc_xC_n^+$ (x=1–3). The SiC_n^+ clusters form virtually identical clusters like C_n^+ except that one carbon atom is replaced by a silicon atom [124]. The $Nb_xC_{60}^+$ clusters are composed of a fullerene C_{60} cage coated by a Nb_x cluster, which can be annealed to $Nb_{x-1}(Nb@C_{60})^+$, a cluster with one Nb atom in the interior of the C_{60} cage [125]. In the ScC_n^+ (n~80) fullerene clusters Sc sits in the interior of the C_n cage [126]. The $Sc_2C_n^+$ clusters show two different structures, one where Sc_2 is inside the C_n cage and one where scandium carbide Sc_2C_2 is inside a C_{n-2} cage. Similarly, the $Sc_3C_n^+$ mobilities agree with a $(Sc_3C_2)@C_{n-2}^+$ structure.

Clusters of the semiconductor elements Si and Ge are much more dense than carbon clusters, but they are not spherical, either, as expected for closed packed atomic spheres. Si_n^+ and Ge_n^+ clusters are prolate with geometries based on stacked tricapped trigonal prisms [127–131]. At a certain cluster size (n~25 for Si_n^+) a structural transition occurs from prolate (n<25) to near spherical (n>25). Interestingly, clusters of tin, which is a metal at room temperature, exhibit very similar structures as Si and Ge, indicating that they are semiconductors as well [131, 132]. Bulk Sn does have a semiconductor form (α-tin), that has the same diamond lattice as Si and Ge. Typical metal clusters appear to pack as tightly as possible and exhibit near spherical shapes as observed for the lead [131, 133], indium [134], and gold [135] clusters. However, the smaller gold clusters Au_n^+ (n≤7) are completely planar [135].

Another class of systems studied by ion mobility is salt clusters, $(NaCl)_nCl^-$ (n<50) [136, 137]. Steps in the ion mobility values as a function of n indicate completion of the 4×3×3, 4×4×3, 4×4×4, and 5×4×4 cuboids. For n>30 several isomers are observed at room temperature, which can be assigned to near cubic and more planar or elongated geometries. At 70°C the less cubic geometries anneal into cuboids with activation barriers of <0.6 eV.

Recent hydration studies of small $(NaI)_nNa^+$ clusters (n=1, 2) indicate that dissociation of $[(NaI)_nNa^+](H_2O)_x$ into $Na^+(H_2O)_y$ ("dissolution") occurs readily at room temperature for x≥6 when n=1 and x≥1 when n=2 [138]. For $[(NaI)Na^+](H_2O)_x$ theory indicates that at least four water molecules are required to solvate one Na^+ ion and form a contact ion pair (i.e., loosen it from the remaining NaI molecule). However, two additional water molecules (six in total) are required to make the $Na^+(H_2O)_4{\cdot}NaI(H_2O)_{x-4}$ cluster a candidate for dissociation under typical experimental conditions [138].

Finally, some ion mobility data is available for a number of clusters composed of an ion and small neutral molecules. Such clusters include $H_3O^+(H_2O)_3$, $NH_4^+(NH_3)_n$, n=1–3, $NO^+(CH_3COCH_3)_n$, n=2, 3 [139–141], complexes between protonated amines and polyethers [21], and ligated transition metal ions

[142–150]. The most unusual geometries among these types of clusters are clearly the multiple-decker sandwich structures observed for vanadium ion-benzene clusters $V_n(C_6H_6)_m^+$ [142, 151–157]. One interesting finding is that "early" transition metals all appear to form sandwich compounds with benzene but for the late metals the benzene adducts coat a central metal cluster [158].

Ligated transition metal ions have been extensively studied by Bowers and coworkers by measuring ligand – metal binding energies ΔH° and entropies ΔS° (as in Sect. 4.3) [143–149]. If ΔH° and ΔS° values are studied as a function of number of ligands added, important information about the electronic structure of the metal center, the arrangement of the ligands, and ligand sigma bond activation can be obtained. For instance, the Fe^+–(CH_4) bond is relatively weak (18 kcal mol^{-1}) compared to the bond energy of the second methane in the $Fe^+(CH_4)_2$ cluster (26 kcal mol^{-1}), indicating that there is a spin change in the Fe^+ ion upon addition of the first ligand requiring a substantial promotion energy [143]. Large binding entropies (~20 cal mol^{-1} K^{-1}) for the first four CH_4 ligands of Fe^+ and Ni^+ and small values (~10 cal mol^{-1} K^{-1}) for the fifth and sixth ligand give information about how the ligands are arranged: four ligands in the first solvation shell and the fifth and higher in a second shell [143]. This conclusion based on entropy data is consistent with very small binding energies measured for the fifth and sixth ligands (~2 kcal mol^{-1}).

Sigma bond activation is observed in several cases for ligands bound to transition metals. For instance, on the basis of binding energy patterns and supporting calculations it was found that H_2 bound to Zr^+ has an H–Zr^+–H structure, where the Zr^+ ion inserted into the H–H sigma bond [149]. The following two H_2 ligands in (H–Zr^+–H)$(H_2)_2$ bind as H_2 molecules. Interestingly uninsertion is observed when more ligands are added yielding $Zr^+(H_2)_n$ clusters for n>4. Sigma bond activation is also observed for hydrocarbon molecules bound to transition metals as in $Co^+(CH_4)_n$, $Ti^+(CH_4)_n$, $Ti^+(C_2H_4)$, $V^+(C_2H_4)$, $Ti^+(C_3H_8)$, $V^+(C_3H_8)$ [159–163], and interestingly for clusters of B^+ [164] and possibly Al^+ [165].

6 Conclusions

We have shown that combining ion mobility spectrometry (IMS) equipment with mass spectrometry (MS) provides a powerful tool to examine the three-dimensional structure of polyatomic ions by measuring collision cross sections of mass identified ions. The technique is particularly useful in conjunction with molecular modeling or electronic structure calculations. Further, we have reviewed applications where the IMS-MS equipment is used to obtain kinetic and thermochemical data of ions.

IMS-MS applications published in the literature can be grouped into two categories. The first contains studies of conformations of flexible molecules. Such flexible molecules include synthetic polymers and biopolymers such as peptides, proteins, and oligonucleotides. The studies of the second category deal with the geometry of cluster ions such as carbon clusters, semiconductor clusters, metal clusters, salt clusters, ion-ligand clusters. The major conclusions regarding structure of these systems are reviewed in Sect. 5.

Acknowledgements. The support of the Air Force Office of Scientific Research under grant F49620-99-1-0048 and the support of the National Science Foundation under grant CHE-0140215 are gratefully acknowledged. The authors would also like to thank their coworkers at UCSB: Dr. P.R. Kemper, Dr. M. Witt, Dr. P. Barran, Mr. Q. Zhang, Ms. S. Bernstein, and Mr. D.F. Liu for supplying unpublished data presented in this review.

7
References

1. Von Helden G, Hsu MT, Kemper PR, Bowers MT (1991) J Chem Phys 95:3835
2. Von Helden G, Hsu MT, Kemper PR, Bowers MT (1992) Novel forms of carbon. MRS Symposium Series. 270:117
3. Von Helden G, Kemper PR, Gotts NG, Bowers MT (1993) Science 259:1300
4. Von Helden G, Hsu MT, Gotts NG, Kemper PR, Bowers MT (1993) Chem Phys Lett 204:15
5. Von Helden G, Gotts NG, Bowers MT (1993) J Am Chem Soc 115:4363
6. Jarrold MF, Constant VA (1991) Phys Rev Lett 67:2994
7. Jarrold MF, Bower JE (1992) J Chem Phys 96:9180
8. Clemmer DE, Jarrold MF (1997) J Mass Spectrom 32:577
9. Mason EA, McDaniel EW (1988) Transport properties of ions in gases. Wiley, New York
10. Wyttenbach T, von Helden G, Batka JJ, Carlat D, Bowers MT (1997) J Am Soc Mass Spectrom 8:275
11. Wyttenbach T, Witt M, Bowers MT (2000) J Am Chem Soc 122:3458
12. Gidden J, Jackson AT, Scrivens JH, Bowers MT (1999) Int J Mass Spectrom 188:121
13. Mesleh MF, Hunter JM, Shvartsburg AA, Schatz GC, Jarrold MF (1996) J Phys Chem 100:16,082
14. Shvartsburg AA, Jarrold MF (1996) Chem Phys Lett 261:86
15. Shvartsburg AA, Liu B, Jarrold MF, Ho KM (2000) J Chem Phys 112:4517
16. Case DA, Pearlman DA, Caldwell JW, Cheatham TE III, Ross WS, Simmerling CL, Darden TA, Merz KM, Stanton RV, Cheng AL, Vincent JJ, Crowley M, Tsui V, Radmer RJ, Duan Y, Pitera J, Massova I, Seibel GL, Singh UC, Weiner PK, Kollman PA (1999) AMBER 6. University of California, San Francisco
17. Gidden J, Kemper PR, Shammel E, Fee DP, Anderson S, Bowers MT (2003) Int J Mass Spectrom 222:63
18. Shammel E, Gidden J, Fee DP, Kemper PR, Anderson S, Bowers MT (2003) Int J Mass Spectrom 222:63
19. Hoaglund-Hyzer CS, Clemmer DE (2001) Anal Chem 73:177
20. Hoaglund CS, Valentine SJ, Clemmer DE (1997) Anal Chem 69:4156
21. Creaser CS, Griffiths JR, Stockton BM (2000) Eur J Mass Spectrom 6:213; Creaser CS, Benyezzar M, Griffiths JR, Stygall JW (2000) Anal Chem 72:2724
22. Wyttenbach T, Kemper PR, Bowers MT (2001) Int J Mass Spectrom 212:13
23. Nestler V, Betz B, Warneck P (1977) Ber Bunsen Phys Chem 81:13
24. Thomas R, Barassin A, Burke RR (1978) Int J Mass Spectrom 28:275
25. Johnson R, Biondi MA, Hayashi M (1982) J Chem Phys 77:2545
26. Bohringer H, Arnold F (1983) Int J Mass Spectrom 49:61
27. Kaneko Y, Megill LR, Hasted JB (1966) J Chem Phys 45:3741
28. Federer W, Ramler H, Villinger H, Lindinger W (1985) Phys Rev Lett 54:540
29. Twiddy ND, Mohebati A, Tichy M (1986) Int J Mass Spectrom 74:251
30. Adams NG, Smith D (1988) Adv At Mol Phys 23:1
31. Viggiano AA, Morris RA, Dale F, Paulson JF, Giles K, Smith D, Su T (1990) J Chem Phys 93:1149
32. McEwan MJ (1992) In: Adams NG, Babcock LM (eds) Advances in gas phase ion chemistry. JAI, Greenwich, CT, vol 1, p 1
33. Krishnamurthy M, De Gouw JA, Bierbaum VM, Leone SR (1996) J Phys Chem 100:14,908
34. McFarland M, Albritton DL, Fehsenfeld FC, Ferguson EE, Schmeltekopf AL (1973) J Chem Phys 59:6610

35. Wu C, Siems WF, Asbury GR, Hill HH (1998) Anal Chem 70:4929
36. Dugourd P, Hudgins RR, Clemmer DE, Jarrold MF (1997) Rev Sci Instr 68:1122
37. Kemper PR, Bowers MT (1990) J Am Soc Mass Spectrom 1:197
38. Bluhm BK, Gillig KJ, Russell DH (2000) Rev Sci Instr 71:4078
39. Hoaglund CS, Valentine SJ, Sporleder CR, Reilly JP, Clemmer DE (1998) Anal Chem 70:2236
40. Valentine SJ, Kulchania M, Barnes CAS, Clemmer DE (2001) Int J Mass Spectrom 212:97
41. Steiner WE, Clowers BH, Fuhrer K, Gonin M, Matz LM, Siems WF, Schultz AJ, Hill HH (2001) Rapid Commun Mass Spectrom 15:2221
42. Gillig KJ, Ruotolo B, Stone EG, Russell DH, Fuhrer K, Gonin M, Schultz AJ (2000) Anal Chem 72:3965
43. Stone EG, Gillig KJ, Ruotolo BT, Russell DH (2001) Int J Mass Spectrom 212:519
44. Woods AS, Koomen JM, Ruotolo BT, Gillig KJ, Russel DH, Fuhrer K, Gonin M, Egan TF, Schultz JA (2002) J Am Soc Mass Spectrom 13:166
45. Kemper PR, Bowers MT (2002) Instrumental paper on TOF-drift cell-quad setup (to be published), see also [17, 18]
46. Kemper PR, Bowers MT (1990) J Am Chem Soc 112:3231
47. Matz LM, Asbury GR, Hill HH (2002) Rapid Commun Mass Spectrom 16:670
48. Witt M, Bowers MT. Unpublished results; experimental details as in [11]
49. Gatland IR (1974) In: McDaniel EW, McDowell MRC (eds) Case studies in atomic collision physics. North-Holland, Amsterdam, vol 4, p 369
50. Koch KJ, Gozzo FC, Zhang DX, Eberlin MN, Cooks RG (2001) Chem Commun 1854
51. Cooks RG, Zhang DX, Koch KJ, Gozzo FC, Eberlin MN (2001) Anal Chem 73:3646
52. Counterman AE, Clemmer DE (2001) J Phys Chem B 105:8092
53. Julian RR, Hodyss R, Kinnear B, Jarrold MF, Beauchamp JL (2002) J Phys Chem B 106:1219
54. Wyttenbach T, Barran P, Bowers MT. Unpublished results
55. Counterman AE, Hilderbrand AE, Barnes CAS, Clemmer DE (2001) J Am Soc Mass Spectrom 12:1020
56. Counterman AE, Valentine SJ, Srebalus CA, Henderson SC, Hoaglund CS, Clemmer DE (1998) J Am Soc Mass Spectrom 9:743
57. Lee SW, Beauchamp JL (1999) J Am Soc Mass Spectrom 10:347
58. Hudgins RR, Jarrold MF (1999) J Am Chem Soc 121:3494
59. Kaleta DT, Jarrold MF (2002) J Am Chem Soc 124:1154
60. Gidden J, Wyttenbach T, Batka JJ, Weis P, Jackson AT, Scrivens JH, Bowers MT (1999) J Am Soc Mass Spectrom 10:883
61. Gidden J, Wyttenbach T, Batka JJ, Weis P, Jackson AT, Scrivens JH, Bowers MT (1999) J Am Chem Soc 121:1421
62. Wyttenbach T, Bowers MT. Unpublished results
63. Meot-Ner M, Field FH (1974) J Am Chem Soc 96:3168
64. Blades AT, Klassen JS, Kebarle P (1996) J Am Chem Soc 118:12,437
65. Gidden J, Bushnell JE, Bowers MT (2001) J Am Chem Soc 123:5610
66. von Helden G, Wyttenbach T, Bowers MT (1995) Science 267:1483
67. von Helden G, Wyttenbach T, Bowers MT (1995) Int J Mass Spectrom Ion Proc 146:349
68. Wyttenbach T, von Helden G, Bowers MT (1996) J Am Chem Soc 118:8355
69. Wyttenbach T, von Helden G, Bowers MT (1997) Int J Mass Spectrom Ion Proc165:377
70. Gidden J, Wyttenbach T, Jackson AT, Scrivens JH, Bowers MT (2000) J Am Chem Soc 122:4692
71. Wyttenbach T, Bushnell JE, Bowers MT (1998) J Am Chem Soc 120:5098
72. Hudgins RR, Jarrold MF (2000) J Phys Chem B 104:2154
73. Gidden J, Bowers MT, Jackson AT, Scrivens JH (2002) J Am Soc Mass Spectrom 13:499
74. Hudgins RR, Ratner MA, Jarrold MF (1998) J Am Chem Soc 120:12,974
75. Kohtani M, Kinnear BS, Jarrold MF (2000) J Am Chem Soc 122:12,377
76. Kinnear BS, Jarrold MF (2001) J Am Chem Soc 123:7907
77. Kinnear BS, Hartings MR, Jarrold MF (2002) J Am Chem Soc 124:4422

78. Kinnear BS, Hartings MR, Jarrold MF (2001) J Am Chem Soc 123:5660
79. Kaleta DT, Jarrold MF (2001) J Phys Chem B 105:4436
80. Counterman AE, Clemmer DE (2001) J Am Chem Soc 123:1490
81. Wyttenbach T, Witt M, Bowers MT (1999) Int J Mass Spectrom 183:243
82. Wyttenbach T, Bowers MT (1999) J Am Soc Mass Spectrom 10:9
83. Jockusch RA, Lemoff AS, Williams ER (2001) J Phys Chem A 105:10,929
84. Strittmatter EF, Lemoff AS, Williams ER (2000) J Phys Chem A 104:9793
85. Jockusch RA, Price WD, Williams ER (1999) J Phys Chem A 103:9266
86. Price WD, Jockusch RA, Williams ER (1997) J Am Chem Soc 119:11,988
87. Cerda BA, Wesdemiotis C (2000) Analyst 125:657
88. Talley JM, Cerda BA, Ohanessian G, Wesdemiotis C (2002) Chem Eur J 8:1377
89. Julian RR, Beauchamp JL, Goddard WA (2002) J Phys Chem A 106:32
90. Julian RR, Hodyss R, Beauchamp JL (2001) J Am Chem Soc 123:3577
91. Wyttenbach T, Bushnell JE, Bowers MT (1998) J Am Chem Soc 120:5098
92. Mao Y, Ratner MA, Jarrold MF (1999) J Phys Chem B 103:10,017
93. Mao Y, Woenckhaus J, Kolafa J, Ratner MA, Jarrold MF (1999) J Am Chem Soc 121:2712
94. Valentine SJ, Clemmer DE (1997) J Am Chem Soc 119:3558
95. Shelimov KB, Clemmer DE, Hudgins RR, Jarrold MF (1997) J Am Chem Soc 119:2240
96. Shelimov KB, Jarrold MF (1997) J Am Chem Soc 119:2987
97. Hoaglund-Hyzer CS, Counterman AE, Clemmer DE (1999) Chem Rev 99:3037
98. Valentine SJ, Anderson JG, Ellington AD, Clemmer DE (1997) J Phys Chem B 101:3891
99. Valentine SJ, Counterman AE, Clemmer DE (1997) J Am Soc Mass Spectrom 8:954
100. Li JW, Taraszka JA, Counterman AE, Clemmer DE (1999) Int J Mass Spectrom 187:37
101. Badman ER, Hoaglund-Hyzer CS, Clemmer DE (2002) J Am Soc Mass Spectrom 13 :719
102. Barran P, Wyttenbach T, Bowers MT. Unpublished results
103. Mao Y, Ratner MA, Jarrold MF (2001) J Am Chem Soc 123:6503
104. Fye JL, Woenckhaus J, Jarrold MF (1998) J Am Chem Soc 120:1327
105. Woenckhaus J, Mao Y, Jarrold MF (1997) J Phys Chem B 101:847
106. Woenckhaus J, Hudgins RR, Jarrold MF (1997) J Am Chem Soc 119:9586
107. Mao Y, Ratner MA, Jarrold MF (2000) J Am Chem Soc 122:2950
108. Barran P, Wyttenbach T, Bernstein S, Liu DF, Bowers MT. To be published; Liu D, Wyttenbach T, Barran PE, Bowers MT (2003) J Am Chem Soc (submitted)
109. Schlosser M, Mongin F, Porwisiak J, Dmowski W, Buker HH, Nibbering NMM (1998) Chem Eur J 4:1281
110. Nourse BD, Hettich RL, Buchanan MV (1993) J Am Soc Mass Spectrom 4:296
111. Lee SW, Lee HN, Kim HS, Beauchamp JL (1998) J Am Chem Soc 120:5800
112. Winger BE, Lightwahl KJ, Rockwood AL, Smith RD (1992) J Am Chem Soc 114:5897
113. Freitas MA, Marshall AG (2001) J Am Soc Mass Spectrom 12:780
114. McLafferty FW, Guan ZQ, Haupts U, Wood TD, Kelleher NL (1998) J Am Chem Soc 120:4732
115. Green MK, Lebrilla CB (1998) Int J Mass Spectrom Ion Proc 175:15
116. Zhang X, Ewing NP, Cassady CJ (1998) Int J Mass Spectrom Ion Proc 175:159
117. Valentine SJ, Clemmer DE (2002) J Am Soc Mass Spectrom 13:506
118. Dugourd P, Hudgins RR, Tenenbaum JM, Jarrold MF (1998) Phys Rev Lett 80:4197
119. Shvartsburg AA, Schatz GC, Jarrold MF (1998) J Chem Phys 108:2416
120. Shvartsburg AA, Hudgins RR, Dugourd P, Gutierrez R, Frauenheim T, Jarrold MF (2000) Phys Rev Lett 84:2421
121. Shvartsburg AA, Hudgins RR, Gutierrez R, Jungnickel G, Frauenheim T, Jackson KA, Jarrold MF (1999) J Phys Chem A 103:5275
122. Shvartsburg AA, Pederson LA, Hudgins RR, Schatz GC, Jarrold MF (1998) J Phys Chem A 102:7919
123. Shvartsburg AA, Hudgins RR, Dugourd P, Jarrold MF (1997) J Phys Chem A 101:1684
124. Fye JL, Jarrold MF (1997) J Phys Chem A 101:1836
125. Fye JL, Jarrold MF (1999) Int J Mass Spectrom 187:507

126. Sugai T, Inakuma M, Hudgins R, Dugourd P, Fye JL, Jarrold MF, Shinohara H (2001) J Am Chem Soc 123:6427
127. Hudgins RR, Imai M, Jarrold MF, Dugourd P (1999) J Chem Phys 111:7865
128. Shvartsburg AA, Liu B, Lu ZY, Wang CZ, Jarrold MF, Ho KM (1999) Phys Rev Lett 83:2167
129. Liu B, Lu ZY, Pan BC, Wang CZ, Ho KM, Shvartsburg AA, Jarrold MF (1998) J Chem Phys 109:9401
130. Ho KM, Shvartsburg AA, Pan BC, Lu ZY, Wang CZ, Wacker JG, Fye JL, Jarrold MF (1998) Nature 392:582
131. Shvartsburg AA, Hudgins RR, Dugourd P, Jarrold MF (2001) Chem Soc Rev 30:26
132. Shvartsburg AA, Jarrold MF (1999) Phys Rev A 60:1235
133. Shvartsburg AA, Jarrold MF (2000) Chem Phys Lett 317:615
134. Lerme J, Dugourd P, Hudgins RR, Jarrold MF (1999) Chem Phys Lett 304:19
135. Gilb S, Weis P, Furche F, Ahlrichs R, Kappes MM (2002) J Chem Phys 116:4094
136. Hudgins RR, Dugourd P, Tenenbaum JM, Jarrold MF (1997) Phys Rev Lett 78:4213
137. Dugourd P, Hudgins RR, Jarrold MF (1997) Chem Phys Lett 267:186
138. Zhang Q, Carpenter CJ, Kemper PR, Bowers MT (2003) J Am Chem Soc (in press)
139. Krishnamurthy M, deGouw JA, Ding LN, Bierbaum VM, Leone SR (1997) J Chem Phys 106:530
140. de Gouw JA, Krishnamurthy M, Bierbaum VM, Leone SR (1997) Int J Mass Spectrom Ion Proc 167:281
141. de Gouw JA, Ding LN, Krishnamurthy M, Lee HS, Anthony EB, Bierbaum VM, Leone SR (1996) J Chem Phys 105:10,398
142. Weis P, Kemper PR, Bowers MT (1997) J Phys Chem A 101:8207
143. Zhang Q, Kemper PR, Bowers MT (2001) Int J Mass Spectrom 210:265
144. Kemper PR, Weis P, Bowers MT, Maitre P (1998) J Am Chem Soc 120:13,494
145. Kemper PR, Weis P, Bowers MT (1998) Chem Phys Lett 293:503
146. Bushnell JE, Maitre P, Kemper PR, Bowers MT (1997) J Chem Phys 106:10,153
147. Kemper PR, Weis P, Bowers MT (1997) Int J Mass Spectrom Ion Proc 160:17
148. Weis P, Kemper PR, Bowers MT (1997) J Phys Chem A 101:2809
149. Bushnell JE, Kemper PR, van Koppen P, Bowers MT (2001) J Phys Chem A 105:2216
150. Leavell MD, Gaucher SP, Leary JA, Taraszka JA, Clemmer DE (2002) J Am Soc Mass Spectrom 13:284
151. Judai K, Sera K, Amatsutsumi S, Yagi K, Yasuike T, Yabushita S, Nakajima A, Kaya K (2001) Chem Phys Lett 334:277
152. Kurikawa T, Takeda H, Hirano M, Judai K, Arita T, Nagao S, Nakajima A, Kaya K (1999) Organometallics 18:1430
153. Yasuike T, Nakajima A, Yabushita S, Kaya K (1997) J Phys Chem A 101:5360
154. Kurikawa T, Takeda H, Nakajima A, Kaya K (1997) Z Phys D Atom Mol Clusters 40:65
155. Hoshino K, Kurikawa T, Taked AH, Nakajima A, Kaya K (1996) Surf Rev Lett 3:183
156. Kurikawa T, Hirano M, Taked AH, Yagi K, Hoshino K, Nakajima A, Kaya K (1995) J Phys Chem 99:16,248
157. Hoshino K, Kurikawa T, Taked AH, Nakajima A, Kaya K (1995) J Phys Chem 99:3053
158. Kemper PR, Bowers MT. Unpublished results
159. Zhang Q, Kemper PR, Shin SK, Bowers MT (2001) Int J Mass Spectrom 204:281
160. Carpenter CJ, van Koppen PAM, Bowers MT, Perry JK (2000) J Am Chem Soc 122:392
161. van Koppen PAM, Perry JK, Kemper PR, Bushnell JE, Bowers MT (1999) Int J Mass Spectrom 187:989
162. van Koppen PAM, Bowers MT, Haynes CL, Armentrout PB (1998) J Am Chem Soc 120:5704
163. Gidden J, van Koppen PAM, Bowers MT (1997) J Am Chem Soc 119:3935
164. Kemper PR, Bushnell JE, Weis P, Bowers MT (1998) J Am Chem Soc 120:7577
165. Kemper PR, Bushnell J, Bowers MT, Gellene GI (1998) J Phys Chem A 102:8590

Top Curr Chem (2003) 225: 233–262
DOI 10.1007/b10468

Threshold Collision-Induced Dissociations for the Determination of Accurate Gas-Phase Binding Energies and Reaction Barriers

P. B. Armentrout

Chemistry Department, University of Utah, 315 S. 1400 E. Rm 2020, Salt Lake City, UT 84112 USA. *E-mail: armentrout@chem.utah.edu*; Fax: 801-581-8433

Over the past two decades, an analysis of the kinetic energy dependence of ion molecule reactions has proven to yield accurate thermochemistry for a wide range of interesting chemical species. In this article, the experimental methods required to collect the data, most notably guided ion beam mass spectrometry, are reviewed along with a comprehensive overview of the details of the requisite data analysis. Although the emphasis of the article is on collision-induced dissociation (CID) reactions, similar methods can also be applied to endothermic bimolecular reactions. For CID studies, the systems are generally chosen such that the threshold corresponds to the bond energy desired, but studies in which reaction barriers have been measured are also reviewed. The range of chemical systems accessible by these techniques is wide and includes small covalently bound molecules (both inorganic and organic), small non-covalently bound complexes, metal ligand complexes, metal clusters, isomeric forms, and systems involving biologically relevant molecules. A fairly comprehensive review of various studies of such systems is included.

Keywords. Bond dissociation energies, Collision-induced dissociation, Guided ion beams, Reaction barriers, Thermochemistry

List of Abbreviations

BDE	Bond dissociation energy
BIRD	Blackbody infrared dissociation
CID	Collision-induced dissociation
CM	Center-of-mass
dc	direct current
eV	Electron volts
GIBMS	Guided ion beam tandem mass spectrometer
lab	Laboratory
MAD	mean absolute deviation
PSL	Phase space limit
rf	Radiofrequency
RRKM	Rice-Ramsperger-Kassel-Marcus
TCID	Threshold collision-induced dissociation
TTS	Tight transition state

1 Introduction

The determination of thermochemistry using threshold collision-induced dissociation (TCID) relies on a deceptively simple experiment: determine the energy threshold for the endothermic reaction (Eq. 1), the dissociation of the molecular ion AB^+:

$$AB^+ + Xe \rightarrow A^+ + B + Xe \tag{1}$$

This CID reaction is intrinsically endothermic. Figure 1 shows a typical example of the data for such a process. At the lowest collision energies, no reaction is observed (any signal observed below about 0.4 eV is equivalent to the noise level of this experiment). This is because insufficient energy is available to allow the $Cr^+(CO)_6$ reactant molecule to decompose. As the energy is increased, first one,

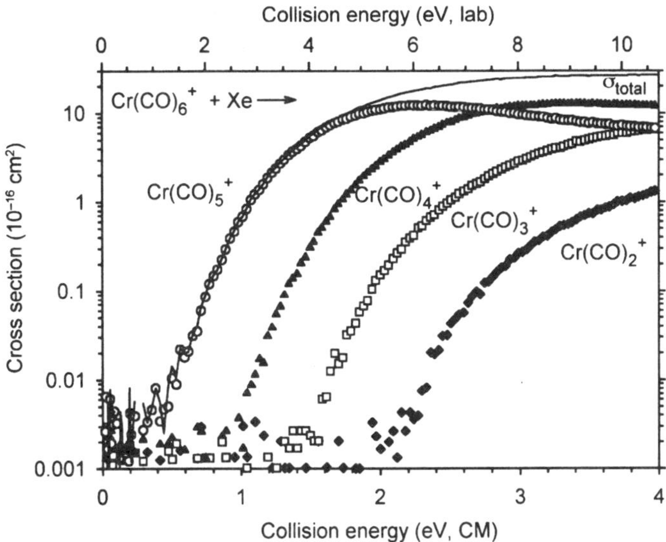

Fig. 1. Integral cross sections for the CID of $Cr(CO)_6^+$ with Xe at a pressure of 0.1 mTorr as a function of collision energy in the center-of-mass frame (*lower x-axis*) and laboratory frame (*upper x-axis*). The *symbols* represent the product cross sections: $Cr(CO)_5^+$ (*open circles*), $Cr(CO)_4^+$ (*solid triangles*), $Cr(CO)_3^+$ (*open squares*), and $Cr(CO)_2^+$ (*solid diamonds*). The *solid line* represents the total cross section. Adapted from [9]

then a second, then a third and fourth carbonyl ligand are lost from the complex. Note that the total cross section varies smoothly with energy, reaching a nearly constant plateau at the highest energies shown. Thus, the CO losses are clearly sequential, as further indicated by the observation that the cross section for the $Cr^+(CO)_5$ primary product declines when the $Cr^+(CO)_4$ secondary product begins to have an appreciable intensity.

The threshold measured for the reaction in Eq. (1), E_0, corresponds to the highest energy along the reaction coordinate for dissociation, i.e., the activation barrier, which occurs at the transition state for the reaction. Such a transition state is termed "tight" if it corresponds to a specific molecular conformation; see Fig. 2a. Interestingly, in the dissociation of many ions, there is no activation barrier for the reverse process (the association reaction), such that the activation energy for dissociation equals the overall reaction endothermicity for the process in Eq. (1); see Fig. 2b. In such a case, the measured threshold is equivalent to the bond dissociation energy (BDE) of AB^+ (Eq. 2):

$$D_0(A^+ - B) = E_0 \tag{2}$$

Now the transition state, termed "loose", is ill-defined and product-like. For ion-molecule reactions, reverse activation barriers are often absent because of the strong long-range ion-induced dipole or ion-dipole potential [1]. Further, for many systems of interest, the cleavage of the AB^+ bond is heterolytic (i.e., the pair of electrons forming the bond is removed by one of the fragments). Quan-

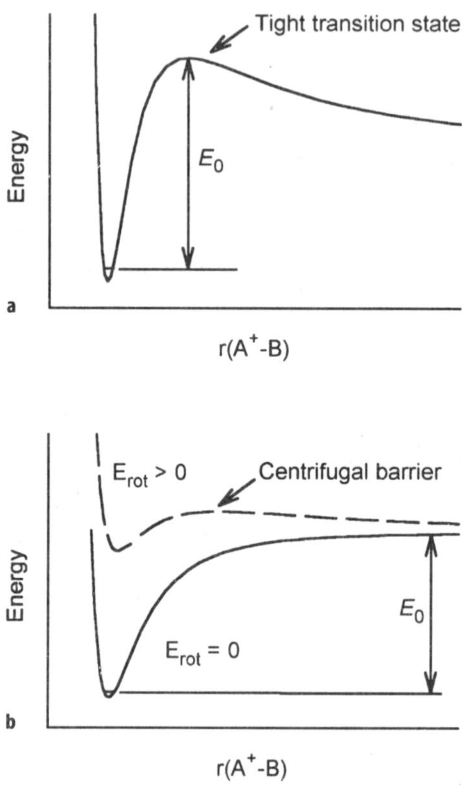

Fig. 2a, b. Potential energy surfaces for dissociation of the AB^+ molecular ion involving: **a** a tight transition state; **b** a loose transition state. The threshold energy, E_0, is shown in both cases. Part b also shows the effects of rotational energy on the dissociation

tum mechanics shows that the electronic surfaces for heterolytic bond dissociations have no barriers [2], although this conclusion may be altered if there are curve crossings and efficient coupling with surfaces of different spin. Overall, it is well-known that exothermic ion-molecule reactions often proceed without an activation energy, and thus the converse should ordinarily be true, i.e., endothermic ion-molecule reactions should proceed once the available energy exceeds the thermodynamic threshold. This assumption has been tested by examining several systems where the thermochemistry is well established, e.g., see [3–7].

To verify that the measured thresholds really do correspond to the thermodynamic limit, it is desirable to measure a particular BDE using more than one reaction system. However, such multiple determinations are often not possible, especially for systems involving noncovalent bond cleavages. In such cases, comparison with values from other experiments (notably equilibrium measurements) and ab initio theory can be used to verify the accuracy of the BDEs obtained.

2 Experimental Methods

2.1 Guided Ion Beam Mass Spectrometry

Although any tandem mass spectrometer can be used to examine the CID reaction (Eq. 1), the accurate measurement of an absolute threshold energy requires much more rigorous experimental instrumentation. A guided ion beam tandem mass spectrometer (GIBMS) is specifically designed for such experiments, and it has proven its ability to provide high quality threshold data that can be analyzed to provide accurate thermodynamic information as described below. Qualitatively, the distinguishing feature of a GIBMS instrument is that the reaction region is larger than the mass-selective regions, in contrast to most commercial mass spectrometers. Details of the GIBMS have been discussed before [3, 8, 9]. One of our instruments is shown in Fig. 3.

In a guided ion beam experiment, reactant ions are created in the source region, mass selected by a mass spectrometer (a magnetic sector in our instruments), decelerated to a desired kinetic energy, and injected into a radiofrequency (rf) octopole ion beam "guide" [10, 11]. This device comprises eight rods cylindrically surrounding the ion beam path. Opposite phases of an rf potential

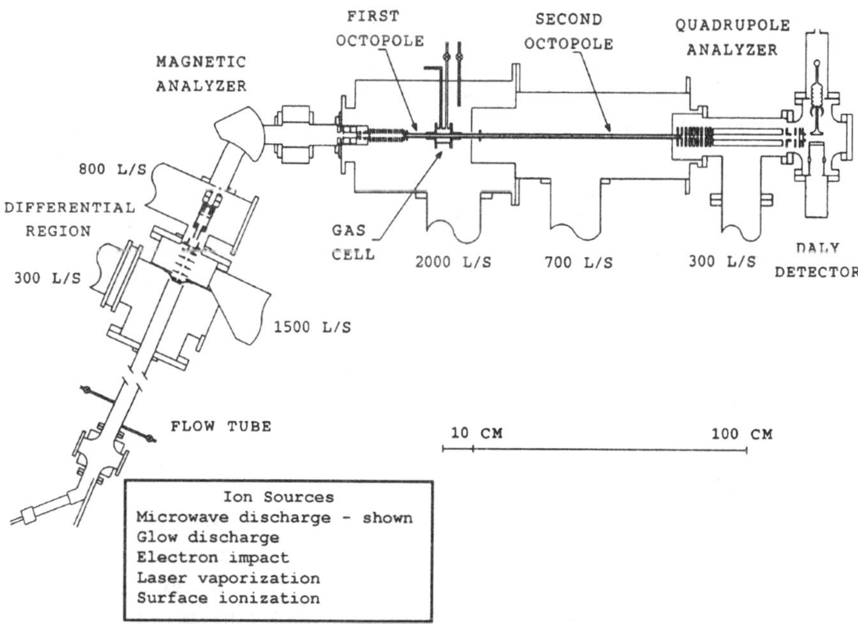

Fig. 3. Schematic diagram of the guided ion beam tandem mass spectrometer, double octopole configuration. Pumping speeds are shown. Ion sources available are listed and the microwave discharge source is shown. Adapted from [9]

are applied to alternate rods to create a radial trap (having an r^6 dependence where r is the distance away from the central axis of the guide) for charged particles, without affecting the ion motion along the axis of the octopole. This trap prevents the loss of ions at low energies resulting from space charge effects and ensures efficient collection of product ions. Compared to an rf-only quadrupole (which has a quadratic r^2 potential), an octopole provides a more homogeneous potential, less perturbation of the kinetic energy distribution of the reactant ions, and thus much better control of the energy.

In our instrument, the octopole passes through a collision cell (which has a length less than half that of the octopole) where the neutral reagent is introduced. (In most triple quad instruments the collision cell extends to the end of the central rf-only quadrupole such that reactive collisions occur in the regions between the quadrupoles where the ion energy is poorly defined.) Reactions take place inside the collision cell (where the pressure is well-defined) and in the regions leading to the cell where the pressure is still appreciable. The effective length of the collision cell is needed for calculation of absolute cross sections and this length can be determined either by calculation or calibration with previous results. It is important to keep the pressure of the neutral reagent sufficiently low that multiple collisions between the ion and neutral reactants are unlikely. In truth, the effects of multiple collisions cannot be avoided completely in any real experiment; however, such effects can be removed easily by performing studies at several different neutral pressures and extrapolating to zero neutral pressure, rigorously single collision conditions [3, 12]. This pressure extrapolation has been found to be particularly important in the analysis of CID reactions because multiple collisions can induce dissociation at energies *below* the threshold associated with a single collision. Without removing the contribution from multiple collisions, the measured threshold will be too small.

After passing through the collision cell, remaining reactant and newly formed product ions drift to the end of the octopole. There, all ions are extracted, focused into a second mass spectrometer (quadrupole mass filters in our instruments) for mass separation, and detected using an ion counter [13]. Our instruments use a Daly detector, which utilizes a high voltage (28 kV) first dynode coupled with a secondary electron scintillator and photomultiplier, thereby providing very high sensitivity (detection efficiencies near 100%).

Acquisition of data involves the measurement of reactant and product ion intensities as a function of ion kinetic energy in the laboratory frame. The latter can be varied from essentially room temperature to hundreds of electron volts by changing the bias potential on the octopole and collision cell relative to the ion source (see below). The ability to vary easily the kinetic energy of the ion through the range of chemical interest is the key capability of the GIBMS. A nuance in the data acquisition for threshold studies is to measure ion intensities with the neutral reagent directed to the collision cell and, on alternate energy scans, to the background of the chamber. This background signal (which includes all random noise as well as reactions taking place outside of the collision chamber) provides a direct measurement of the absolute zero of the cross sections for all product ions formed inside the collision cell. Apparatus-independent results are obtained by converting the raw data to absolute reaction cross sections as a function of the relative kinetic energy, or energy in the center-of-mass (CM) frame, $\sigma(E)$. The cross section is a direct measure of the probability of reaction and can be thought of as a microscopic rate constant because $k=\sigma v$, where v is the relative velocity of the reactants. The center-of-mass energy is used because it excludes the kinetic energy of the entire reaction system moving through the laboratory (which must be conserved and therefore is unavailable for driving chemical reactions). Ion intensities are converted to cross sections using a Beer's law type of formula [8]. Laboratory energies are converted to relative energies using a simple stationary target approximation, $E(\text{CM})=\{E(\text{lab})-E_z(\text{lab})\}m/(m+M)$, where m and M are the masses of the neutral and ionic reactants, respectively, and E_z is the absolute zero of energy. The distribution of ion kinetic energies and E_z are directly measured in each experiment by a retarding potential analysis that is made particularly accurate by the use of the octopole ion guide. The accuracy of this retarding analysis has been verified by time-of-flight measurements [8, 9] and comparisons with theoretical cross sections [14]. In the low energy region where the ion beam begins to be truncated (because not all ions have sufficient energy to be transmitted by the octopole), the conversion from lab to CM frame explicitly accounts for this change in the ion energy distribution [8].

2.2
Generation of Ions

To acquire accurate thermochemical data from a TCID experiment, the energy available to the reactants must be characterized in detail. The kinetic energy distributions of the ionic reactant are measured directly, as outlined in the previous section, and those of the neutral reagents are described by a Maxwell-Boltzmann distribution at the temperature of the collision cell (~300 K). This same temperature characterizes the internal energy of the neutral reagent (although for the monatomic gas often used in TCID studies, this point is moot). The internal energies of the ions are the most difficult quantities to determine because ionization is generally an energetic process that leaves ions with poorly characterized vibrational and rotational energies. Although a number of ionization methods can be used to minimize the internal excitation, most of these methods are still poorly characterized with regard to the extent of internal excitation in all degrees of freedom.

To overcome these limitations, we utilize a flowing afterglow ion source [15–17] in which ions undergo about 10^5 collisions with a flowing He bath gas (present at a pressure of ~0.5 Torr and a temperature of ~300 K). The collisions thermalize the ions rotationally and vibrationally no matter how they are formed. For many systems of interest in our laboratory, we generate metal-ligand complexes by condensing the ligand on the bare metal ion (three-body complexation that is enabled by the high pressure bath gas). Metal ions are generated in a d.c.-discharge where the metal of interest is the cathode [3]. Alternatively, we can start with a stable organometallic or organic precursor and ionize it by electron impact ionization, charge transfer from He^+, or Penning ionization with He*. Having carefully thermalized the ions, it is also critical to remove them from the high-pressure flow tube without collisional excitation altering the internal energy distribution of the ions. Differential pumping and low voltage lenses provide this capability in our apparatus [3, 18].

Ions produced in the flow tube ion source should have internal energies characterized by a Maxwell-Boltzmann distribution at 300 K. No direct measure of the internal temperature of the ions is available, but extensive studies of systems for which thermodynamic information is available in the literature provide some confidence that this assumption is valid. These studies include TCID measurements on N_4^+ [18], $Fe(CO)_x^+$ (x=1–5) [3], $Cr(CO)_x^+$ (x=1–6) [5,9], SiF_x^+ (x=1–4) [19], $H_3O^+(H_2O)_x$ (x=1–5) [4], $Cu^+(H_2O)_x$ (x=1–4) [20], $Li^+(H_2O)_x$ (x=1–6) [7], and $K^+(NH_3)_x$ (x=1–4) [21].

3 Data Analysis

Having obtained reaction cross sections as a function of relative kinetic energy, the data must be analyzed to extract the onset for the endothermic process of interest. Because of the extensive distributions of energies, accurate location of this onset is a difficult process. In our work, we choose to model the experimental cross sections with a mathematical expression that is theoretically justified [22–24], Eq. (3). In addition, this expression has proven to yield accurate thermochemistry in a number of systems, providing a reasonable empirical justification for its use:

$$\sigma(E) = \sigma_0 \sum g_i \, (E + E_i - E_0)^n / E^m \tag{3}$$

In Eq. (3), σ_0 is a scaling factor, E is the relative (CM) kinetic energy, E_0 is the threshold energy, and n and m are adjustable parameters. The sum is over individual reactant states (vibrational, rotational, and/or electronic), denoted by i, with energies E_i and populations g_i (Σg_i=1). An exposition of theoretically justified values of n and m has been provided recently [24]. Empirically, experimental results for TCID processes can generally be described using m=1, as predicted for translationally driven reactions [25]. Recent work has shown that when the reaction of interest involves a spin change, it may be more appropriate to use m=3/2 in order to account for the $E^{-1/2}$ dependence of crossing between surfaces [26]. A detailed understanding of the value of n has not been developed, although a value near unity is common (the so-called line-of-centers model). Generally, n increases as the complexity of the system increases and decreases as the threshold decreases. In TCID studies, recent work has shown that the value of n also describes the distribution of energy transferred in the collision referred to in Eq. (1) [9], as discussed further below.

However, Eq. (3) is not yet complete as it does not include the distributions of the kinetic energy of the neutral and ionic reactants. These are included as first described by Chantry [27] and later developed elsewhere [8, 28]. Once these distributions are included, the σ_0, n, and E_0 parameters of Eq. (3) are optimized using a non-linear least squares analysis to give the best reproduction of the data. Because all sources of energy are included in the analysis of the data, the threshold obtained is believed to correspond to a BDE at 0 K, Eq. (2) [4, 29].

Uncertainties in E_0 are calculated from the range of threshold values obtained for different data sets with different values of n coupled with the error in the absolute energy scale (0.05 eV lab). If the internal energy of the reactant ion is non-negligible and the vibrational frequencies are not well known, the uncertainty in E_0 includes variations in the calculated internal energy distribution of Eq. (3). Nowadays, vibrational frequencies and rotational constants of the reactant ion complexes are generally taken from ab initio calculations. Precise values are not required as it is most important that the correct number of low frequency vibrations be correctly identified. The accuracy of the thermochemistry obtained by this modeling procedure depends on a number of experimental parameters discussed in detail elsewhere and reviewed below [29–31]. It might be noted that this modeling procedure is not limited to TCID reactions but can also be used to measure the thresholds for chemical exchange reactions, $A^+ + BC \rightarrow AB^+ + C$, in which covalent bonds are made and broken [29].

An example of this modeling procedure is shown in Fig. 4 for the total dissociation cross section of the $Cr^+(CO)_6$ complex [9]. The energy at which the $Cr^+(CO)_5$ product signal first deviates from zero clearly depends on the sensitivity of the experiment, but certainly rises above the background at about 0.5 eV (Fig. 1). This apparent threshold energy differs appreciably from the E_0 value obtained from modeling with Eq. (3), 1.59±0.09 eV, as indicated by where the dashed line deviates from zero in Fig. 4. The convolution of Eq. (3) over the

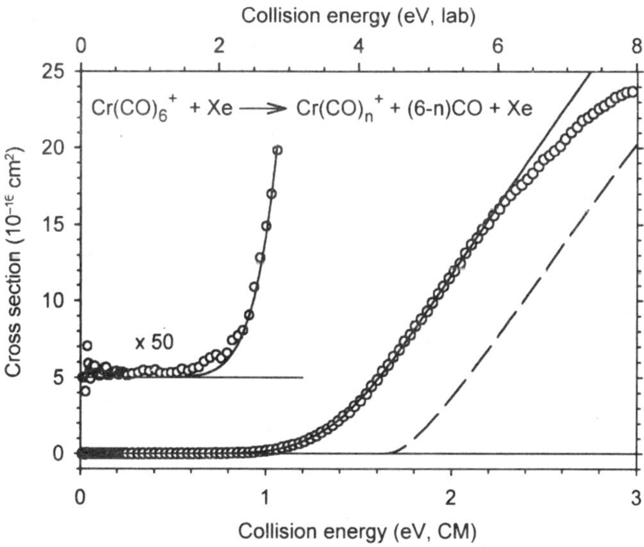

Fig. 4. Total cross section for the CID of $Cr(CO)_6^+$ with Xe, extrapolated to zero pressure, as a function of collision energy in the center-of-mass frame (*lower x-axis*) and laboratory frame (*upper x-axis*). The *solid line* shows a representative fit to the data using the model of Eq. (4) convoluted over the energy distributions of the two reactants. The *dashed line* shows the same model in the absence of energy convolution, for reactants with an internal temperature of 0 K. A 50× magnification of the threshold region of the cross section is presented in the *upper left side* of the figure. Adapted from [9]

internal and kinetic energy distributions is shown by the full line, which reproduces the data extremely well over a wide range of magnitudes and energies. To assess the accuracy of this modeling, the threshold can be compared with literature values; however, the effect of kinetic shifts needs to be included before this comparison is meaningful.

3.1 Kinetic Shifts

An energized molecule AB^+ that has enough energy to dissociate has a lifetime that depends on the available energy and the complexity of the molecule. Small molecules dissociate rapidly because it is straightforward for the energy to find its way into the vibration corresponding to the reaction coordinate. For the dissociation of large molecules having many degrees of freedom, it is possible to have enough internal energy to allow dissociation, but the average time needed for the decomposition exceeds the flight time of the ion from the collision region to the analysis mass spectrometer, τ. In the instrument shown in Fig. 3, this time scale is about 5×10^{-4} s. Thus, as the lifetime of the energized molecule approaches this limit, the *apparent* threshold for reaction shifts to higher translational energies, resulting in a "kinetic" shift. The magnitude of this shift is instrument dependent and varies with the sensitivity of the apparatus and the experimental time scale available. The lifetime for dissociation of an energized molecule can be quantified using statistical kinetic theories, such as Rice-Ramsperger-Kassel-Marcus (RRKM) theory [32], for predicting the unimolecular dissociation rate constant, $k(E^*)$, at an available energy, E^*. The specifics of how RRKM theory is incorporated into Eq. (3) (for m=1) has evolved [5, 33, 34], but is presently described by Eq. (4):

$$\sigma(E) = (n\sigma_0/E) \sum_i g_i \int_{E_0-E_i}^{E} [1 - e^{-k(\varepsilon+E_i)\tau}] (E - \varepsilon)^{n-1} \, d\varepsilon \tag{4}$$

In this equation, all quantities are as described above and ε is the energy deposited in the complex upon collision with the neutral reagent. The energy deposition distribution has been measured recently by measuring the kinetic energy of the reactants after collision, specifically for the $Cr^+(CO)_6$ complex discussed above [9]. This work finds that the distribution can be parameterized using n, Eq. (5), where n is the same parameter used in Eq. (3):

$$P(\varepsilon) = \sigma_0 n(E - \varepsilon)^{n-1}/E \tag{5}$$

These two forms are self-consistent in that the same value of n can describe the energy dependence of the cross section and the distribution of deposited energies. In cases where the decomposition rate is fast compared to τ, the integration over ε in Eq. (4) recovers Eq. (3).

As for most applications of statistical rate theory, the most difficult aspect is the characterization of the transition state. As discussed above, if the E_0 value determined from a TCID experiment corresponds to the bond energy of AB^+, then the transition state must be a loose one. In such a case, it is generally appropriate to utilize a "phase space limit" (PSL), which assumes that the transition state is located at the centrifugal barrier for dissociation. Then, the TS is product-like, such that the molecular parameters for the transition state are easily calculated [34]. In this analysis of the data, the uncertainties in the threshold now include uncertainties associated with the RRKM analysis of lifetime effects. This generally involves ±10% uncertainties in the vibrational frequencies and an uncertainty of a factor of two in the time available for dissociation.

The importance of including such kinetic shifts in the modeling is now absolutely clear, and the accuracy of this PSL model for estimating kinetic shifts has now been verified in several applications [6, 34–41]. An extreme case is illustrated by the dissociation of Na^+(18-crown-6). When modeled without consideration of the lifetime for dissociation, a threshold of 7.37±0.24 eV is obtained, but when lifetime effects are included using the PSL model, a threshold of 3.07±0.20 eV is found [38]. The accuracy of the latter value is confirmed by high-level ab initio calculations that predict a bond energy at 0 K of 3.44 eV [42]. It generally appears that lifetime effects begin to be important when the total number of heavy atoms in the complex exceeds five and when the BDE for the bond being broken is greater than about 1 eV. Lifetime effects become more significant as the complexity of the system increases and as the bond strength increases (and less significant for weaker bonds).

We now return to our example of $Cr^+(CO)_6$ dissociation, which exhibits a threshold energy of 1.59±0.09 eV without consideration of lifetime effects and 1.43±0.09 eV when these are included. As discussed elsewhere [5], the most reliable values from the literature for this bond energy come from appearance energy measurements of Michels et al. [43] (1.38±0.04 eV) and the threshold photoelectron-photoion coincidence measurements of Meisels and co-workers [44] (1.44±0.25 eV). Another check of the thermochemistry is to compare the sum of the individual BDEs of all six $Cr^+(CO)_x$ (x=1–6) systems with the known value for the enthalpy of reaction for losing all six CO ligands from $Cr^+(CO)_6$ at 298 K, 5.23±0.09 eV [5]. When all six individual bond energies are measured using TCID methods, we obtain an experimental sum of 5.25±0.09 eV (5.41±0.09 eV if lifetime effects are excluded) [5, 9]. These comparisons demonstrate that the agreement between the literature and TCID values with lifetime effects included is very good.

3.2
Competitive Shifts

In addition to kinetic shifts, the threshold for a CID process can be delayed by competition with another decomposition channel. Consider the bis-ligated sodium cation, $Na^+(H_2O)(NH_3)$, which can decompose by losing either ligand, Eq. (6):

$$Na^+(H_2O)(NH_3) + Xe \rightarrow Na^+(NH_3) + H_2O + Xe \tag{6a}$$

$$\rightarrow Na^+(H_2O) + NH_3 + Xe \tag{6b}$$

Although the lowest energy dissociation pathway can be described accurately using Eqs. (3) or (4) up to the energy where the second channel starts, the higher

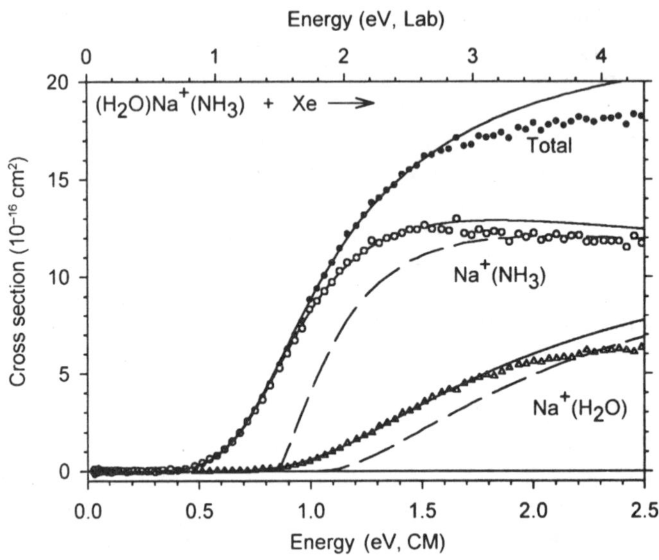

Fig. 5. Zero pressure extrapolated cross sections for the competitive collision-induced dissociation processes of $(H_2O)Na^+(NH_3)$ with xenon in the threshold region as a function of kinetic energy in the center-of-mass frame (*lower axis*) and laboratory frame (*upper axis*). *Solid lines* show the best fits to the data using the model of Eq. (7) convoluted over the neutral and ion kinetic energies and the internal energies of the reactants, using common scaling factors. *Dashed lines* show the model cross sections in the absence of experimental energy broadening for reactants with an internal energy of 0 K. Adapted from [45]

energy process is inhibited by the competition with the energetically more favorable dissociation pathway. As is evident in Fig. 5 [45], the lower energy channel has a cross section that rises rapidly from threshold, whereas the higher energy channel increases more slowly because of the competition. The consequences of this competition are made evident when it is realized that the true difference in thresholds between the two channels is only 0.15±0.03 eV, whereas the apparent difference is several tenths of an electron volt.

The effects of competition can be estimated by extending the statistical method introduced above for handling kinetic shifts [46]. This is accomplished by introducing a branching ratio for multiple channels indexed using j (Eq. 7):

$$\sigma_j(E) = (n\sigma_0/E) \sum_i g_i \int_{E_0-E_i}^{E} (k_j/k_{tot}) \left[1 - e^{-k_{tot}(\varepsilon+E_i)\tau}\right] (E-\varepsilon)^{n-1}\, d\varepsilon \tag{7}$$

where $k_{tot}(\varepsilon+E_i)=\Sigma k_j(\varepsilon+E_i)$. Significantly, this method allows simultaneous determination of reaction thermochemistry for multiple dissociation channels, as illustrated in Fig. 5. Because both channels are constrained to use the same value of n, the precision of the method is actually higher than for single channel CID processes. However, various entropic factors can influence the relative magnitudes of the two channels, and these need to be considered carefully [45]. This is both a difficulty (because the data cannot be reproduced accurately unless the character and molecular parameters of the transition state are properly chosen) and a benefit (for instance, it is straightforward to identify loose vs tight transition states). Equation (7) was originally developed for competitive pathways that dissociate by loose (PSL) transition states [46] and has proven useful in several additional studies [45, 47, 48]. It has also been shown to be effective in systems of mixed transition states (loose and tight) [49], and it should be applicable in systems that have only tight transition states (although any applications involving tight transition states rely on determining accurate molecular parameters). The method has been applied to two and three competing channels, but the methodology should be applicable to systems having even more competitive processes. Finally, although precise relative thresholds are determined, much like in equilibrium studies, competitive TCID studies also yield the absolute BDEs for both ligands bound to a solvated ion.

3.3 Theoretical Support

Theoretical calculations augment experimental TCID work in several ways, although they are never necessary for the accurate measurement of CID thresholds. Most importantly, ab initio calculations can provide the vibrational frequencies and rotational constants needed to evaluate the internal energy of the reactant complex and required for the statistical calculations of rate constants used to estimate kinetic and competitive shifts. Such calculations are needed for the dissociating complex (AB^+) and the products (A^+ and B). For these purposes, low level calculations are sufficient because there are accurate scaling procedures that have been developed [50–52]. Calculations are also used to envision the structure of the dissociating molecule, thereby making sure that the dissociation is unlikely to involve a reverse activation barrier. Finally, calculated energies double check the accuracy of the experimentally measured BDEs and can identify ambiguities associated with electronic or structural isomers. Such theoretical energetics need to involve calculations at a fairly high level of theory. Thus, structures need to be calculated using methods that include electron correlation, such as MPx [53] or density functional methods [54–58] and a double-zeta basis set like 6-31G* that includes polarization. For final energetics, single point energy calculations using a larger basis set (e.g., triple-zeta 6-311+G(2d,2p), that includes more polarization and diffuse functions) are needed. It is critical to correct for zero point energies (ZPE), and even though it is still a point of debate, basis set superposition errors (BSSE) are also usually corrected for in the full counterpoise approximation [59, 60].

An example of the utility of such calculations involves the bis-ligated sodium cation complexes $Na^+(H_2O)(NH_3)$ and $Na^+(H_2O)(C_2H_5OH)$, which are calculated to have the structures shown in Fig. 6 [45]. In both cases, the two ligands are well separated such that there should be little interaction between them upon dissociation. However, in the $Na^+(H_2O)(C_2H_5OH)$ complex, the calculations show that the ethanol ligand, which binds to Na^+ primarily through the lone pair electrons on oxygen, also has a weak interaction using the methyl

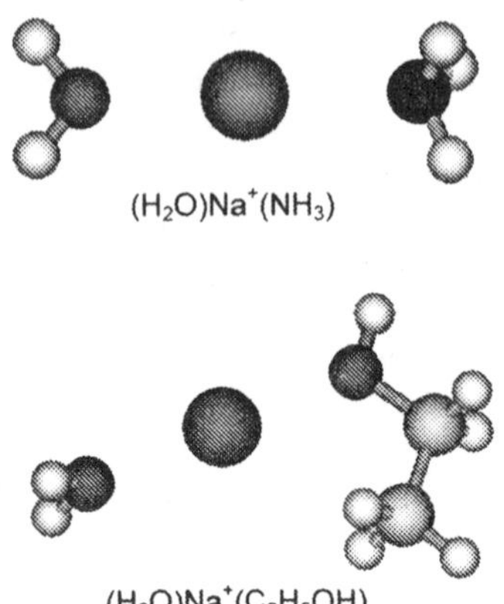

Fig. 6. Ground state geometries of $(H_2O)Na^+(NH_3)$ and $(H_2O)Na^+(C_2H_5OH)$ complexes optimized at the MP2(full)/6-31G* level of theory. Adapted from [45]

group. This greatly hinders the rotation about the C-C bond such that the entropy for dissociation of the ethanol ligand is particularly large, favoring this channel appreciably. The effects of this are easily seen in the competitive dissociation of the $Na^+(H_2O)(C_2H_5OH)$ complex [45].

Typical agreement between experimental and theoretical bond energies can also be illustrated by these complexes, as shown in Table 1 for each ligand alone and their sum. It can be seen that the values generally agree within experimental error. A more comprehensive comparison between eight different levels of theory and TCID experimental bond energies for 18 different Na^+(ligand) com-

Table 1. Experimental and theoretical absolute Na^+–L bond dissociation energies at 0 K (in kJ mol^{-1})

Bond	Experiment		Theory[b]			
	TCID[a]	Literature	MP2	B3LYP	G2	CBS-Q
Na^+-C_2H_5OH	110.0 (5.5)	110.1 (5.1)[c]	108.9	114.7	107.6	104.4
Na^+-NH_3	102.2 (5.4)	102.6 (4.0)[c] 103.1 (0.8)[d]	102.5	108.7	102.2	96.7
Na^+-OH_2	87.8 (6.0)	89.4 (5.2)[c]	89.2	94.6	88.8	88.8
Na^+-(OH_2+NH_3)	184.2 (7.5)		179.6			
Na^+-$(OH_2+C_2H_5OH)$	187.2 (8.9)		184.7			

[a] [45].
[b] [39] and [45]; MP2=MP2(full)/6-311+G(2d,2p)//MP2(full)/6-31G*, B3LYP=B3LYP/6-311+G(2d,2p)//B3LYP/6-31G*.
[c] Calculated from experimental ΔG_{298} values reported by McMahon and Ohanessian [199] and enthalpy and entropy corrections determined at the MP2(full)/6-31G* level, [39].
[d] [70].

plexes is also available [39]. Mean absolute deviations (MADs) between experiment and theory are generally around 5 kJ mol^{-1}, except for B3LYP/6-311+G(2d,2p)//B3LYP/6-31G* and G3 [61] calculations, which deviate by averages of 8 and 12 kJ mol^{-1}, respectively.

4 Necessities for Accurate Thermochemistry

Although the idea of measuring a threshold from an endothermic cross section appears deceptively simple, the discussion above should make it clear that accurate thermochemistry requires attention to a number of specific steps that need to be addressed in both the experiment and the data analysis. Systems presently amenable to study could not have been accurately treated a decade ago, since which time our current understanding of the importance of these various steps has evolved. The TCID methods are seeing increased application using both equipment designed for such studies (such as the GIBMS) and instrumentation that has been developed primarily for other tasks (e.g., triple quads, ion cyclotron resonance mass spectrometers, and ion traps). Here, the specific details that are crucial to measurements of accurate thermodynamics using TCID methods are outlined briefly:

- Collision energies must be well defined. In our apparatus this definition is achieved using an octopole ion beam guide that is substantially longer than the collision cell. This provides a homogeneous electronic environment with no end effects in regions where the ion energy is poorly controlled.
- Ions of interest must have well-characterized internal energies. This is generally achieved in my laboratory by using a flow tube ion source, where ions are thoroughly thermalized by about 10^5 collisions with room temperature bath gases. (One might question whether the use of a supersonic jet expansion source might provide colder ions, thereby lessening the effects of internal energy. Although such a source would yield ions having lower internal energies, the energy distribution can change from translation to rotation to individual vibrational modes, such that the overall ion temperature is poorly characterized.) Other types of ion sources could certainly be used in TCID experiments given that the internal temperature is characterized.
- The absolute zero of the cross-section scale must be accurately measured. Fully 50% of our data acquisition time is spent collecting background signals.
- The collision gas should provide efficient kinetic to internal energy transfer. Several studies [30, 62, 63] convincingly demonstrate that Xe is a preferred collision gas for TCID studies. Because Xe is atomic, it cannot carry away energy in internal modes. Further it is heavy, so the interaction time with the ion is long, and it is polarizable, making the collision "sticky". The latter two attributes mean that the transient $ABXe^+$ complex behaves more statistically, which maximizes the energy transfer (as little energy is generally released into kinetic energy in statistical processes).
- Cross sections corresponding to single collision conditions should be analyzed. As noted above, this requires that cross sections be measured at several

neutral reactant pressures and then extrapolated to zero pressure. This procedure analytically eliminates the effects of secondary collisions that can induce dissociation below the threshold for single collision CID. Although time-consuming, we always measure the pressure dependence of TCID cross sections.

- Analysis of the energy dependent cross sections must explicitly include all sources of energy (internal and kinetic) and their distributions. Consideration of these factors is achieved using the model in Eq. (3) and its derivative forms, Eqs. (4) and (7), and includes convolution over the kinetic energy distributions. A computer program that facilitates this analysis, CRUNCH, is available by contacting the author.
- For larger ions (approximately six heavy atoms or more), the kinetics of the dissociation process must be included in the data analysis. This requires the use of Eqs. (4) or (7), a capability included in the CRUNCH program.
- Direct competition between parallel reaction channels should be explicitly considered in the systems where multiple dissociation channels exist. Again, CRUNCH includes this capability.

5
Comparison with Other Methods

Of course, thermochemistry in the form of bond energies can be measured by a variety of other experimental techniques. Key among these are temperature-dependent equilibrium methods [64–74]. More recently, absolute BDEs have also been measured using blackbody infrared radiative dissociation (BIRD) [75–81] and radiative association [82–84] methods. In addition, precise relative BDEs can be determined using equilibrium [85, 86] and kinetic method procedures [87–90]. Direct comparisons between results of TCID methods and those of these alternative techniques provide some confidence that all these various methods, when adequately interpreted, do yield accurate thermochemistry.

Some distinctions and similarities between these various techniques are worth mentioning. Threshold CID has proven to be a fairly robust method capable of measuring absolute bond energies over a very wide range of values, ~10–400 kJ mol^{-1}. Because they are dependent on establishing constant temperature conditions, equilibrium, BIRD, and radiative association methods are generally limited to a smaller range, ~40–200 kJ mol^{-1}. Likewise, a larger range appears to be available to relative measurements made by competitive CID methods, up to about 30 kJ mol^{-1} [46], whereas other methods are often limited to differences of less than 10 kJ mol^{-1}. In addition, competitive CID studies also provide absolute BDEs for two ligands simultaneously along with a precise measure of the relative difference between them. A primary limitation of the threshold CID method is that it is model dependent, as is obvious from the exposition above, although efforts are underway to provide a more quantitative basis for these models. Likewise, BIRD, radiative association, and kinetic method approaches share a dependence on models. In contrast, the underpinnings of equilibrium methods are a very well established part of thermodynamics.

6 Thermodynamics of Reaction Barriers and Bimolecular Reactions

6.1 Reaction Barriers

As noted above, most systems chosen for study by TCID are designed such that there is no reverse activation barrier, hence the threshold corresponds to the BDE of the bond being broken. In some systems, however, there is an activation barrier that must be overcome for the reaction to rearrange to the observed products. The earliest examination of such a system involved threshold collisional activation of $Fe^+(C_3H_8)$ [91], which has the energetics shown in Fig. 7. Three major product channels are observed at lower collision energies (below 2 eV): $Fe^+(C_2H_4)+CH_4$, $Fe^+(C_3H_6)+H_2$, and $Fe^++C_3H_8$. The threshold for the latter channel, simple CID, showed that the complex was bound by 0.82±0.07 eV. On the basis of other related studies [29], thermodynamic thresholds for the first two channels are approximately 0.1 and 0.5 eV, respectively. In contrast, dissociation of $Fe^+(C_3H_8)$ to form these two product channels both exhibited thresholds of 0.47±0.12 eV. Thus, the thresholds correspond to a barrier attributed to a tight transition state for C-H bond activation, which can lead to both products. This is consistent with density functional calculations on the related $Co^+(C_3H_8)$ system [92], although these calculations also suggest the possibility that the rate limiting tight transition states are associated with the β-H transfer steps in which the CH_4 and H_2 molecules are formed. (See [93] for a complete discussion of the mechanisms of alkane activation by transition metal ions.)

More recently, studies that combine TCID studies with high level computations allow for accurate assessment of tight transition states and the competition between these and entropically favored, but higher energy, decomposition pathways. This work includes case studies of the $(OCS{\cdot}C_2H_2)^+$ complex [49] and the dissociation of phenol cation (which forms $C_5H_6^+$+CO) [94]. Similar studies that combine TCID work with theory have been performed for protonated biological molecules, which have only covalent bonds and hence their dissociation pathways generally involve activation barriers [95, 96].

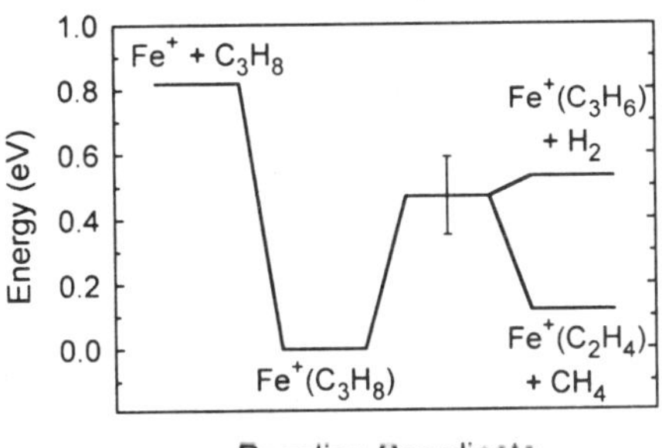

Fig. 7. Energetics of the $Fe^+(C_3H_8)$ system. The uncertainty in the determination of the reaction barrier is shown by the *error bar* at the transition state. Adapted from [91]

6.2
Bimolecular Reactions

The methods introduced above for measuring the kinetic energy dependence of CID reactions can also be applied with success to bimolecular reactions such as

$$A^+ + BC \rightarrow AB^+ + C \tag{8}$$

Indeed, these methods were largely developed for such systems and then applied to CID processes. One key advantage of bimolecular reactions is the absence of a third body, such that the energy available to the reaction intermediate, ABC^+, is the entire kinetic energy of the collision. In such a case, the parameter n no longer describes the energy deposition process, but rather the efficiency of the reaction, such that Eq. (3) is still an appropriate means to determine the threshold for such reactions.

This is illustrated in Fig. 8 which shows the reaction of platinum cations with dihydrogen [97], where Pt^+ has been formed in its ground electronic state, $^2D_{5/2}$. The dashed line shows Eq. (3) with n=1.1, m=1.0, and E_0=1.68±0.06 eV, whereas results on the analogous D_2 system find E_0=1.71±0.05 eV. The full line has been convoluted with the kinetic energy distributions and can be seen to reproduce the data very nicely. At high energies, above the bond energy of H_2 at 4.475 eV, a model for the subsequent dissociation of the PtH^+ product to Pt^++H is included [98]. The threshold values for the H_2 and D_2 systems should differ by the zero-point energy differences in the reactions. Calculations indicate that the threshold for reaction with H_2 should be 0.04 eV lower than that of the D_2 reaction,

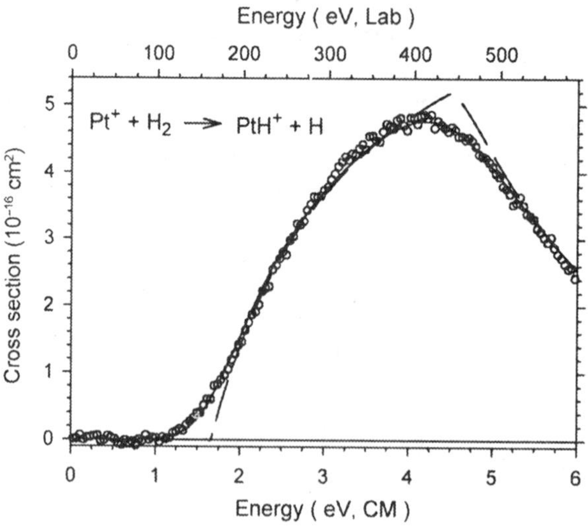

Fig. 8. Cross sections for reaction of $Pt^+(^2D_{5/2})$ with H_2 as a function of kinetic energy in the center-of-mass frame (*lower axis*) and laboratory frame (*upper axis*). The best fit of Eq. (3) with parameters given in the text to the data is shown as a *dashed line*. The *solid line* shows this model convoluted over the kinetic and internal energy distributions of the reactant neutral and ion. Adapted from [97]

consistent with the experimental results and showing that they exhibit very good precision. In this simple reaction, the H_2 (D_2) bond is broken and the PtH^+ (PtD^+) bond is made such that the threshold energy is simply the difference between these two bond energies. This fact is used to extract an average Pt^+-H bond energy of 2.81±0.05 eV (271±5 kJ mol^{-1}).

This methodology, which allows the measurement of covalent bond energies, has been applied to a wide variety of systems. Many of these have been reviewed elsewhere and include atomic transition metal cations bonded to H, C, N, O, CH, NH, OH, CH_2, NH_2, CH_3 [29, 93], S, and 2S [99]; transition metal cluster cations bound to D, O, [100], S [101], CH, CH_2, and CH_3 [102]; and various inorganic species relevant to plasma chemistry [103].

7 Range of Systems Amenable to TCID Studies

7.1 Small Molecules: Covalently Bound

7.1.1 *Inorganic Cations*

The thermochemistry of a number of inorganic cations has been examined using TCID methods. These include UO^+ and UO_2^+ [104], VO^+ [62], SiD_3^+ [105], CS_2^+ [106], SF_x^+ (x=1–5) [107], SiF_x^+ (x=1–4) [19], N_3^+ [108], $Si(CH_3)_x^+$ (x=1–4) [109], and the acids $H_2NO_3^+$ [110] and $H_3SO_4^+$ [48]. The VO^+ work examined CID with all five rare gases and demonstrated that the efficiency of energy transfer was best with Xe, as discussed above. The study of CS_2^+ was interesting as both CS^++S and S^++CS products were observed, thereby allowing the heat of formation of the CS radical to be measured. The $Si(CH_3)_4^+$ work provided a favorable comparison between bond energies measured using TCID methods and BIRD.

However, studies of small, strongly bound inorganic ions have highlighted one of the potential difficulties with TCID techniques, namely, that energy is inefficiently transferred from the kinetic energy of the collision to the internal energy of the ion. Certainly the smaller the ion and the more strongly bound it is, the less likely the collision is to behave statistically. Indeed, TCID studies of diatomic transition metal oxide cations routinely provide only upper limits to the true bond energies, as measured using bimolecular reactions. The accuracy of CID thermochemistry in such cases can be illustrated using two examples that provide extremes in the results. Studies of the CID of SiF_x^+ (x=1–4) [19] provided bond energies for the cations and complementary studies of related bimolecular reactions and charge transfer processes also yielded thermochemistry for the neutral SiF_x (x=1–4) molecules. The results for the cations (measured directly by TCID) are shown in Table 2 along with subsequent high level, CCSD(T), calculations [111]. It can be seen that the agreement is quite satisfying with an average absolute deviation of 0.09±0.07 eV, comparable to the experimental error. In contrast to these results are those for SF_x^+, also listed in Table 2 along with the results of G2 [112] and G2(MP2) [113] calculations from two separate inves-

Table 2. Bond dissociation energies (eV) at 0 K of SiF_x^+ (x=1–4) and SF_x^+ (x=1–5)

Bond	Experiment (TCID)	Theory
Si^+-F	7.04±0.06[a]	6.90[b]
SiF^+-F	3.18±0.04[a]	3.34[b]
SiF_2^+-F	6.29±0.10[a]	6.34[b]
SiF_3^+-F	0.85±0.16[a]	0.83[b]
S^+-F	3.56±0.05[c]	3.80±0.08[d], 3.35[e]
SF^+-F	4.17±0.10[c]	3.86±0.06[d], 4.09[e]
SF_2^+-F	4.54±0.08[c]	4.15±0.06[d], 4.16[e], 4.30[e]
SF_3^+-F	0.36±0.05[c]	0.58±0.07[d], 0.56[e]
SF_4^+-F	4.60±0.10[c]	3.84±0.07[d], 3.87[e]

[a] [19].
[b] CCSD(T) results from [111].
[c] [107].
[d] G2 results from [114].
[e] G2 and G2(MP2) results from [115].

tigations [114, 115]. Oddly, the theoretical values for x=1 and 2 do not agree, but the experimental values for x=1–3 lie about 0.2±0.1 eV above the theoretical values from Cheung et al. This is satisfactory agreement, especially given the uncertainties in the theoretical values, but the systematically larger experimental values could indicate some inefficiencies in energy transfer. A similar discrepancy of 0.2 eV is observed for the bond energy of SF_4^+, but here the experimental value lies below theory. It is possible that incomplete thermalization of this ion left it with excess internal energy leading to a slightly lower threshold. The bond energy for SF_5^+ measured by TCID lies well above the theoretical predictions, which have now been confirmed within about 0.1 eV by other experiments [116]. Collisional dissociation of SF_5^+ is complicated because the initially formed SF_4^+ product rapidly dissociates to SF_3^+ at slightly collision higher energies because of the small SF_3^+-F bond energy (Table 2). This complication coupled with the possibility that energy transfer is inefficient may have led to the large discrepancy with theory, although some sort of activation barrier might also be imagined. Fortunately, in the multitude of cases where comparisons can be made, such large deviations are quite uncommon. Nevertheless, the measurement of strong bonds of small molecules by TCID can be problematic and caution should be exercised in the interpretation of the data.

7.1.2
Inorganic Anions

TCID methods have been employed, primarily by Sunderlin and coworkers, to study a number of inorganic anions, generally involving halogens. These include the polyhalide anions: F_2^-, F_3^-, Cl_3^-, Br_3^-, Br_5^-; group 15 halides: ACl_4^- (A=P, As, Sb, Bi), PBr_4^-, and PI_4^-; sulfur and sulfoxy halides: SCl_3^-, SF_5^-, SF_6^-, $SOCl_3^-$, and SO_2F^-; and other species such as $N_3O_2^-$ and $X^-(HCOOH)_x$ (x=1, 2 and X=Cl, Br, I). This work has been reviewed elsewhere [117].

7.1.3 *Organic Ions*

Squires pioneered the use of CID methods to elucidate the thermochemistry of a wide range of organic species. The work of his group includes carboxylates [118], carbenes (from CID of carbene-X^- where X is a halogen atom) [119–123], carbynes [124, 125], benzynes [126–129], and radical anions [130–133]. Contributions from other groups include the gas-phase acidity and C-H bond energy of diacetylene [134], gas-phase acidities for a series of alcohols [47], investigations of isonitrile [135], *m*-xylylene [136], triradicals [137, 138], phenyltrifluorosilane [139], and diazeniumdiolates [140].

7.2 Small Molecules: Non-Covalent

A number of small non-covalently bound complexes have also been examined using CID methods including $(N_2)_2^+$ [18], $(N_2O{\cdot}H_2O)^+$ [141], $(OCS{\cdot}C_2H_2)^+$ [49], and protonated water clusters, $(H_3O)^+(H_2O)_x$ (x=1–5) [4]. The latter study provided an excellent test of the absolute accuracy of TCID experiments, as thermochemistry for these clusters is available from a number of equilibrium studies. Compared to results of Meot-Ner and Speller [142], the TCID bond energies had an average deviation of 2.9±4.0 kJ mol^{-1} (MAD=4.4±1.4 kJ mol^{-1}), and compared to results of Kebarle and coworkers [143, 144], the average deviation was 1.0±3.7 kJ mol^{-1} (MAD=3.0±1.9 kJ mol^{-1}). Considering that the uncertainties are ±6 kJ mol^{-1} in the TCID measurements and ±4 kJ mol^{-1} in the equilibrium studies, this is excellent agreement.

7.3 Metal Ligand Complexes and Ion Solvation

A great deal of work utilizing CID has involved the interactions of metal ions with various ligands. Molecules studied include the binding of metal ions to rare gases (Ar, Kr, Xe), diatomics (H_2, CO, N_2, NO, CS), triatomics (H_2O, CO_2, CS_2), simple solvents (H_2O, NH_3, acetonitrile, acetone), hydrocarbons (alkanes, alkenes, benzene), substituted benzenes (C_6H_5X where X=F, OH, NH_2, OCH_3, NH_2, CH_3), alcohols (methanol–butanols), ethers (dimethyl ether, dimethoxy ethane, and cyclic crown ethers), aldehydes and ketones, azoles (five-membered ring heterocycles containing N including imidazole, triazoles, and tetrazoles), and azines (six-membered ring heterocycles containing N including pyridine, pyridazine, pyrimidine, pyrazine, *s*-triazine). Metals in this work are generally singly charged cations and include the alkali metals (Li^+, Na^+, K^+, Rb^+, and Cs^+), Mg^+, and Al^+, the first-row transition metals (Sc^+–Zn^+), and several second-row and third-row transition metals. Some anionic studies are also available [145, 146]. A comprehensive listing of most of these metal-ligand bond dissociation energies as determined by threshold CID has been provided [147]. This review includes a discussion of the trends in these values, thereby elucidating the importance of ion-dipole and ion-induced dipole interactions, chelation, different conformers and tautomers, steric interactions, solvation phenomena, and electronic effects such as hybridization and promotion.

Work performed subsequent to this review includes a systematic evaluation of the binding of alkali metal ions with one and two ligands such as substituted pyridines (c-C_5H_4NX, X=H [40], CH_3 [148], NH_2 [149]), substituted benzenes (c-C_6H_5X, X=CH_3 [150], NH_2 [151], OH [152], OCH_3 [153], F [154]), and azoles [155]. Periodic trends in the binding of first-row transition metal cations with pyridine [156], pyrimidine [157], and adenine [158] have been studied. Solvation of Na^+ by acetonitrile [159], K^+ by ammonia [21], and Cu^+ by NO [160], acetonitrile [161], acetone [162], dimethyl ether [163], and dimethoxyethane [164] have been examined. Several studies of Ag^+ bound to organonitriles [165], small oxygen- [166] and nitrogen-containing ligands [167] have been published. Two TCID studies involving Pt^+ include the sequential binding energies of CO [168] and interactions with methane [169].

As alluded to above, some of these studies include an examination of solvation from a very fundamental level, i.e., the measurement of sequential bond energies up to (and sometimes past) the first solvent shell. Notable among such work are TCID studies involving the ligands of H_2O, CO, NH_3, and dimethyl ether which have been performed for a range of different metal ions [147, 170]. An example of such data for the case of the ubiquitous H_2O molecule is given in Table 3. In addition to these ligands, solvation of Mg^+ by CO_2 and CH_3OH [41] and Fe^+ by N_2 [171], CH_4 [172], CH_2O [171], and CS_2 [173] have been studied. Mono and bis complexes of several metal ions with C_2H_4 [174], C_6H_6 [174], substituted benzenes [150–154], and dimethoxyethane ($CH_3OCH_2CH_2OCH_3$) [164, 175] have been examined. The trends in these sequential bond energies can be very interesting and illuminating, as illustrated in Fig. 9 which shows hydration

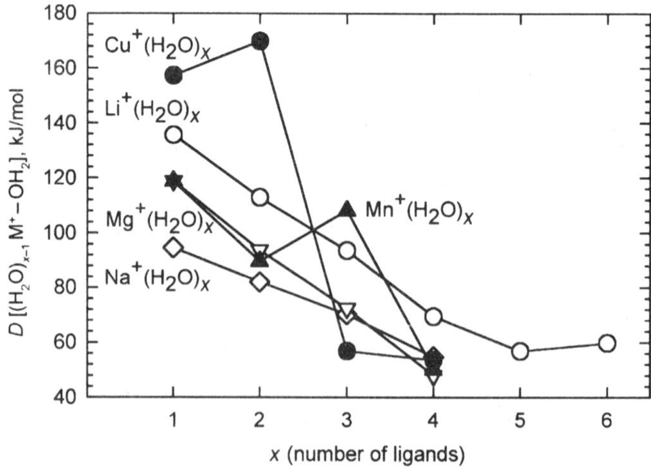

Fig. 9. Sequential hydration energies in kJ/mol of the singly charged alkali ions, lithium (*open circles*) and sodium (*open diamonds*), the alkaline earth, magnesium (*open triangles*), and two first row transition metal ions, manganese (*solid triangles*) and copper (closed circles). All data taken from Table 3

Table 3. Sequential bond energies of $(H_2O)_{x-1}M^+–H_2O$ (kJ/mol)[a]

M	x=1	x=2	x=3	x=4	x=5	x=6
Li[b]	135 (8)[c]	113 (10)	94 (4)	70 (5)	57 (4)	60 (5)
Na[d]	95 (8)	82 (6)	70 (6)	55 (6)		
Mg[d]	119 (13)	94 (7)	72 (9)	48 (9)		
Al[d]	104 (15)	67 (5)	64 (8)	52 (6)		
Ti[e]	154 (6)	136 (5)	67 (7)	84 (8)		
V[e]	147 (5)	151 (10)	68 (5)	68 (8)		
Cr[e]	129 (9)	142 (6)	50 (5)	51 (7)		
Mn[e]	119 (6)	90 (5)	108 (6)	50 (5)		
Fe[f]	128 (5)	164 (4)	76 (4)	50 (7)		
Co[e]	161 (6)	162 (7)	65 (5)	58 (6)		
Ni[e]	180 (3)	168 (8)	68 (6)	52 (6)		
Cu[e]	157 (8)	170 (7)	57 (8)	54 (4)		

[a] All values are at 0 K with uncertainties listed in parentheses.
[b] [7].
[c] Reevaluated in [46].
[d] [200].
[e] [20].
[f] [172], reevaluated in [20].

of five different metal ions. Note that the BDEs decrease with increasing solvation for Li^+, Na^+, and Mg^+, the trend that might be expected for purely electrostatic bonding. However, the BDEs for Na^+ and Mg^+ cross one another, i.e., those for $x=1–3$ are stronger for Mg^+ compared to Na^+, but weaker for $x=4$. In more striking contrast to the monotonic trends in the BDEs for these three metals are the trends for Mn^+ and Cu^+. The explanation of the trends in all five cases, which involve hybridization (sp for Mg^+ and sd for Mn^+ and Cu^+) and changes in spin state (Mn^+), are straightforward and can be found in detail elsewhere [147, 170]. These expositions also demonstrate that the trends are sensitive to the nature of the ligand, e.g., whether it is a π-donor or π-acceptor.

7.4 Metal Clusters

CID methods have proven to be very useful in measuring the stabilities of clusters of bare transition metal atoms, providing many more thermochemical values than photodissociation methods. In our laboratory, we have used CID to study the cationic clusters of ten different transition metal elements, including Ti_x^+ ($x=2–22$), V_x^+ ($x=2–20$), Cr_x^+ ($x=2–21$), Mn_2^+, Fe_x^+ ($x=2–19$), Co_x^+ ($x=2–18$), Ni_x^+ ($x=2–18$), and Cu_2^+; the second row transition metal clusters of Nb_x^+ ($x=2–11$); and the third row transition metal clusters of Ta_x^+ ($x=2–4$). These results have been summarized and trends analyzed previously [176, 177]. CID methods have also been used by Ervin et al. to measure the stabilities of anionic clusters of the coinage metals: Cu_x^- ($x=2–8$) [178], Ag_x^- ($x=2–11$) [179], and Au_x^- ($x=2–7$) [180]; and group 10 metals: Pd_3^- [181] and Pt_x^- ($x=3–6$) [182]. A multiple collision-induced dissociation experiment in an ICR trap has recently been used to measure dissociation energies for Ag_x^+ (x=2–25) [183] and Ag_x^{2+} (x=9–25) [184]. The various CID results compare well with values obtained from photodissociation methods for a much more limited number of small cluster cations, a more comprehensive set of Ag_x^+ (x=8–21) clusters [185], and several cluster anions: Ag_x^- (x=7–11) [186], and Au_x^- (x=6, 7) [180]. Such comparisons have been reviewed elsewhere [100, 176].

TCID experiments have also been used to examine the binding of small molecules to metal clusters. An extensive set of data for transition metal cluster carbonyls has been reviewed by Ervin [187]. Kinetic energy dependent bimolecular reactions have also been used to provide binding energies of atoms (D, C, O) to metal clusters as well, as reviewed elsewhere [100]. Sulfur atoms have recently been included in such studies, partly because of their biological interest [101]. Further such work has been extended to molecular fragments such as CH, CH_2, and CH_3 [102], which is particularly interesting as no comparable information is available for the binding of such fragments to metal surfaces.

7.5 Isomers

Collision-induced dissociation techniques have long been utilized in mass spectrometry as a tool for structural elucidation. Especially at high collision energies, the connectivity of the molecular ion is reflected in the fragment ions that appear. TCID provides another view of such structural assignments because different isomers of the molecular ion should have different energy content, which can be reflected in the threshold observed. Such energetic information can sometimes be correlated with fragmentation patterns for structural identification, but in other cases the threshold process (the lowest energy decomposition pathway) will lead to the same fragment ions as this is the lowest energy asymptote no matter what structure the molecular ion starts with.

One example of such a study involved an examination of the potential energy surface for carbon-dioxide activation by V^+ [188]. Here, complexes of $V^+(CO_2)$ and $OV^+(CO)$ were independently generated and collisionally activated. At low collision energies these complexes dissociate exclusively to form V^++CO_2 and VO^++CO, respectively, such that the results are characteristic of the structure of the complexes and the V^+-CO_2 and OV^+-CO bond energies can be measured. However, bimolecular reactions of V^++CO_2 and VO^++CO demonstrate that these two complexes share common potential energy surfaces, but do not interconvert readily because their electronic ground states have different spin and there is a potential barrier separating them. Related studies of MCO_2^+ species have provided similar results for M=Y, Zr, Nb, and Mo [189–192]. Other studies have examined isomers such as $Fe^+(C_2H_6)$ vs $Fe(CH_3)_2^+$ [193], metallacycle and metal-alkene isomers of $MC_3H_6^+$ where M=Fe [194] and Co [195]. In these cases, the structures were distinguished largely on the basis of the means used to form the initial complexes and confirmed by different energetics. In a recent example [169], the complex of $Pt^+(CH_4)$ was collisionally activated and found to form appreciable amounts of Pt^+ and PtH^+, suggesting that the complex actually had the H-Pt^+-CH_3 structure. This suggestion was confirmed by good agreement between the thermochemistry obtained from the thresholds for both product channels and high level ab initio calculations, which show that the hydrido methyl complex is the global minimum.

7.6 Biomolecules

One of the exciting new applications of TCID methods is in the area of biomolecules. The earliest work is that of Klassen et al. [196] who used TCID methods to study the binding of Na^+ and K^+ to acetamide, *N*-methylacetamide, *N,N*-dimethylacetamide, glycine, and glycylglycine. This has been followed by another measurement of the Na^+(glycine) bond energy, which was then compared to those of Na^+ binding to related molecules having single (1-propanol, 1-propylamine, methyl ethyl ketone) and double (propionic acid and 1-amino-2-ethanol) functional groups [197]. This permits the details of the metal ion-glycine interaction to be more clearly understood. Rodgers and Armentrout have measured the binding energies of several nucleic acid bases (adenine, uracil, and thymine) to Li^+, Na^+, and K^+ [198] and the first-row transition metal ions (V^+–Zn^+) to adenine [158]. In the latter study, the early transition metal ions (Sc^+, Ti^+, and V^+) are observed to activate covalent bonds in the adenine ligand, making thermodynamic information impossible to obtain in the first two metal systems. Rodgers and coworkers have also provided a series of TCID studies on the interactions of the alkali metal ions (Li^+, Na^+, and K^+) with various substituted benzenes [150–154]. This work is designed to elucidate the systematic influences on cation π-interactions, which are of potential interest in many biological systems, and in particular, for metal ion selective channels. Finally, the energetics for fragmentation of protonated glycine, glycinamide, and some related small peptides and peptide amino amides [95] and the $(Gly)_3$ tripeptide [96] have been elucidated by using TCID in comparison with theory. Although most TCID studies in the biological area to date have been confined to model systems and small biologically relevant molecules, the prospects for continued work on increasingly large systems are very good.

Acknowledgements. The work described in this article could not have been accomplished without the dedication and insight of a great group of students, whose names are listed throughout the references. Profs. K.M. Ervin and M.T. Rodgers have been instrumental in the development of the analysis tools (CRUNCH) described here. The work has been funded by the National Science Foundation and the Department of Energy, Office of Basic Energy Sciences.

8 References

1. Talrose VL, Vinogradov PS, Larin IK (1979) In: Bowers MT (ed) Gas phase ion chemistry, vol 1. Academic Press, New York, p 305
2. Armentrout PB, Simons J (1992) J Am Chem Soc 114:8627
3. Schultz RH, Crellin KC, Armentrout PB (1991) J Am Chem Soc 113:8590
4. Dalleska NF, Honma K, Armentrout PB (1993) J Am Chem Soc 115:12,125

5. Khan FA, Clemmer DC, Schultz RH, Armentrout PB (1993) J Phys Chem 97:7978
6. Rodgers MT, Armentrout PB (1997) J Phys Chem A 101:2614
7. Rodgers MT, Armentrout PB (1997) J Phys Chem A 101:1238
8. Ervin KM, Armentrout PB (1985) J Chem Phys 83:166
9. Muntean F, Armentrout PB (2001) J Chem Phys 115:1213
10. Teloy E, Gerlich D (1974) Chem Phys 4:417
11. Gerlich D (1992) Adv Chem Phys 82:1
12. Hales DA, Lian L, Armentrout PB (1990) Int J Mass Spectrom Ion Processes 102:269
13. Daly NR (1960) Rev Sci Instrum 31:264
14. Burley JD, Ervin KM, Armentrout PB (1987) Int J Mass Spectrom Ion Processes 80:153
15. Leopold DG, Murray KK, Miller AES, Lineberger WC (1985) J Chem Phys 83:4849
16. Leopold DG, Ho J, Lineberger WC (1987) J Chem Phys 86:1715
17. Graul ST, Squires RR (1988) Mass Spectrom Rev 7:263
18. Schultz RH, Armentrout PB (1991) Int J Mass Spectrom Ion Processes 107:29
19. Fisher ER, Kickel BL, Armentrout PB (1993) J Phys Chem 97:10,204
20. Dalleska NF, Honma K, Sunderlin LS, Armentrout PB (1994) J Am Chem Soc 116:3519
21. Iceman C, Armentrout PB (2003) Int J Mass Spectrom 222:329
22. Armentrout PB, Beauchamp JL (1981) J Chem Phys 74:2819
23. Aristov N, Armentrout PB (1986) J Am Chem Soc 108:1806
24. Armentrout PB (2000) Int J Mass Spectrom Ion Processes 200:219
25. Chesnavich WJ, Bowers MT (1979) J Phys Chem 83:900
26. Rue C, Armentrout PB, Kretzschmar I, Schröder D, Harvey JN, Schwarz H (1999) J Chem Phys 110:7858
27. Chantry PJ (1971) J Chem Phys 55:2746
28. Lifshitz C, Wu RLC, Tiernan TO, Terwilliger DT (1978) J Chem Phys 68:247
29. Armentrout PB, Kickel BL (1996) In: Freiser BS (ed) Organometallic ion chemistry. Kluwer, Dordrecht, p 1
30. Armentrout PB (1992) In: Adams NG, Babcock LM (eds) Advances in gas phase ion chemistry, vol 1. JAI Press, Greenwich, p 83
31. Armentrout PB (2002) J Am Soc Mass Spectrom 13:419
32. Holbrook KA, Pilling MJ, Robertson SH (1996) Unimolecular reactions. Wiley, New York
33. Loh SK, Hales DA, Lian L, Armentrout PB (1989) J Chem Phys 90:5466
34. Rodgers MT, Ervin KM, Armentrout PB (1997) J Chem Phys 106:4499
35. More MB, Ray D, Armentrout PB (1997) J Phys Chem A 101:831
36. Rodgers MT, Armentrout PB (1999) J Phys Chem A 103:4955
37. Rodgers MT, Armentrout PB (1999) Int J Mass Spectrom 185/187:359
38. More MB, Ray D, Armentrout PB (1999) J Am Chem Soc 121:417
39. Rodgers MT, Armentrout PB (2000) J Phys Chem A 104:2238
40. Amunugama R, Rodgers MT (2000) Int J Mass Spectrom 195/196:439
41. Andersen A, Muntean F, Walter D, Rue C, Armentrout PB (2000) J Phys Chem A 104:692
42. Glendening ED, Feller D, Thompson MA (1994) J Am Chem Soc 116:10,657
43. Michels GD, Flesch GD, Svec HJ (1980) Inorg Chem 19:479
44. Das PR, Nishimura T, Meisels GG (1985) J Phys Chem 89:2808
45. Amicangelo JC, Armentrout PB (2001) Int J Mass Spectrom 212:301
46. Rodgers MT, Armentrout PB (1998) J Chem Phys 109:1787
47. DeTuri VF, Ervin KM (1999) J Phys Chem A 103:6911
48. Pommerening CA, Bachrach SM, Sunderlin LS (1999) J Phys Chem A 103:1214
49. Muntean F, Armentrout PB (2000) Z Phys Chem 214:1035
50. Pople JA, Schlegel HB, Raghavachari K, DeFrees DJ, Binkley JF, Frisch MJ, Whitesides RF, Hout RF, Hehre WJ (1981) Int J Quantum Chem Symp 15:269
51. DeFrees DJ, McLean AD (1985) J Chem Phys 82:333
52. Seeger DM, Korzeniewski C, Kowalchyk W (1991) J Phys Chem 95:6871
53. Moller C, Plesset MS (1934) Phys Rev 46:618
54. Hohenberg P, Kohn W (1964) Phys Rev 136:B864
55. Kohn W, Sham LJ (1965) Phys Rev 140:A1133

56. Lee C, Yang W, Parr RG (1988) Phys Rev B 37:785
57. Parr RG, Yang W (1989) Density functional theory of atoms and molecules. Oxford University Press, Oxford
58. Perdew JP, Yang W (1992) Phys Rev B 45:13,244
59. Boys SF, Bernardi R (1970) Mol Phys 19:553
60. van Duijneveldt FB, van Duijneveldt de Rijdt JGCM, van Lenthe JH (1994) Chem Rev 94:1873
61. Curtiss LA, Raghavachari K, Redfern PC, Rassolov V, Pople JA (1998) J Chem Phys 109:7764
62. Aristov N, Armentrout PB (1986) J Phys Chem 90:5135
63. Hales DA, Armentrout PB (1990) J Cluster Sci 1:127
64. Dzidic I, Kebarle P (1970) J Phys Chem 74:1466
65. Davidson WR, Kebarle P (1976) J Am Chem Soc 98:6133
66. Peschke M, Blades AT, Kebarle P (1998) J Phys Chem A 102:9978
67. Kemper PR, Bushnell J, van Koppen P, Bowers MT (1993) J Phys Chem 97:1810
68. Bushnell JE, Kemper PR, Bowers MT (1995) J Phys Chem 99:15,602
69. Bowers MT, Kemper PR, van Koppen P, Wyttenbach T, Carpenter CJ, Weis P, Gidden J (1999) In: Minas de Piedade ME (ed) Energetics of stable molecules and reactive intermediates. Kluwer, Dordrecht, p 235
70. Hoyau S, Norrman K, McMahon TB, Ohanessian G (1999) J Am Chem Soc 121:8864
71. Bouchard F, Hepburn JW, McMahon TB (1990) J Am Chem Soc 111:8934
72. Holland PM, Castleman AW (1980) J Am Chem Soc 102:6174
73. Holland PM, Castleman AW (1982) J Chem Phys 76:4195
74. Castleman AW, Peterson KI, Upschulte BL, Schelling FJ (1983) Int J Mass Spectrom Ion Phys 47:203
75. Dunbar RC, McMahon TB (1998) Science 279:194
76. Schnier PD, Price WD, Jockusch RA, Williams ER (1996) J Am Chem Soc 118:7178
77. Schnier PD, Klassen JS, Strittmatter EF, Williams ER (1998) J Am Chem Soc 120:9605
78. Jockusch RA, Williams ER (1998) J Phys Chem 102:4543
79. Klassen JS, Schnier PD, Williams ER (1998) J Am Soc Mass Spectrom 9:1117
80. Price WD, Jockusch RA, Williams ER (1998) J Am Chem Soc 120:3474
81. Strittmatter EF, Schnier PD, Klassen JS, Williams ER (1999) J Am Soc Mass Spectrom 10:1095
82. Pozniak BP, Dunbar RC (1997) J Am Chem Soc 119:10,439
83. Ho Y-P, Yang Y-C, Klippenstein SJ, Dunbar RC (1997) J Phys Chem A 101:3338
84. Ryzhov V, Dunbar RC (1999) J Am Chem Soc 121:2259
85. Woodin RL, Beauchamp JL (1978) J Am Chem Soc 100:501
86. Taft RW, Anvia F, Gal J-F, Walsh S, Capon M, Holmes MC, Hosn K, Oloumi G, Vasanwala R, Yazdani S (1990) Pure Appl Chem 62:17
87. Cooks RG, Wong PH (1998) Acc Chem Res 31:379
88. Cooks RG, Koskinen JT, Thomas PD (1999) J Mass Spectrom 34:85
89. Armentrout PB (1999) J Mass Spectrom 34:74
90. Armentrout PB (2000) J Am Soc Mass Spectrom 11:371
91. Schultz RH, Armentrout PB (1991) J Am Chem Soc 113:729
92. Reichert EL, Weisshaar JC (2002) J Phys Chem A 106:5563
93. Armentrout PB (1999) In: Brown JM, Hofmann P (eds) Topics in organometallic chemistry, vol 4. Springer, Berlin Heidelberg New York, p 1
94. Muntean F, Armentrout PB (2002) J Phys Chem B 106:8117
95. Klassen JS, Kebarle P (1997) J Am Chem Soc 119:6552
96. El Aribi H, Rodriquez CF, Almeida D, Ling Y, Mak WW-N, Hopkinson AC, Siu KWM work in progress
97. Zhang X-G, Armentrout PB (2002) J Chem Phys 116:5565
98. Weber ME, Elkind JL, Armentrout PB (1986) J Chem Phys 84:1521
99. Kretzschmar I, Schröder D, Schwarz H, Armentrout PB (2001) In: Duncan MA (ed) Advances in metal and semiconductor clusters, vol 5, p 347

100. Armentrout PB (2001) Ann Rev Phys Chem 52:423
101. Koszinowski K, Schröder D, Schwarz H, Liyanage R, Armentrout PB (2002) J Chem Phys 117:10.039
102. Liyanage R, Zhang X-G, Armentrout PB (2001) J Chem Phys 115:9747
103. Armentrout PB (2000) In: Inokuti M, Becker K (eds) Fundamentals of plasma chemistry (advanced atomic, molecular, and optical physics), vol 43. Academic Press, New York, p 187
104. Armentrout PB, Beauchamp JL (1980) Chem Phys 50:21
105. Boo BH, Armentrout PB (1987) J Phys Chem 91:5777
106. Prinslow DA, Armentrout PB (1991) J Chem Phys 94:3563
107. Fisher ER, Kickel BL, Armentrout PB (1992) J Chem Phys 97:4859
108. Haynes CL, Freysinger W, Armentrout PB (1995) Int J Mass Spectrom Ion Processes 149/150:267
109. Lin C-Y, Dunbar RC, Haynes CL, Armentrout PB, Tonner DS, McMahon TB (1996) J Phys Chem 100:19,659
110. Sunderlin LS, Squires RR (1993) Chem Phys Lett. 212:307
111. Ricca A, Bauschlicher CW (1998) J Phys Chem A 102:876
112. Curtiss LA, Raghavachari K, Trucks GW, Pople JA (1991) J Chem Phys 94:7221
113. Curtiss LA, Raghavachari K, Pople JA (1993) J Chem Phys 98:1293
114. Irikura KK (1995) J Chem Phys 102:5357
115. Cheung Y-S, Chen Y-J, Ng CY, Chiu S-W, Li W-K (1995) J Am Chem Soc 117:9725
116. Evans M, Ng CY, Hsu C-W, Heimann P (1997) J Chem Phys 106:978
117. Sunderlin LS (2001) In: Babcock LM, Adams N (eds) Advances in gas phase ion chemistry, vol 4. JAI/Elsevier Science, Amsterdam, p 49
118. Graul ST, Squires RR (1990) J Am Chem Soc 112:2517
119. Paulino JA, Squires RR (1991) J Am Chem Soc 113:5573
120. Poutsma JC, Paulino JA, Squires RR (1997) J Phys Chem A 101:5327
121. Poutsma JC, Nash JJ, Squires RR (1997) J Am Chem Soc 119:4686
122. Seburg RA, Hill BT, Squires RR (1999) J Chem Soc Perkin Trans 11:2249
123. Poutsma JC, Upshaw SD, Squires RR, Wenthold PG (2002) J Phys Chem A 106:1067
124. Seburg RA, Hill BT, Squires RR (1999) J Am Chem Soc 121:63
125. Jesinger RA, Squires RR (1999) Int J Mass Spectrom 185/187:745
126. Wenthold PG, Hu J, Squires RR (1994) J Am Chem Soc 116:6961
127. Wenthold PG, Squires RR (1994) J Am Chem Soc 116:6401
128. Wenthold PG, Wierschke SG, Squires RR (1994) J Am Chem Soc 116:7378
129. Wenthold PG, Hu J, Squires RR (1996) J Am Chem Soc 118:11,865
130. Wenthold PG, Squires RR (1994) J Am Chem Soc 116:11,890
131. Hu J, Squires RR, Jotacs R (1996) J Am Chem Soc 118:5816
132. Hill BT, Squires RR (1998) J Chem Soc Perkin Trans 5:1027
133. Hill BT, Poutsma JC, Squires RR (1999) J Am Soc Mass Spectrom 10:896
134. Shi Y, Ervin KM (2000) Chem Phys Lett 318:129
135. Wenthold PG (2000) J Phys Chem A 104:5612
136. Hammad LA, Wenthold PG (2000) J Am Chem Soc 122:11,203
137. Hammad LA, Wenthold PG (2001) J Am Chem Soc 123:12,311
138. Lardin HA, Nash JJ, Wenthold PG (2002) J Am Chem Soc 124:12,612
139. Krouse IH, Lardin HA, Wenthold PG (2002) Int J Mass Spectrom (in press)
140. Hammad LA, Ruane PH, Kumar NA, Toscano JP, Wenthold PG (2002) Int J Mass Spectrom (submitted)
141. Bastian MJ, Dressler RA, Levandier DJ, Murad E, Muntean F, Armentrout PB (1997) J Chem Phys 106:9570
142. Meot-Ner M, Speller CV (1986) J Phys Chem 90:6616
143. Cunningham AJ, Payzant AD, Kebarle P (1972) J Am Chem Soc 94:7627
144. Lau YK, Ikuta S, Kebarle P (1982) J Am Chem Soc 104:1462
145. Sunderlin LS, Wang D, Squires RR (1993) J Am Chem Soc 115:12,060
146. Sunderlin LS, Squires RR (1999) Int J Mass Spectrom 182:149

147. Rodgers MT, Armentrout PB (2000) Mass Spectrom Rev 19:215
148. Rodgers MT (2001) J Phys Chem A 105:2374
149. Rodgers MT (2001) J Phys Chem A 105:8145
150. Amunugama R, Rodgers MT (2002) J Phys Chem A 106:5529
151. Amunugama R, Rodgers MT (2002) Int J Mass Spectrom (accepted)
152. Amunugama R, Rodgers MT (2002) J Phys Chem A 106:9718
153. Amunugama R, Rodgers MT (2003) Int J Mass Spectrom 222:431
154. Amunugama R, Rodgers MT (2002) J Phys Chem A 106:9092
155. Huang H, Rodgers MT (2002) J Phys Chem A 106:4277
156. Stanley JR, Amunugama R, Rodgers MT (2000) J Am Chem Soc 122:10,969
157. Amunugama R, Rodgers MT (2001) J Phys Chem A 105:9883
158. Rodgers MT, Armentrout PB (2002) J Am Chem Soc 124:2678
159. Valina AB, Amunugama R, Huang H, Rodgers MT (2001) J Phys Chem A 105:11,057
160. Koszinowski K, Schröder D, Schwarz H, Holthausen MC, Sauer J, Koizumi H, Armentrout PB (2002) Inorg Chem 41:5882
161. Vitale G, Valina AB, Huang H, Amunugama R, Rodgers MT (2001) J Phys Chem A 105:11,351
162. Chu Y, Yang Z, Rodgers MT (2002) J Am Soc Mass Spectrom 13:453
163. Koizumi H, Zhang X-G, Armentrout PB (2001) J Phys Chem A 105:2444
164. Koizumi H, Armentrout PB (2001) J Am Soc Mass Spectrom 12:480
165. Shoeib T, El Aribi H, Siu KWM, Hopkinson AC (2001) J Phys Chem A 105:710
166. El Aribi H, Shoeib T, Ling Y, Rodriquez CF, Hopkinson AC, Siu KWM (2002) J Phys Chem A 106:2908
167. El Aribi H, Rodriquez CF, Shoeib T, Ling Y, Hopkinson AC, Siu KWM (2002) J Phys Chem A 106:8798
168. Zhang X-G, Armentrout PB (2001) Organometallics 20:4266
169. Zhang X-G, Liyanage R, Armentrout PB (2001) J Am Chem Soc 123:5563
170. Armentrout PB (1995) Acc Chem Res 28:430
171. Tjelta BL, Armentrout PB (1997) J Phys Chem A 101:2064
172. Schultz RH, Armentrout PB (1993) J Phys Chem 97:596
173. Capron L, Feng WY, Lifshitz C, Tjelta BL, Armentrout PB (1996) J Phys Chem 100:16,571
174. Meyer F, Khan FA, Armentrout PB (1995) J Am Chem Soc 117:9740
175. Armentrout PB (1999) Int J Mass Spectrom 193:227
176. Armentrout PB, Hales DA, Lian L (1994) In: Duncan MA (ed) Advances in metal and semiconductor clusters, vol 2. JAI, Greenwich, p 1
177. Armentrout PB (1996) In: Russo N, Salahub DR (eds) Metal-ligand interactions – structure and reactivity. Kluwer, Dordrecht, p 23
178. Spasov VA, Lee T H, Ervin KM (2000) J Chem Phys 112:1713
179. Spasov VA, Lee TH, Maberry JP, Ervin KM (1999) J Chem Phys 110:5208
180. Spasov VA, Shi Y, Ervin KM (2000) Chem Phys 262:75
181. Spasov VA, Ervin KM (1998) J Chem Phys 109:5344
182. Grushow A, Ervin KM (1997) J Chem Phys 106:9580
183. Krückeberg S, Dietrich G, Lützenkirchen K, Schweikhard L, Walther C, Ziegler J (1999) J Chem Phys 110:7216
184. Krückeberg S, Dietrich G, Lützenkirchen K, Schweikhard L, Ziegler J (1999) Phys Rev A 60:1251
185. Hild U, Dietrich G, Krückeberg S, Lindinger M, Lützenkirchen K, Schweikhard L, Walther C, Ziegler J (1998) Phys Rev A 57:2786
186. Shi Y, Spasov VA, Ervin KM (1999) J Chem Phys 111:938
187. Ervin KM (2001) Int Rev Phys Chem 20:127
188. Sievers MR, Armentrout PB (1995) J Chem Phys 102:754
189. Sievers MR, Armentrout PB (1998) Int J Mass Spectrom 179/180:103
190. Sievers MR, Armentrout PB (1998) J Phys Chem A 102:10,754
191. Sievers MR, Armentrout PB (1999) Inorg Chem 38:397
192. Sievers MR, Armentrout PB (1999) Int J Mass Spectrom 185/187:117

193. Schultz RH, Armentrout PB (1992) J Phys Chem 96:1662
194. Schultz RH, Armentrout PB (1992) Organometallics 11:828
195. Haynes CL, Armentrout PB (1994) Organometallics 13:3480
196. Klassen JS, Anderson SG, Blades AT, Kebarle P (1996) J Phys Chem 100:14,218
197. Moision RM, Armentrout PB (2002) J Phys Chem A 106:10,350
198. Rodgers MT, Armentrout PB (2000) J Am Chem Soc 122:8548
199. McMahon TB, Ohanessian G (2000) Chem Eur J 6:2931
200. Dalleska NF, Tjelta BL, Armentrout PB (1994) J Phys Chem 98:419

IV
Medicinal Chemistry

Top Curr Chem (2003) 225: 265–282
DOI 10.1007/b10476

Investigating Viral Proteins and Intact Viruses with Mass Spectrometry

Sunia A. Trauger · Teri Junker · Gary Siuzdak

The Scripps Center for Mass Spectrometry, The Scripps Research Institute, 10550 North Torrey Pines Road, La Jolla, California 92037, USA. *E-mail: suizdak@scripps.edu*

Mass spectrometry has become a fundamental technology for virology and an important tool in probing the structure and function of viruses. It has been used to identify viral capsid proteins, detect viral mutants, characterize post-translational modifications, and measure intact viruses. Mass-based analysis techniques including time resolved proteolysis, tandem mass spectrometry, and computer-assisted database searching are also providing new insights into viral structure and the dynamic changes in the capsid domains at the onset of infection. This capsid mobility is allowing researchers to develop and evaluate new drugs for viral inactivation. More recently, new mass spectrometry approaches have been used to detect intact viruses. Overall, its broad application to local and global viral structure is providing unique insight into the solution and gas-phase properties of viruses.

Keywords. Virus, Viral, Viral structure, Capsid, Dynamic, Capsid mobility, Whole virus, Intact virus, MALDI, Electrospray, Proteins, Inactivation

1 Introduction

Mass spectrometry is an emerging technology and now a core resource in all areas of viral research [1, 2]. It is becoming a fundamental tool for work in areas such as protein characterization [3–6], structural virology [7–12], drug discovery [12, 13], and clinical chemistry [14]. Most of the increased interest in this technique has been due to the development of matrix-assisted laser desorption/ionization [15] (MALDI) and electrospray ionization (ESI) [16], which are highly sensitive and "soft" ionization techniques.

Electrospray ionization involves the introduction of a liquid solution directly into the atmospheric pressure source through an emitter. The liquid forms a droplet at the end of the emitter, where it is exposed to a high electrical field (Fig. 1). This results in a buildup of multiple charges on the surface of the droplet. The coulombic forces from these charges ultimately result in the droplet's expulsion from the surface. The ions produced in the ion source are then extracted into the mass analyzer. ESI is now widely used for identifying small molecules, proteins, studying large non-covalent complexes, structural analysis, and as a detector for separation methods such as HPLC and capillary

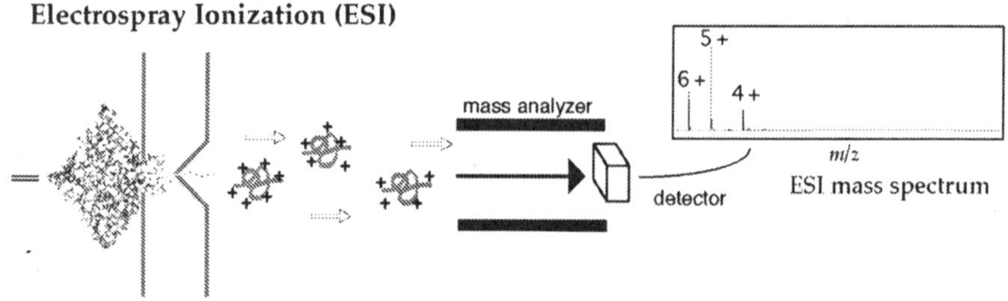

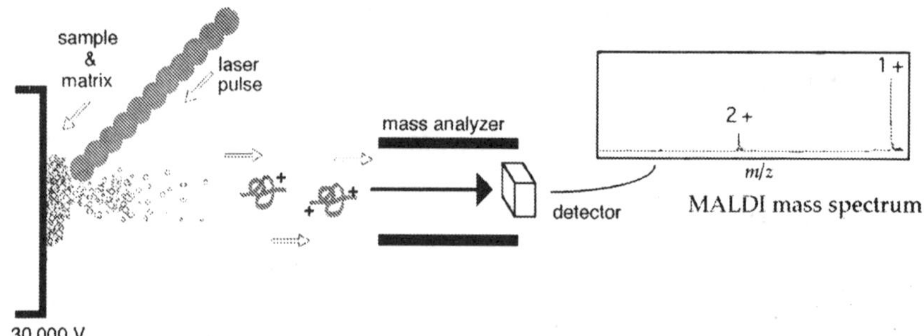

Fig. 1. MALDI and electrospray processes are represented schematically. Electrospray ionization involves the formation of charged droplets, while MALDI occurs through the laser evaporation from a crystallized sample/matrix mixture. In both techniques, the sample ions are extracted and transmitted through electrostatic lenses to a mass analyzer

zone electrophoresis (CZE). The low tolerance of ESI to salt (<10 mmol/l) has been overcome by coupling liquid chromatography to ESI-MS. Nanoliter/min low flow nanoESI-MS/MS has extended the application of ESI to highly sensitive means of protein identification.

MALDI MS has also emerged as an extremely useful method in biological analysis. A pulsed laser is used to irradiate a sample that is co-crystallized on a surface along with a matrix (e.g., 2,5-dihydroxybenzoic acid). The matrix absorbs the excess laser energy and facilitates ionization of the sample as the sample/matrix mixture is vaporized (Fig. 1). The matrix acts as a receptacle for the laser energy, therefore minimizing the decomposition of the analyte ion that would normally occur from direct laser desorption. The ions are then directed into the mass analyzer using a combination of electrostatic lens elements. While MALDI is not amenable to direct LC/MS applications, MALDI is more tolerant to salts and also better suited for direct complex mixture analysis than ESI. For this reason MALDI is very useful in the direct analysis of proteolytic digests with reported sensitivities in the low femtomole to attomole level.

ESI and MALDI-MS are routinely used for both accurate mass information on intact proteins and their proteolytic digests. As a result these methods have already helped detect viral mutants, identify capsid proteins, and post-translational modifications. Recent work has also included the detection of the first intact viral particles as well as a viral protein capsid. Other mass-based approaches like time-resolved proteolysis is giving new insight into the dynamics of viral capsid proteins in solution. This information, when combined with complementary information from X-ray crystallography studies is leading to a better understanding of viral structure and function.

2 Identification of Viruses and Viral Proteins

Mass spectrometry has been highly successful for identifying all classes of proteins including those originating from viral capsids and membrane proteins from enveloped viruses using existing genomic databases. Protein identification generally involves purification (commonly performing using gel electrophoresis), proteolytic digestion, and mass analysis. Trypsin is most often used to generate fragments since it cleaves at Lys and Arg, two amino acids which are in high abundance and usually situated in accessible regions in most proteins, therefore resulting in a significant number of peptides. Computer-aided data analysis conducted by comparing observed mass of proteolytic fragments against the predicted mass can be used to generate a list of potential matches. The computer-based analysis also provides confidence levels for each match that helps in distinguishing between statistically random and significant hits. This approach relies on the availability of the genomic sequences within the database. However, because of the small size of viral genomes, their genomes are readily available.

As a demonstration of the power of mass mapping for viral identification MALDI-MS and LC/MS/MS experiments were performed on Hong Kong 97 (HK97) icosohedral virus capsid [17]. HK97 presents an interesting system for these analyses because it has a highly stable interlinked "chain-mail" capsid that

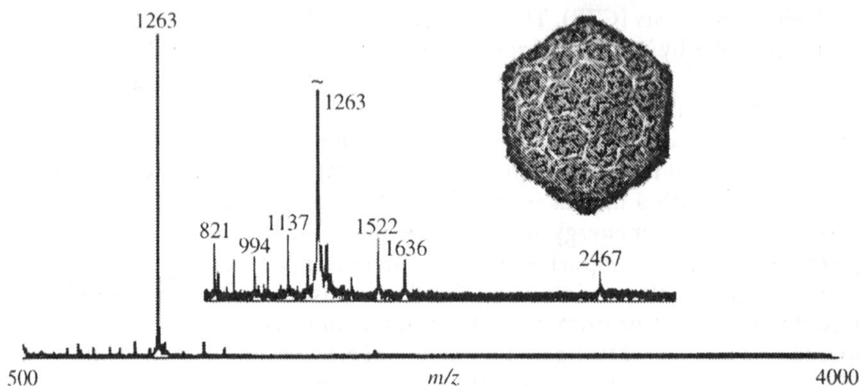

Fig. 2. Thermolysin rapidly cleaves the highly interlinked capsid from Hong Kong 97 virus at 65°C. Cleavage reaction was performed for 2 h in 100 mmol/l Tris-HCl, 10 mmol/l $CaCl_2$. Trypsin cleavage showed no significant fragmentation within this time. *Inset*: an image of HK97

is relatively impervious to enzyme digestion. Attempts at trypsin digestion of this capsid generated minimal peptide cleavage, even after 24 h. However, high temperature digestion with thermolysin generated significant cleavage within 2 h (Fig. 2). The 1263 (1263.66) mass peptide fragment appears most rapidly during the digestion process. This peptide corresponds to residues 172–181 or 173–182 in the capsid sequence. These peptides are situated at a beta sheet interface between capsid proteins and demonstrate this to be the first region of the virus coat to be susceptible to denaturation.

To establish whether thermolysin was a viable enzyme for identification using mass mapping, BSA and HK97 virus were digested and further analyzed. The molecular masses of the fragments were obtained and searches were undertaken using Profound. Using only F, L, I, and V as amino acid cleavage sites, we obtained an unequivocal identification of both BSA and HK97 virus capsid proteins. These identifications were made using the molecular weight range of 0–100 kDa and the database searches were constrained to mammals for BSA and viruses for HK97. Expanding the searches to include *all taxa* in the mass range of 0–500 kDa did not alter the identification significantly. This data indicates that thermolysin allows for identification of proteins that would be difficult or impossible to identify otherwise.

3 Identification of Mutations and Post-Translational Modifications

DNA sequencing methods have been traditionally used to identify viral mutants yet peptide mass mapping has recently been successfully used to localize mutations. By comparing proteolytic fragments from wild-type viruses and genetic mutants their characterization has been successful. The mass mapping strategy schematically, represented in Fig. 3, was first shown for identifying the mutants of the human rhinovirus (HRV) and the tobacco mosaic virus (TMV) [10].

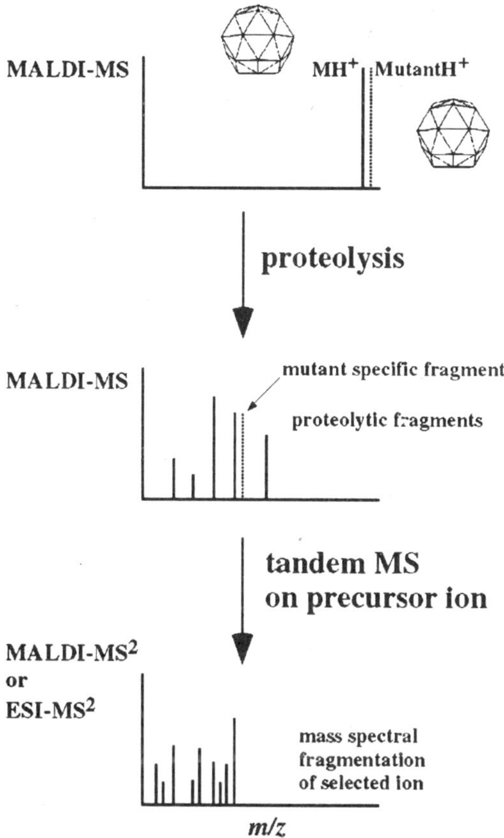

Fig. 3. The strategy for detecting viral mutations is represented. Differences between two strains of a virus are characterized by comparing mass spectra of proteolytic fragments. The specific peptide ion modification can then be localized by tandem mass spectrometry in combination with MALDI or ESI

Trypsin digests of both wild type HRV virus and the mutant were analyzed using MALDI-TOF and MALDI Fourier transform mass spectrometry (FTMS). For HRV, the mass spectra for both wild-type and mutant were identical except for one peptide occurring at *m/z* 4700. This corresponds to residues 187–227 in the wild type sequence. The corresponding peak in the mutant mass spectrum occurs at 4783.5 (Fig. 4, inset). This mass difference of 83 Da corresponds exactly to a mutation of a Cys to Trp residue and there are no other possible mutations that would be separated by 83 Da. Since there is only one Cys in the peptide 187–227 at position 199, the mutant can be localized as HRV14-Cys199Trp, which contains a Trp at position 199 instead of Cys in the wild type.

Another mutant for the tobacco mosaic virus (TMV Asp77Arg) (Fig. 4), was identified using the same approach. MALDI mass spectra of the digests of the capsid protein were identical except for the fragment occurring at 2051.4 Da in the wild type and 2091.8 Da in the mutant. The mass of the peptide for the wild

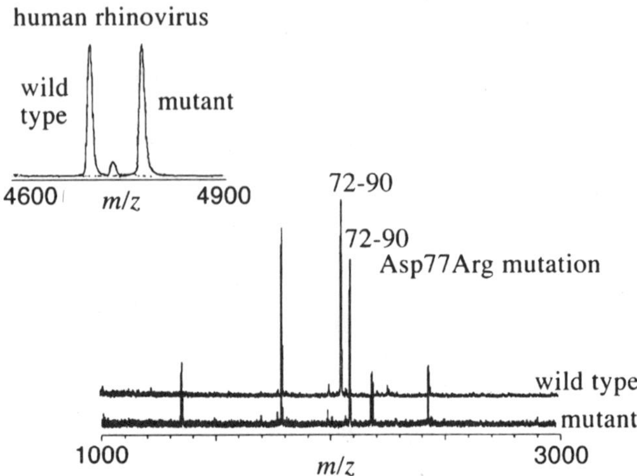

Fig. 4. The detection of mutations is shown for two viruses, HRV 14 and tobacco mosaic virus. The 83 Da mass difference for the HRV14 mutant helps identify the mutation as Cys199Tyr as the only possible mutation. For the tobacco mosaic virus, a mutation at Asp77Arg is identified similarly by comparison of mass difference with the known sequence for the tryptic fragment

type corresponds to the amino acids 72–90 in the sequence. Of the three possible mutations that cause an increase in the mass of the mutant by 40 Da, only one was possible in residues 72–90. The mutation at residue 77 replaces the Asp with Arg. Other mutations that differed by as little as 1 Da could be identified by a combination of high resolution/accurate mass measurement, tandem mass spectrometry, and genomic sequencing. In cases where genomic sequencing was required, the region to be sequenced was greatly reduced by using mass spectrometry.

Another wide application of mass spectrometry is the detection and characterization of post-translational modifications such as myristoylation, phosphorylation, disulfide bridging, etc. The detection and localization of post-translational modifications has been a rapidly developing area of mass spectrometry due to the functional importance of these modifications in biological systems. An example of this was recently shown for the case of the human rhinovirus HRV14 [10]. Electron density maps from crystallography data indicated a myristoylation of VP4. Mass analysis of VP4 also indicated a mass difference of 212 Da (consistent with myristoylation of VP4). Additional experiments with proteolytic digestion and tandem mass spectrometry were able to localize the modification to the N-terminus of VP4.

MALDI quadrupole ion trap mass spectrometry has also been used to localize and identify the post-translational modifications on the Sindai virus [18]. The polymerase associated protein (P protein) from this virus is reported in the literature to be highly phosphorylated. In vitro studies have detected phosphorylation in different regions of the protein, while a single phosphorylation site was found in the in vivo studies. Mass spectral data, along with computer-aided analysis, enabled the identification and localization of two phosphorylation sites.

Mass spectrometry was also used to localize disulfide bridging in the human respiratory syncitial virus (RSV) [19]. The attachment protein was digested with trypsin and separated using reversed phase high-performance liquid chromatography (HPLC). One tryptic peptide in the digest was detected using MALDI time-of-flight mass spectrometry that corresponded to residues 152–187 in the sequence, with four Cys residues in a disulfide linkage. Further selective digestion with thermolysin and pepsin was able to identify the linkages between Cys 173 and Cys 186 and between Cys 176 and Cys 182.

4 Peptide Mass Mapping for Protein Structure

The same proteolytic digestion methods used to analyze primary protein structure can be applied to the study of higher order structure of macromolecular assemblies. This is because the degree of proteolytic degradation for a protein complex will depend on the exposure of accessible regions on its surface. Enzymes such as trypsin when used under native conditions will be limited to surface accessible regions of the protein [4, 20–22]. The choice of proteases selected to probe the structure of the complex depends on the distribution of amino acids within the protein. Trypsin and V8 are frequently employed since these have a cleavage specificity for the hydrophilic amino acids, generally found on the surface of native proteins (R, K and D, E). The basic strategy for mass mapping for protein complexes involves the contrasting of associated and unassociated states after proteolytic digestion. The adduction of a macromolecule on the surface of the protein protects the surface in such a way as to reduce access of the protease to potential cleavage sites. This effect results in a dramatic reduction in proteolysis in the region of complexation. The basic experiment is shown in Fig. 5. Arrows mark susceptible cleavage sites on the surface and flexible loop regions. Only some of the possible tryptic fragments are generated due to presence of inaccessible regions on the protein surface.

Mass mapping for protein structure has been utilized for studying protein-protein [4, 15] and protein-DNA complexes [23]. An example of this is shown in Fig. 6 for cell cycle regulatory proteins Cdk2 and p21. A comparison of the proteolysis products of p21B for the associated and unassociated states was used to identify a 24 amino acid region on p21B as the binding site. The relative intensity of the proteolytic fragments was reduced for the p21/Cdk2 complex. Due to limited mass accuracy of the available equipment at the time and the general complexity of the mass spectra, ^{15}N labeling was used to identify unambiguously peaks originating from the p21B subunit. With the availability of more accurate MALDI-TOF mass spectrometers that facilitate ppm mass accuracy, the interpretation of mass spectra obtained from the digestion of complex assemblies is becoming easier without the need for additional steps taken in this experiment. However, it is important to note that information on the binding site of these two proteins was obtained in a matter of minutes in a mass spectrum without the need for additional separation and purification steps such as SDS/PAGE and reverse phase chromatography. This work was later corroborated by Pavletich and co-workers [24]. The structures of more complex viral protein assemblies have

in theory - proteolysis of a protein and a protein/protein complex

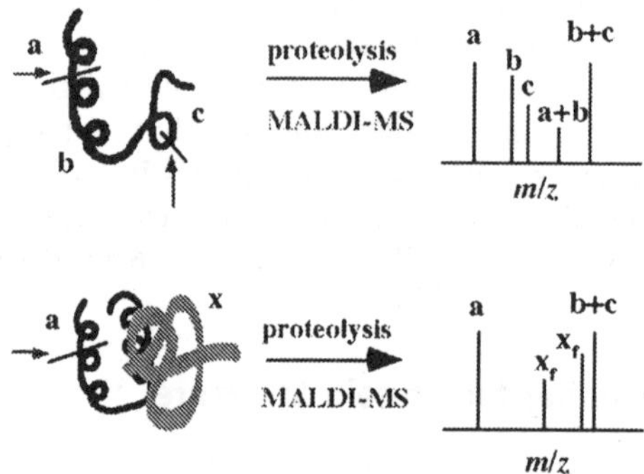

in practice - limited proteolysis of a p21-B/Cdk2 complex

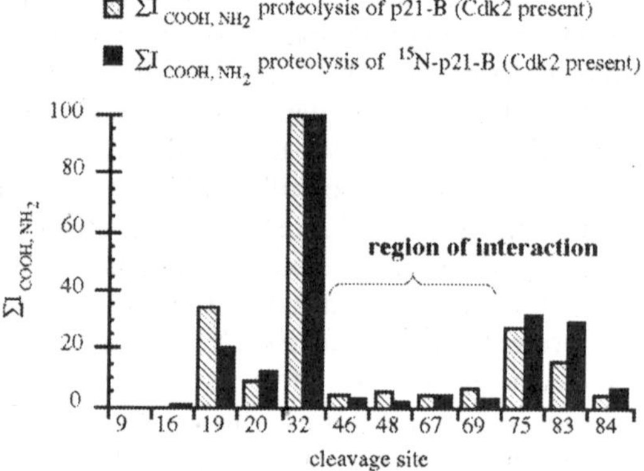

Fig. 5. The approach to probing higher order protein structure by mass spectrometry is shown. The cleavage of a hypothetical protein generates fragments (a, b, c, a+b, and b+c) that can be monitored by MALDI-MS. Here, *arrows* denote possible cleavage sites for proteolysis. The adduction of this protein with another to form a complex leads to a suppression of proteolysis in the region of association as well as fragments from the other protein (x_f). An example is shown for the limited proteolysis of p21-B/Cdk2 complex. The suppression of proteolysis in a 24 residue region of the sequence helps identify this area of association

since been explored using protein mass mapping and building on the success of the earlier work described above [8, 9, 11, 25].

5 Probing Viral Capsid Dynamics

A natural extension of protein mass mapping is getting a greater understanding of the local and global viral capsid structure. The viral capsid consists of a protein coat that can be mapped using limited and selective proteolytic digestion, since the proteins lying exterior to the viral surface are expected to be more susceptible to cleavage. Because proteolytic digestion is performed in solution, this technique is capable of detecting multiple conformers of the viral capsid and thereby extending our understanding of the dynamic domains within the virus structure.

Limited proteolysis was used on the human rhinovirus 14 (HRV14) and the Flock House virus (FHV) [9]. Four proteins, VP1, VP2, VP3, and VP4 comprise the protein coat of the HRV. According to crystal structure, VP4 lies inside the virus at the RNA/capsid interface. FHV has a protein coat which is made up of a single protein that autocatalytically dissociates to β-protein and γ-peptide during maturation. These, according to the crystal structure, have regions which lie internally and externally on the surface (Fig. 6). Time resolved proteolysis, in combination with MALDI MS analysis, was used to map the surface accessible proteins in the virus capsid. The experiment involved sequential digestion by trypsin (endoprotease), followed by carboxypeptidase Y (exoprotease). In both HRV and FHV, the analysis was able to identify surface proteins. However, fragments resulting from the digestion of internal regions, as determined by crystallography data, were also observed very early on in the time course of the proteolysis experiments. Although these results were initially perplexing, additional studies along with crystallography data suggest that the internal proteins within the capsid are highly mobile. The domains lying internally can be transiently exposed on the surface of the virus in solution. For FHV, the regions of the β-protein and γ-peptide that were transiently exposed correlate with the regions implicated in the RNA release and delivery. More recently, mass spectrometry, along with proteolytic digestion, were used to distinguish between FHV and virus-like particles (VLP) [26]. These virus-like particles are cystallographically identical to the wild-type virus except that they are generated in a special baculovirus expression system to contain the cellular RNA instead of the viral RNA. The degree of proteolysis was found to be much higher for the VLP, suggesting that the viral capsid mobility and stability in solution is significantly affected by the specific interaction at the RNA-capsid interface.

Another way of examining viral dynamics is through chemical modification. Proteins within the capsid can be exposed to modifying agents. The analysis of

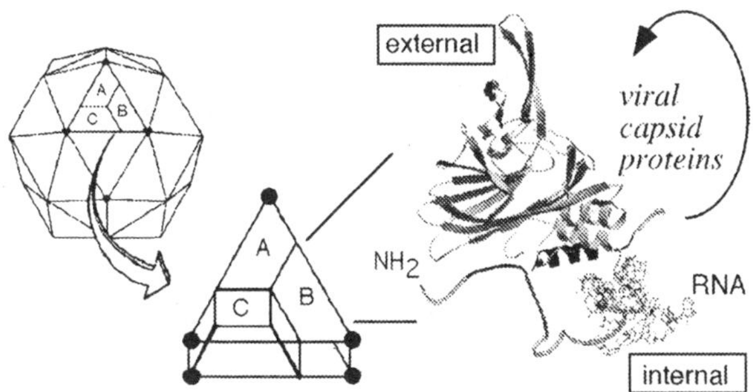

Fig. 6. The Flock House virus has an icosahedral symmetry with the γ-peptide and the C and N termini of the β-protein lying internally and away from the surface. However, time resolved proteolysis data indicates that the viral capsid is highly mobile and that internal domains are transiently exposed on the surface

the products helps identify regions of the proteins that are transiently exposed on the surface. In addition, hydrogen-deuterium (H/D) exchange experiments can yield information on different conformer populations present for a protein. For example, denatured proteins will be more susceptible to H/D exchange than proteins in a more folded state. These experiments can give helpful information on both the degree of folding and the number of conformations present. Two examples of structural characterization of viruses using chemical modification and H/D exchange are shown in Fig. 7. Here, acetylation is using to help localize the surface exposed region of the γ-peptide. An H/D exchange experiment is also shown in which the number of exchangeable hydrogen atoms reveals information about the structure of the capsid protein. Recently, H/D exchange experiments have been used to probe stages of viral capsid assembly for the dsDNA bacteriophage [27]. Here, limited proteolysis and the degree of back exchange of deuterated capsid protein were used to characterize the surface exposed regions of the virus during its dynamic transformation to the mature capsid form. In another study, pH-induced structural changes in the viral capsid that cause destabilization were characterized using mass spectrometry [28]. Therefore, time-

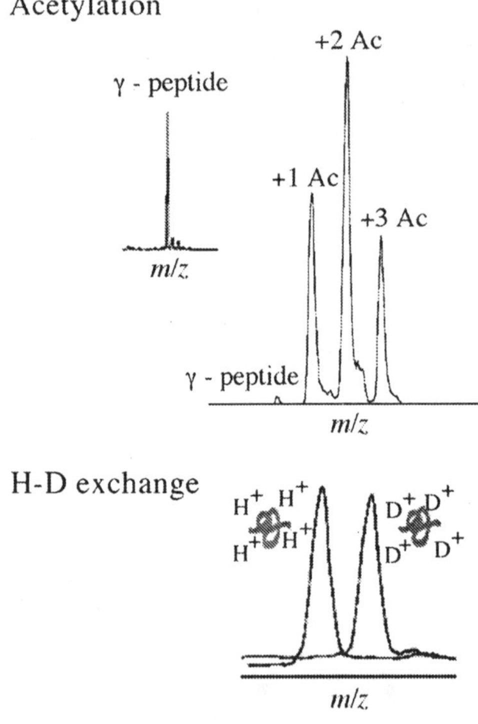

Fig. 7. *Top*: mass spectra of FHV γ-peptide after acetylation. The degree and site of acetylations, determined by tandem mass spectrometry, can be used to characterize surface accessible regions of the peptide. *Bottom*: a viral capsid protein ion generated by ESI undergoes hydrogen-deuterium exchange. The multiple populations of ions generated can help distinguish between multiple conformers present for a capsid protein

course proteolysis and chemical modification methods, when combined with mass spectrometry, provide complimentary information to X-ray crystallography. These, when taken together, can help broaden our understanding of the dynamic domains of the viral capsid.

6 Screening for Antiviral Drugs

There are many human diseases caused by viruses, but not many anti-viral drugs that directly attack the virus. Screening of new and existing compounds for anti-viral activity is a necessary step towards the development of new drugs. Inhibition of viral enzymatic reactions can be monitored with mass spectrometry. These approaches have already been developed for non-viral enzymatic reactions such as glycosylation. In the mass based approach to screening activity, the substrate, inhibitor, product, and internal standard mixture are introduced into a mass spectrometer. The product formation for a reaction can be monitored by mass analysis. This methodology was illustrated for a galactosyltransferase catalyzed reaction (Fig. 8) [13]. The inhibition of this reaction was analyzed using a library of 20 small molecules. The experiment involved monitoring product formation for 22 parallel reactions which included two controls without any inhibitor (Fig. 9). These reactions were quenched by the addition of methanol before their introduction into an ESI mass spectrometer. Three new inhibitors were identified through the screening procedure. Although this example involves a non-viral enzyme system, this technique is a fast and efficient technique that can be used to screen for antiviral drugs that target viral enzymes.

The traditional method for determining anti-viral drug activity involves conducting a plaque assay. A preliminary mass based screening can be helpful in reducing the number of drug candidates that need to be screened by more

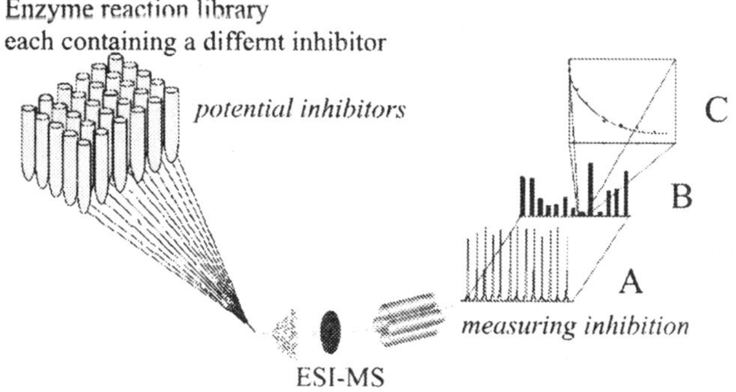

Fig. 8. A quantitative enzyme assay of enzyme inhibition is shown. The ESI/MS analysis can be automated. Samples introduced include enzyme, substrate, product, potential inhibitor and internal standard. (A) Total ion current is measured for each sample. (B) Product ion formation is plotted with respect to internal standard. (C) If an inhibitor is found, the degree of inhibition is derived using ESI-MS

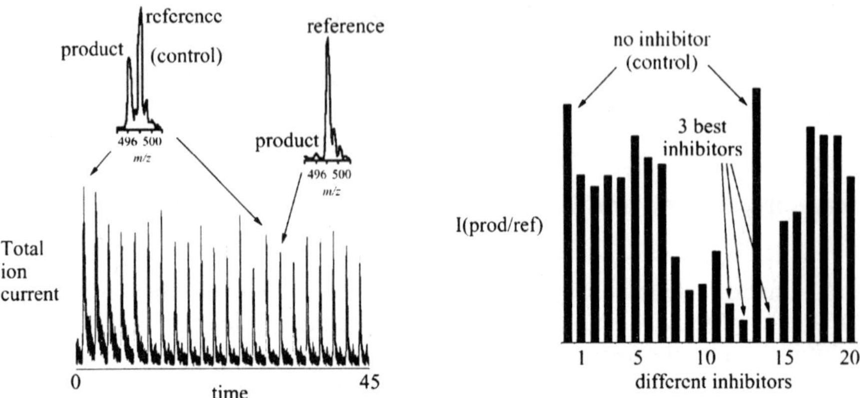

Fig. 9. A Total current observed for all 22 reactions; 20 of these had each a different inhibitor. *Insets* show data for control (*left*) and an effective inhibitor (*right*). **B** A bar graph was plotted of product ion/reference ion intensity ratio to help identify potential inhibitors

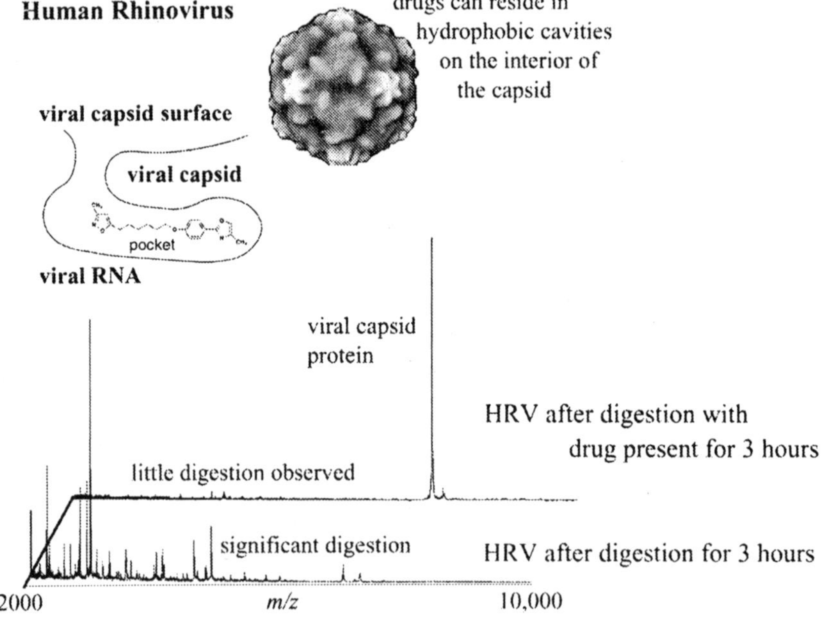

Fig. 10. Proteolysis is used to probe the antiviral activity of WIN52084, an experimental antiviral drug. The binding of WIN52084 to the viral capsid results in a suppression of proteolysis that can be monitored by mass spectrometry

time consuming plaque assays. Most antiviral drugs are targeted at the protein coat of a virus. The picornaviruses are a class of icosahedral viruses of which the human rhinovirus 14 (HRV14) is a member. These are known to have 25-Å canyons on each of the fivefold symmetry axes. The experimental drug WIN52084 is known to bind to a hydrophobic pocket within the canyon. This is of particular interest to researchers since the canyon is also believed to be the site of the cell receptor attachment before endocytosis. The binding of the antiviral drug WIN52084 has been observed to block cell attachment, inhibit uncoating, and make it easier to inactivate the virus thermally. A family of the WIN drugs were analyzed for suppression of capsid proteolysis [29]. This effect, which results from the drug binding to the protein coat, was observed for other drugs, but was most pronounced in WIN52084 (Fig. 10). Other viruses were also analyzed using this method and their capsid proteolysis was found not to be inhibited by WIN52084. This was further evidence of the specificity of the drug for HRV14, and that proteolysis suppression was not an effect resulting from the inhibition of the enzyme. The dramatic reduction of proteolysis for the virus in the presence of WIN52084 is believed to result from the interruption of virus capsid mobility which is essential for the virus to initiate receptor attachment and the ensuing endocytosis.

7 Differentiating Viral Genomes by Mass

The analysis of viral DNA and RNA directly using mass spectrometry has been challenging due to the low levels present in biological samples and the limited mass range offered by most commercial mass spectrometers. However, nested polymerase chain reaction (PCR), along with MALDI/TOF analysis, has been used to detect the presence of low levels of virus in hepatitis B patients [30]. The hepatitis B virus related products were purified, immobilized, denatured, and analyzed using MALDI/TOF mass spectrometry. Very massive single ions of DNA of several megadaltons molecular weight have been observed using ESI Fourier transform mass spectrometry (FTMS) [31] and charge detection mass spectrometry [32, 33]. Both techniques employ the amplification and measurement of a weak image current induced by the ion. In charge detection mass spectrometry, a very sensitive amplifier is used to measure directly the charge and velocity of an ion. The velocity is used to deduce the m/z which can be used along with the measured charge to give molecular weights of very heavy ions. These techniques may be promising for measuring viral DNA directly as a means for identification and characterization of genetic variability. The first observation of an intact virus particle was made by charge detection mass spectrometry as discussed in the section below.

8 Analyzing Whole Viruses

There have been significant developments in the past few years in the area of macromolecular analysis of whole viruses and viral capsids. The first was the

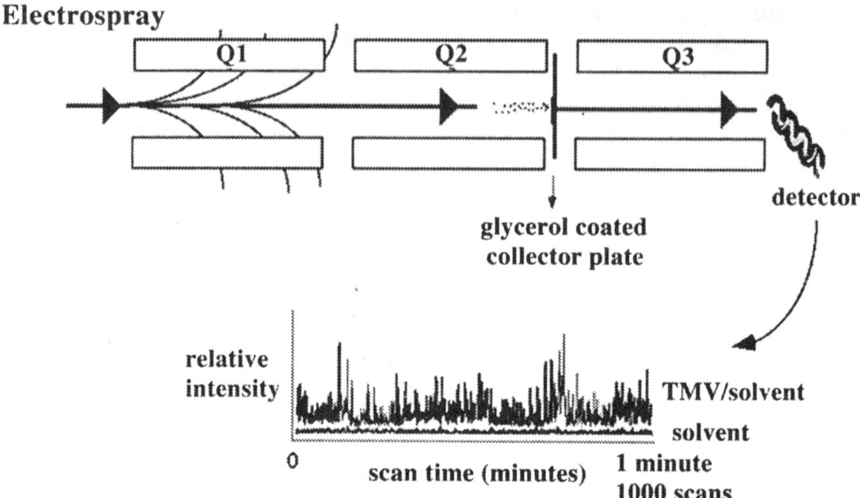

Fig. 11. The experiment of indirectly "detecting" an intact viral particle after its transmission through an electrospray mass spectrometer. The viral particles were observed under an electron microscope after collection since the massive particles exceeded the working range of the electron multiplier normally used with this mass spectrometer

examination of virus viability following transmission of the whole virus through a mass spectrometer [34]. A brass plate was placed between Q2 and Q3 in a triple quadrupole mass spectrometer where viral particles could be collected (Fig. 11). The plates, when examined using electron microscopy, were found to have intact viral particles. The mass analyzed and collected TMV viral particles were successfully used to infect tobacco plants. The surprising results of these studies demonstrated that intact viruses could be transmitted through the ESI mass spectrometer and that the ionization method was gentle enough for the viruses to retain their virulence. More recently, the measurement of an intact viral capsid was accomplished using ESI/TOF mass spectrometry by Robinson and coworkers [35].

Because of the size of viruses the challenge of measuring them intact required the development of new technology. The direct measurement of the charge state of very massive ions has been shown using charge detection TOF/MS through image current measurement. In charge detection mass spectrometry, both the charge and the time-of-flight, yielding *m/z*, are measured one ion at a time. Intact viral particles of the rice yellow mottle virus (RYMV) and the tobacco mosaic virus (TMV) were studied using this methodology [36]. The RYMV is an icosahedral non-enveloped virus consisting of a single strand of RNA and multiple copies of a single protein. The RYMV and TMV have theoretical molecular weights of 6.5×10^6 Da and 40.5×10^6 Da, respectively. The mass spectra obtained for RYMV and TMV are shown in Fig. 12. The broad peaks obtained for each virus correlated with the calculated masses. This work demonstrates the use of mass spectrometry for identifying pathogenic viruses.

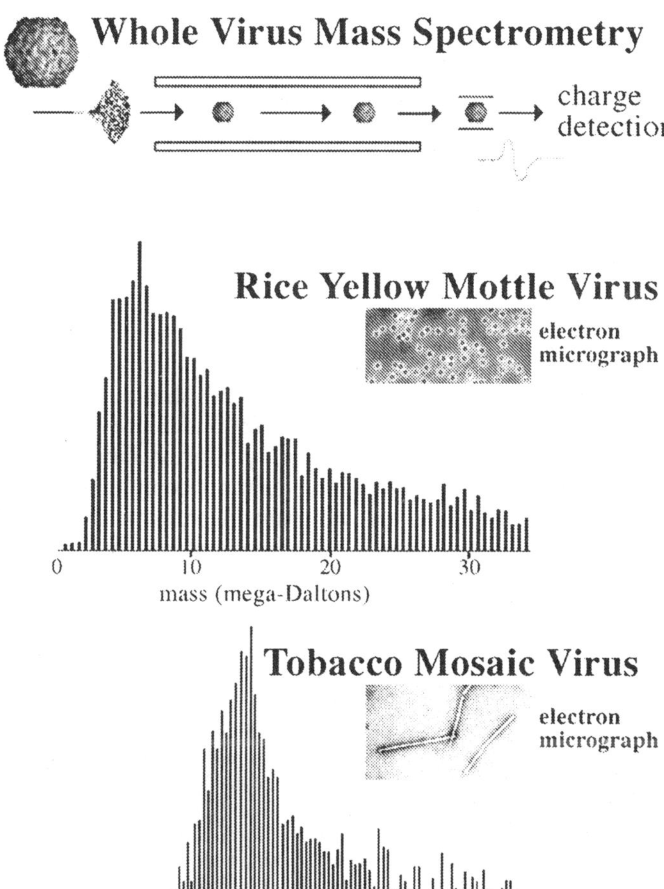

Fig. 12. Mass spectra of Rice Yellow Mottle Virus (RYMV) and Tobacco Mosaic Virus (TMV) particles analyzed with an electrospray ionization charge detection time-of-flight mass spectrometer. *Inset*: electron micrographs of the icosahedral RYMV (diameter 28.8 nm) and the cylindrical TMV (~300 nm long and 17 nm in diameter). The known molecular weight of RYMV and TMV are 6.5×10^6 and 40.5×10^6 Daltons, respectively

9
Future Prospects and Conclusions

Mass spectrometry has opened up new opportunities for understanding viruses. ESI, MALDI, and emerging new technologies such as desorption ionization on silicon (DIOS) [37–39] are in a continual state of flux, with significant improvements accruing yearly. The ultra high accuracy provided by FTMS is being increasingly applied to solve biologically significant problems and promises to become an important tool for probing viruses. The use of proteolysis, with mass based approaches is helping researchers not only to probe the functional and genetic diversity within viral species, but also to understand the dynamic nature of the viral capsid. One of the most exciting developments has been the use of time-resolved proteolysis to study capsid dynamics, and as a means to characterize protein-protein and RNA-capsid interactions. Since proteolysis is performed in solution, it can yield complementary information to crystallography data. These techniques, when extended to even more complex enveloped viruses like HIV and influenza, will surely provide valuable information about these important pathogens, and may lead to the development of new and more effective anti-viral drugs.

10
References

1. Hop CECA, Bakhtiar R (1997) An introduction to electrospray ionization and matrix-assisted laser desorption/ionization mass spectrometry: essential tools in a modern biotechnology environment. Biospectroscopy 3:259–280
2. Siuzdak G (1996) Mass spectrometry for biotechnology. Academic Press, San Diego
3. Shevchenko A, Jensen ON, Podtelejnikov AV, Sagliocco F, Wilm M, Vorm O, Mortensen P, Shevchenko A, Boucherie H, Mann M (1996) Linking genome and proteome by mass spectrometry: large-scale identification of yeast proteins from two dimensional gels. Proc Natl Acad Sci USA 93:14,440–14,445
4. Kriwacki RW, Wu J, Tennant L, Wright PE, Siuzdak G (1997) Probing protein structure using biochemical and biophysical methods – proteolysis, matrix-assisted laser desorption/ionization mass spectrometry, high-performance liquid chromatography and size-exclusion chromatography of p21(Waf1/Cip1/Sdi1). J Chromatogr A 777:23–30
5. Chaurand P, Luetzenkirchen F, Spengler B (1999) Peptide and protein identification by matrix-assisted laser desorption ionization (MALDI) and MALDI-post-source decay time-of-flight mass spectrometry. J Am Soc Mass Spectrom10:91–103
6. Yates JR, Carmack E, Hays L, Link AJ, Eng JK (1999) Automated protein identification using microcolumn liquid chromatography-tandem mass spectrometry. Methods Mol Biol 112:553–569
7. Siuzdak G (1998) Probing viruses with mass spectrometry. J Mass Spectrom 33:203–211
8. Broos K, Wei J, Marshall D, Brown F, Smith TJ, Johnson JE, Schneemann A, Siuzdak G (2001) Viral capsid mobility: a dynamic conduit for inactivation. Proc Natl Acad Sci USA 98:2274–2277
9. Bothner B, Dong X-F, Bibbs L, Johnson JE, Siuzdak G (1998) Evidence of viral capsid dynamics using limited proteolysis and mass spectrometry. J Biol Chem 273:673–676
10. Lewis JK, Bendahmane M, Smith TJ, Beachy RN, Siuzdak G (1998) Identification of viral mutants by mass spectrometry. Proc Natl Acad Sci, USA 95:8596–8601
11. Lewis JK, Bothner B, Smith TJ, Siuzdak G (1998) Antiviral agents block breathing of common cold virus. Proc Natl Acad Sci USA 95:6774–6778

12. Lewis JK, Chiang J, Siuzdak, G (1999) Monitoring protein-drug interactions with mass spectrometry. J Assoc Lab Automat 4:46–48
13. Wu J, Shuichi T, Wong C-H, Siuzdak G (1997) Quantitative electrospray mass spectrometry for the rapid assay of enzyme inhibitors. Chem Biol 4:653–657
14. Chace DH (2001) Mass spectrometry in the clinical laboratory. Chem Rev 101:445–477
15. Hillenkamp F, Karas M, Beavis RC, Chait BT (1991) Matrix-assisted laser desorption ionization mass spectrometry of biopolymers. Anal Chem 63:A1193-A1202
16. Siuzdak G, Bothner B, Yeager M, Brugidou C, Fauquet CM, Hoey K, Chang C-M (1996) Mass spectrometry and viral analysis. Chemi Biol 3:45–48
17. Bark SJ, Muster N, Yates JR, Siuzdak G (2001) High temperature protein mass mapping using a thermophilic protease. J Am Chem Soc 123:1774–1775
18. Jonscher KR, Yates JR (1997) J Biol Chem 272:1735–1741
19. Gorman JJ, Ferguson BL, Speelman D, Mills D (1997) Determination of the disulfide bond arrangement of human respiratory syncytial virus attachment (G) protein by matrix-assisted laser desorption/ionization time-of-flight mass spectrometry. Protein Sci 6:1380
20. Kriwacki RW, Wu J, Siuzdak G, Wright PE (1996) Probing protein/protein interactions with mass spectrometry and isotopic labeling: analysis of the p21/Cdk2 complex. J Am Chem Soc 118:5320–5321
21. Fontana A, Zambonin M, deLaureto PP, DeFilippis V, Clementi A, Scaramella E (1997) Probing the conformational state of apomyoglobin by limited proteolysis. J Mol Biol 266:223–230
22. Fontana A, Polverino de Laureto P, De Filippis V, Scaramella E, Zambonin M (1997) Probing the partly folded states of proteins by limited proteolysis. Folding Des 2:R17–R26
23. Cohen SL, Ferre-D'Amare AR, Burley SK, Chait BT (1995) Probing the solution structure of the DNA-binding protein Max by a combination of proteolysis and mass spectrometry. Protein Sci 4:1088–1099
24. Russo AA, Jeffrey PD, Patten AK, Massague J, Pavletich NP (1996) Crystal structure of the $p27^{kip1}$ cyclin-dependent-kinase inhibitor bound to the cyclin A-Cdk2 complex. Nature 382:325–331
25. Millar AL, Jackson N, Dalton H, Jennings KR, Levi M, Wahren B, Dimmeck N (1998) Rapid analysis of epitope-paratope interactions between HIV-1 and 17 amino-acid neutralizing microantibody by ESI mass spectrometry. Eur J Biochem 258:164–169
26. Bothner B, Schneemann A, Marshall D, Reddy V, Johnson JE, Siuzdak G (1999) Crystallographically identical virus capsids display different properties in solution. Nat Struct Biol 6:114–116
27. Tuma R, Coward LU, Kirk MC, Barnes S, Prevelige PE (2001) Hydrogen-deuterium exchange as a probe of folding and assembly in viral capsids. J Mol Biol 306:389–396
28. Wang L, Lane LC, Smith DL (2001) Detecting structural changes in viral capsids by hydrogen exchange and mass spectrometry. Protein Sci 10:1234–1243
29. Lewis JK, Bothner B, Smith TJ, Siuzdak G (1998) Antiviral agent blocks breathing of common cold virus. Proc Natl Acad Sci USA 95:6774–6778
30. Jurinke C, Zöller B, Feucht H, Van den Boom D, Jacob A, Polywka S, Laufs R, Koster H (1998) Application of nested PCR and mass spectrometry for DNA-based virus detection: HBV-DNA detected in the majority of isolated anti-HBC positive sera. Genet Anal Biomol Eng 14:97–102
31. Chen X, Camp DG II, Wu Q, Bakhtiar R, Springer DL, Morris BJ, Bruce JE, Anderson GA, Edmonds CG, Smith RD (1996) Nucl Acids Res 24:2183–2189
32. Schultz JC, Hack CA, Benner WH (1998) Mass determination of megadalton-DNA electrospray ions using charge detection mass spectrometry. J Am Soc Mass Spectrom 9:305–313
33. Fuerstenau SD, Benner WH (1995) Molecular weight determination of megadalton DNA electrospray ions using charge-detection time-of-flight mass spectrometry. Rapid Commun Mass Spectrom 9:1528–1538
34. Siuzdak G, Bothner B, Yeager M, Brugidou C, Fauquet CM, Hoey K, Chang CM (1996) Mass spectrometry and viral analysis. Chem Biol 3:45–48

35. Tito MA, Tars K, Valegard K, Hajdu J, Robinson CV (2000) Electrospray time-of-flight mass spectrometry of the intact MS2 virus capsid. J Am Chem Soc 122:3550–3551
36. Fuerstenau SD, Benner WH, Thomas JJ, Brugidou C, Bothner B, Siuzdak G (2001) Mass spectrometry of an intact virus. Angew Chem Int Ed 40:542–544
37. Wie J, Buriak JM, Siuzdak G (1999) Desorption-ionization mass spectrometry on porous silicon. Nature 399:243–246
38. Thomas JJ, Shen Z, Crowell JE, Finn MG, Siuzdak G (2001) Desorption/ionization on silicon (DIOS): a diverse mass spectrometry platform for protein characterization. Proc Natl Acad Sci 98:4932–4937
39. Shen Z, Thomas JJ, Averbuj C, Broo KM, Engelhard M, Crowell JE, Finn MG, Siuzdak G (2001) Porous silicon as a versatile platform for laser desorption/ionization mass spectrometry. Anal Chem 73:612–61

Top Curr Chem (2003) 225: 283–302
DOI 10.1007/b10472

High-Throughput Mass Spectrometry for Compound Characterization in Drug Discovery

Mark Brönstrup

Aventis Pharma, Chemistry, Building G878, 65926 Frankfurt, Germany
E-mail: Mark.Broenstrup@aventis.com, Fax: +49 69 305 21618

A key element of modern drug discovery is the efficient parallel synthesis of large numbers of individual compounds, followed by extensive profiling for chemical as well as biological properties. As a consequence, analytical chemistry is continually challenged to provide compound characterization with sufficient throughput and quality. This review covers those techniques using mass spectrometry (MS) to address the three simple questions 'What is it?', 'What is the relative purity?', and 'What is the absolute content?' Today, nominal molecular mass can be rapidly determined by flow injection analysis (FIA) with speeds of less than 5 s per sample and with miniaturized formats with very low sample consumption. In most cases, mass spectrometry is coupled to a separation technique to obtain purity information. Fast serial HPLC/MS, SFC/MS, or parallel HPLC/MS are shown to reach throughputs of less than 1 min per sample. For relative and absolute quantifications, ELSD and CLND detectors have been established along with UV or TIC detection. Structural information is obtainable through automated accurate mass measurements and tandem MS experiments. Finally, software tools that facilitate data processing and interpretation for increased levels of mass spectrometric data are presented.

Keywords. High throughput techniques, Chromatography, Mass spectrometry, Drug discovery Combinatorial chemistry

List of Abbreviations

APCI	Atmospheric pressure chemical ionization
API	Atmospheric pressure ionization
APPI	Atmospheric pressure photoionization
CLND	Chemiluminescent nitrogen detection
DIOS	Desorption ionization on porous silicon
DMPK	Drug metabolism and pharmacokinetics
EI	Electron ionization
ELSD	Evaporative light scattering detection
ESI	Electrospray ionization
FIA	Flow injection analysis
FTICR	Fourier transform ion cyclotron resonance
HPLC	High performance liquid chromatography
HT	High throughput
LIMS	Laboratory information management system
MALDI	Matrix assisted laser desorption ionization
MS	Mass spectrometry
MS/MS	Tandem mass spectrometry
SFC	Supercritical fluid chromatography
TIC	Total ion current
TOF	Time of flight

1 Introduction

In the last decade, key steps in drug discovery have been accelerated by remarkable technology breakthroughs: genomics and proteomics have disclosed a large number of potential targets, which can be rapidly tested for biological activity by high throughput screening techniques. The proliferation of chemical compounds for testing has been speeded up by combinatorial chemistry. The remarkable increase in compound output challenged analytical chemistry to keep pace by providing a) analytical techniques with sufficient throughput and b) tools to interpret the large quantity of generated data. The techniques have to follow changes in combinatorial chemistry strategies, where split-and-mix synthesis of complex mixture libraries has by and large given way to parallel synthesis of single substances in separate vials. In addition, focus has shifted from large, diverse collections for lead identification to the synthesis of smaller compound series directed by medicinal chemistry programs to establish structure-activity relationships. In consequence, the requirements on quality of analytical data has increased, because a correct determination of structural integrity as well as relative and absolute purity are crucial for parallel compound profiling programs. Thorough profiling at early project stages has proven to be cost- and time-saving and includes structural, physicochemical, and DMPK characterization next to biological activity determination. Thus, the review deals with the three simple analytical questions 'What is it?', 'What is the relative purity?', and 'What is the absolute content?'. Ideally, a synergistic combination of complementary analytical techniques is applied to give a thorough answer. Considerable advances in NMR [1–3] or infrared spectroscopy [4, 5] have facilitated this combined approach, which is indispensable for safe confirmation or even elucidation of chemical structures. This contribution is confined to techniques using mass spectrometry. As demonstrated in precedent reviews [6–13], mass spectrometry undoubtedly plays a key role in high throughput compound characterization, mainly since established MS-merits like sensitivity and specificity have been successfully combined with ruggedness, LC-compatibility, and speed. For example, small molecules can routinely be investigated with a sample consumption at the low pmol level. Advances on all instrument components including sample delivery, ion sources, mass analyzers, and software control have enhanced performance and in particular robustness, rendering the open-access operation of instruments by a large number of users without extensive training possible [14–17]. Mixture analysis is readily achieved and can be extended to exceedingly complex compositions. As more than 10,000 compounds have been resolved in a single FTICR mass spectrum [18], and as HPLC is able to separate more than 1000 compounds in time [19], the resolution of $>10^6$ compounds is conceivable by HPLC/MS. For applications of mass spectrometry in closely related drug discovery areas like genomics [20, 21], proteomics [22], library purification [8, 9, 13, 23], or pharmacokinetics [6, 12], the reader is referred to the literature. An innovative approach to lead generation uses mass spectrometry to screen for small molecules that bind to biomolecular targets. Experiments that involve detection of the high-mass target are reviewed elsewhere [24–27]. If small molecules are monitored [27–29], the same MS techniques as described below are relevant. The speed of analysis is a central aspect of this review. Factors contributing to speed are discussed for a suite of analytical steps starting from sample introduction via separation techniques, ionization, mass analysis, to data analysis.

2
Flow Injection Analysis

One of the most basic characteristics of a compound is its molecular mass; in this context, much of today's popularity of MS can be traced back to the simplicity of molecular mass readout from API spectra. A common experiment is flow injection analysis (FIA). Here, an autosampler is used to inject an aliquot of dissolved sample into a liquid stream, which is provided by an LC pump, to the MS-detector. FIA offers the advantages of easy automation and fast cycle times of about 30 s per sample. In order to increase throughput, multiprobe autosam-

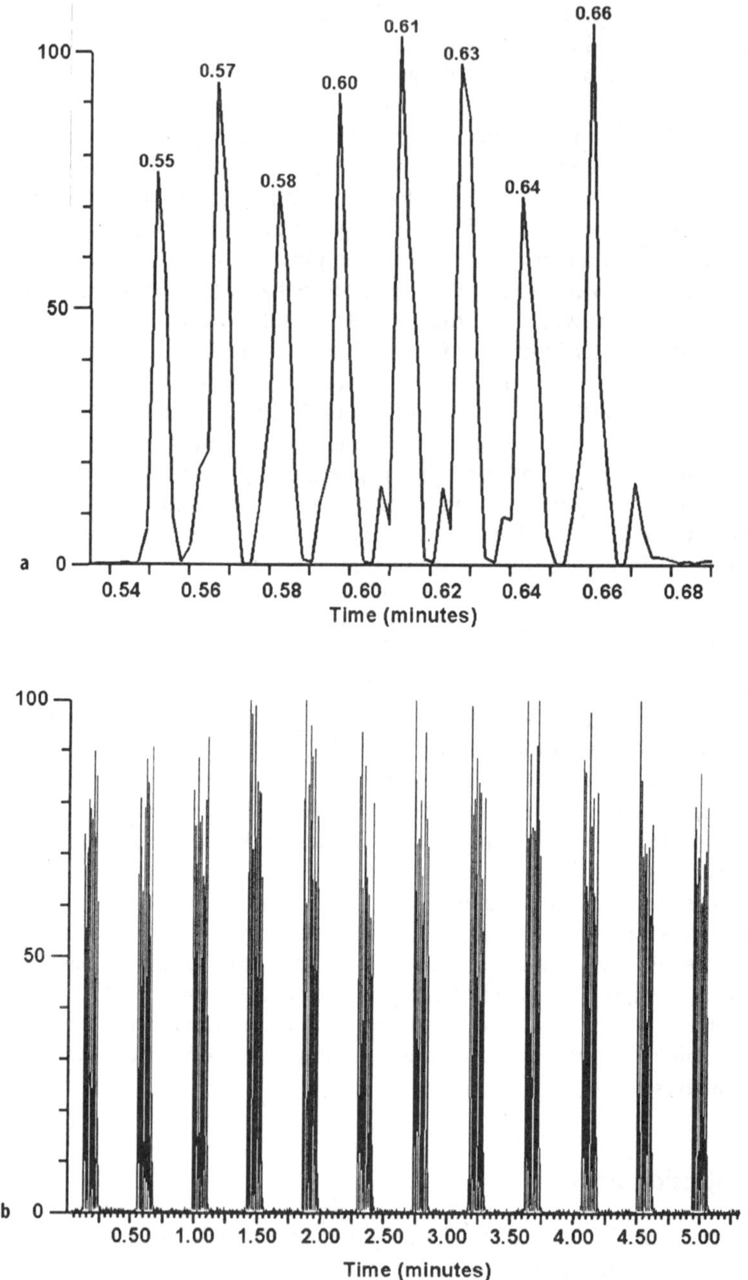

Fig. 1a, b. FIA-MS single ion chromatogram of quinine (MW=324.2, $C_{20}H_{24}N_2O_2$): **a** single set of eight injections; **b** full 96-well microtiter plate, obtained with a flow rate of 2 ml/min, and no split prior to sample introduction to the mass spectrometer. The switching valves were triggered at a rate of ~0.8 s/sample. The peak widths were measured as 0.5 s at fwhm (a) and 0.9 s at the base (b). Reprinted with permission from K. Morand [31]

plers have been applied to inject eight samples to eight injector valves, which are connected sequentially to the detector in short time distances [30, 31]. Morand et al. have reported an optimized setup with average analysis times of 4 s per sample (Fig. 1). As the width of the flow injection peak with baseline resolution is only 0.9 s, further optimization will focus on a reduction of the intraplate cycling time, which accounted for 74% of the total analysis time [31].

Approaches to miniaturize FIA completely omit the pumping unit; sample introduction is achieved with a short connection from the sample vial to the source by either reducing the pressure in the source or by applying headspace gas pressure on the sample. Main advantages of automated nano-ESI systems are high sensitivity, very low sample consumption, and variable analysis times. Felton et al. describe the direct connection of the sample vials of a microtiter plate to the ESI interface with a single ESI-nozzle to achieve a cycle time of about 10 s per sample with a sample volume consumption of 120 nl [32]. An integrated washing device reduces sample carry-over. Carry-over is completely avoided by arrays of ESI capillary tips, rendering a throughput of 5 s per sample possible, as reported by Liu et al. [33]. A fully automated setup has been released commercially by Advion Biosciences recently [34]. The sample is aspirated from a 96-well plate into a conductive pipette, which is then connected to a microfabricated 3.9×3.9 cm chip that holds a 10×10 array of nanoelectrospray nozzles with 10 μm ID (Fig. 2). For each sample, a new pipette tip and a separate nozzle is used. Spray current fluctuations during the spraytime as well as nozzle-to-nozzle variability have been shown to be very small, allowing applications in quantitative MS, proteomics, or noncovalent interactions. Further progression leaps towards the 'lab on a chip' are conceivable when miniaturized sprayers are coupled to microfluidic separation devices [35–37] and, ultimately, to miniaturized mass analyzers [38].

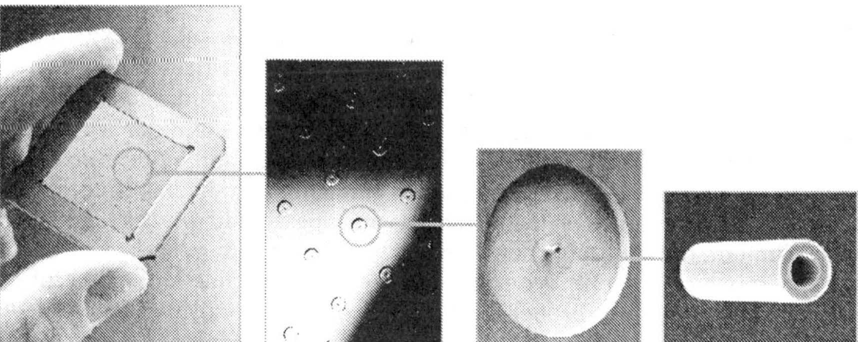

Fig. 2. Advion's ESI Chip is an array of 100 independent nanoelectrospray nozzles etched in silicon. Image 1: the 10×10 array ESI Chip (chip dimensions=3.9×3.9 cm). Image 2: magnification of 10×10 array. Image 3: magnification on one nanoelectrospray device. Image 4: magnification on one nanoelectrospray nozzle. Reprinted with permission from Advion Bio-Sciences, Inc

3 Separations

The most frequently chosen method for compound characterization in the pharmaceutical industry is LC/MS [6]. Replacing flow injection by a chromatographic separation prior to MS analysis offers three major advantages: i) impurities or by-products are separated in time from the product of interest, rendering a purity assessment of the sample possible; ii) ionization suppression of the compound of interest by salts, detergents, or by-products is avoided; iii) the interpretation of mass spectra of 'pure' compounds is much easier than the MS analysis of mixtures. Combinations of separation techniques with mass spectrometry have been reviewed recently by Tomer [39].

Compared to LC/MS, GC/MS methods are applied to a much smaller extent, mainly because many pharmaceutical compounds cannot be eluted from a GC column without prior derivatization steps. However, GC/MS is still the method of choice for the analysis of small molecules that can be evaporated without decomposition due to its unmatched chromatographic resolution and highly reproducible, compound-characteristic EI-MS spectra [40]. Whereas compound-specific derivatization is time-consuming, the GC/MS analysis itself can be very fast, as a review by Leclercq and Cramers demonstrates [41].

3.1 Fast LC/MS

The LC-runtimes can be minimized by employing elements of fast LC/MS like short columns (50–20 mm length, 4.6–2 mm inner diameter) with small particles (<5 μm), high flow rates (2–5 ml min^{-1}) and steep gradients. Fortunately, the gain in speed is associated with only little sacrifice in chromatographic resolution, as proven by several groups [42–46]. Kennedy et al. [47] have comprehensively reviewed recent progress in fast LC. A concise theoretical and experimental overview is given by Cheng et al. [48]; the authors investigated the effect of flow rates, column dimensions, particle size, gradient duration, and solvent pH on peak capacity, and demonstrate that separation times <1 min can be realized. Under their 'ultrafast' conditions, which comprise a 2.1 mm×20 mm column with 2.5 μm particles, 1.5 ml flow rate and a gradient duration of 0.7 min, a mixture of five compounds is eluted within 9 s with a peak width of <1 s (Fig. 3). An upper limit on the applicable flow rate is set by the maximum back-pressure of the system, which is mainly determined by the column. Reduction of back-pressure has been achieved through heating the column (to ca. 40–60°C), thereby lowering the viscosity of the mobile phase [49]. Another approach makes use of monolithic silica columns, which are featured by a sponge-like, continuous silica network structure with high through-pore-size/skeleton-size ratio and high porosity, and a much lower back-pressure compared to particulate column material [50, 51]. For example, flow rates as high as 6 ml min^{-1} at standard column dimensions and pressures (4.6 mm×50 mm, 61 bar) could be applied for the separation of a test mixture in <1 min [52]. Increasing the flow rate from 1 ml min^{-1} to 6 ml min^{-1} did not harm separation efficiency, signal-to-noise ratio, and re-

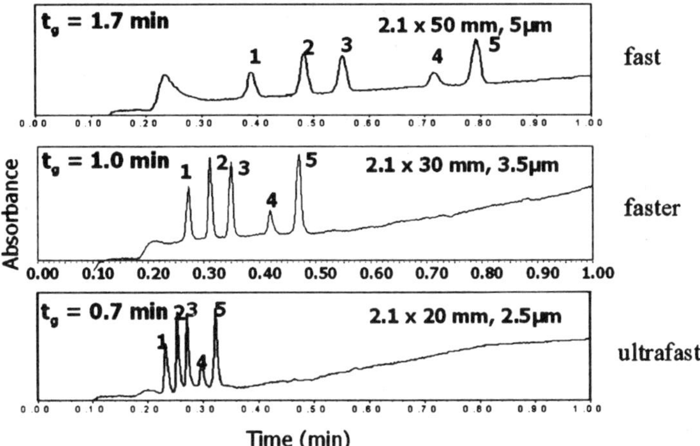

Fig. 3. Strategy to achieve fast, faster, and ultrafast analysis. Both the particle size and the gradient duration are scaled down in proportion to the column length. This ultrafast analysis demonstrates that the smallest particle size (2.5 µm) in the shortest column length (20 mm), together with the shortest gradient duration time (t_g=0.7 min), resulted in a baseline resolution of a wide range of analytes in a 9 s window. The size of the detector cell was 2.6 µl. Reprinted with permission from U. Neue [48]

solution. We note in passing that fast LC also imposes special requirements for MS instrumentation. Small peak widths of 1 s or less require fast acquisition of full-range mass spectra; TOF analyzers that can achieve sampling rates beyond 100 spectra per second are superior to other principles in this respect [53–56]. As today's ESI sources usually operate at flow rates <1 ml min^{-1}, a considerable portion of the analyte has to be split away from the MS detector for applications with mobile phase flows of several milliliters. In order to minimize peak broadening effects caused by splitting, atmospheric pressure ion sources that can handle higher flow rates whilst maintaining sensitivity are desirable.

3.2 SFC/MS

An alternative separation technique that has attracted growing interest during the last years is supercritical fluid chromatography (SFC) [57, 58]. In SFC, the mobile phase is carbon dioxide compressed to its supercritical state. A gradient of a polar, organic modifier like methanol is added in order to improve the solubility of polar analytes in the mobile phase. Being a normal-phase method, the separation characteristics of packed column SFC are complementary to conventional reversed-phase HPLC. As the mobile phase has a lower viscosity and higher diffusion coefficients, the range of optimum linear velocity is increased up to fivefold according to Van Deemter plots, and the number of theoretical plates is up to three times higher compared to HPLC [59]. In consequence, very high flow rates are applicable, which renders the chromatographic resolution or the achievable separation speed higher than in HPLC. Convincing examples for the high resolving power of SFC come, inter alia, from enantiomeric separations [60–62]. Coupling SFC to MS detectors is facilitated by the fact that carbon dioxide is much more volatile than LC solvents; therefore, the effluent from SFC systems can be directly coupled to MS detectors without flow splitting to give TIC traces without any loss in resolution [60]. Its speed has been demonstrated by Ventura et al., who analyzed combinatorial library compounds by SFC/MS using APCI in positive and negative modes [63, 64]. Compared to their standard HPLC method (6.5 min gradient, 2.1 mm×30 mm column with 3 µm particles, 0.55 ml min^{-1} flow), the authors observe enhanced selectivity, more precise quantitation of mixtures, and a threefold reduction in cycle times. Recently, the high throughput capabilities for compound characterization have been fully exploited by Bolanos et al., who reported SFC/ESI-TOF-MS cycle times of 40 s at a flow rate of 10 ml min^{-1} [65], which boosts the maximum capacity of a single instrument in serial mode far beyond 100,000 samples per year. Even faster runtimes are possible in DMPK applications, as demonstrated by Hoke et al. [66]. The group coupled the multiinjector setup described above for FIA-MS [31] with SFC, a short guard column (2×10 mm, 5 µm), and a triple quadrupole mass spectrometer to determine dextromethorphan with a cycle time of 5 s per sample only.

4
Parallel Separations

The answer for high-throughput analytical chemistry to parallel organic synthesis must be parallel compound characterization. In the case of LC/MS, parallel analysis cannot be realized by multiplication of the number of instruments due to limited benchspace, limited number of operators, and limited budgets. Fully indexed, parallel chromatographic systems with either four or eight channels have been developed instead [67–71]. In the eight channel configuration depicted in Fig. 4, the flow from one HPLC pump is equally split to eight columns that are connected to one multichannel UV detector and one TOF mass spectrometer. The system's heart is a multiple electrospray source (MUX, Micromass, Manchester, UK) with a sampling rotor that connects only one spray at a time with the TOF mass analyzer, thereby blocking the seven other sprays (Fig. 5). A cross contamination from the other sprays is practically absent. After an acquisition time of typically 100 msec per channel, the rotor hops to the next spray within ≈50 msec, yielding an overall cycle time of ca. 1.2 s^{-1}. The computer software generates a unique single data file for each channel with the corresponding MS and UV data. Although ultrafast LC methods as described above, that are featured by peak widths of about 1 s, cannot be applied due to the limited MUX duty cycle, the gain in throughput over conventional systems is tremendous. After careful optimization of LC parameters like columns, flow rates, or splitter position, Fang et al. achieve a long-term throughput of 3200 compounds per day [71] with a 3.5 min cycle time for eight samples from combinatorial chemistry. The setup depicted in Fig. 4 with a non-regulated tee-device after the LC pump only provides eight equal flow rates if all channels exhibit identical back-pressures; this condition may not be maintained due to contamination effects. Tolson et al. reported a significantly improved chro-

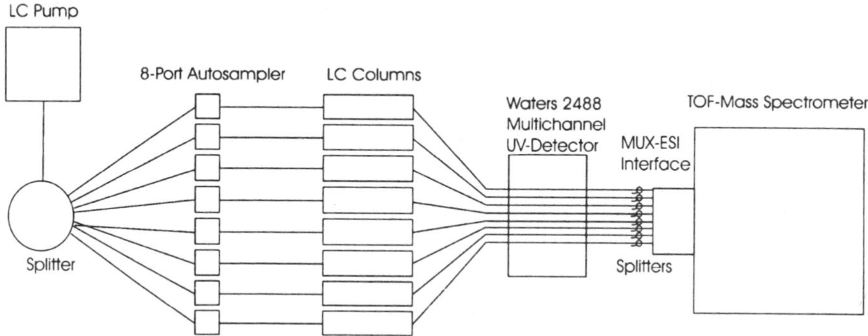

Fig. 4. Schematic representation of a commercial parallel LC/MS system. Liquid flow is delivered to eight columns by a pump and a flowsplitter. Peaks are detected by a multichannel UV detector and a TOF mass spectrometer with a multiplexed electrospray source

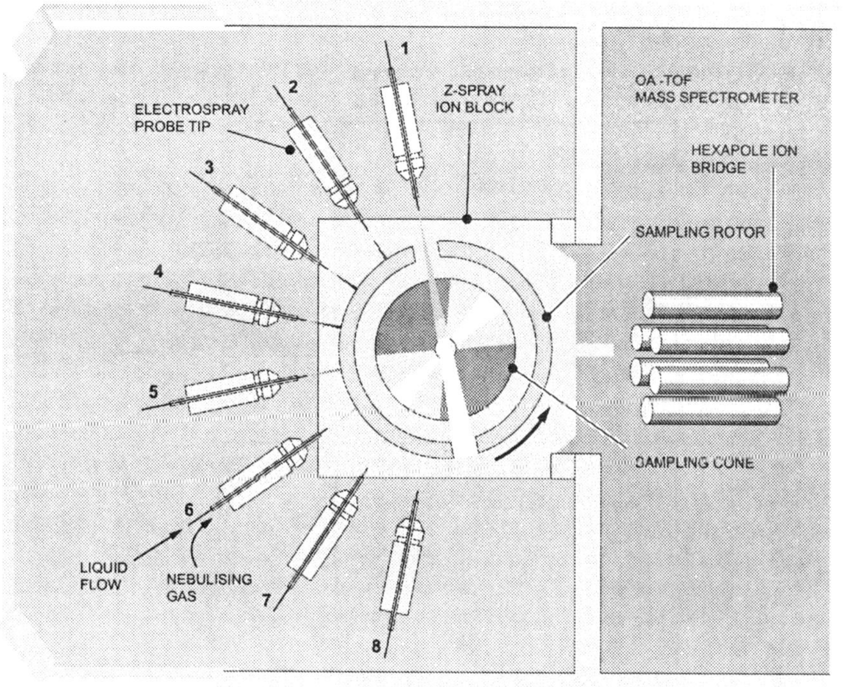

Fig. 5. Schematic diagram of an eight-channel multiplexed electrospray source. At the current representation, only spray 1 can pass through the sampling rotor to the mass analyzer, whereas sprays 2–8 are blocked. Reprinted with permission from Micromass, UK

matographic performance if equal gradient flows for all channels are assured by a system with two pumps for gradient mixing and eight pumps for flow delivery [72]. A related advantage of this setup is the possibility to use different columns on the eight channels in order to optimize LC conditions [73]. The four-channel MUX device has also been coupled to quadrupole mass analyzers for quantification purposes [51, 74–76]. Other applications of MUX include catalyst screening [77], library purification [23], logP measurements [78], or protein analysis [73, 79]. Parallel LC/MS systems with non-indexed dual sprayers have been described by Zeng and Kassel [80] and Hiller et al. [81].

In addition, parallel approaches that apply a single-spray mass spectrometer have been established. They address the fact that in serial LC, a considerable portion of the cycle time has to be devoted to sample loading or column re-equilibration after a gradient run, thereby pausing data acquisition. This problem may be avoided by a switching procedure between two columns. In a setup described by Janiszewski et al., one sample is eluted and analyzed through the first column, while the second column is regenerated by an auxiliary pump and loaded with

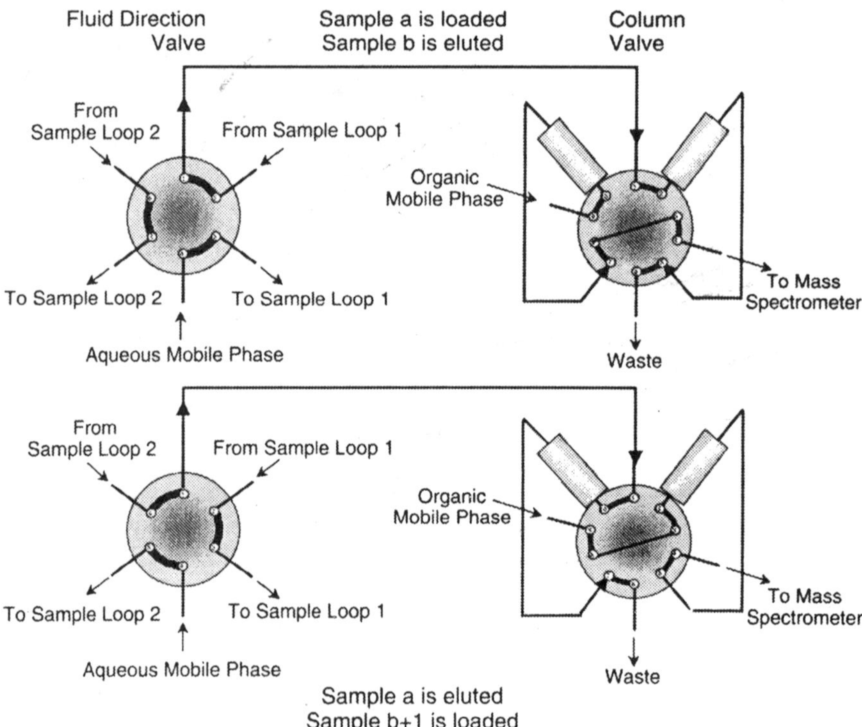

Fig. 6. Schematic representation of a two-column chromatographic system with a six-port fluid direction valve and a ten-port column valve as described in [82]. The column toggles between load and elution modes. Note that the fluid direction valve, and aqueous loading flow, are always directed to waste, and likewise, elution flow is always directed to the detector. Reprinted from [82] with permission from J.S. Janiszewski

the next sample [82]. In this manner, capacities for 2000 DMPK samples per day per instrument have been provided using conventional LC equipment. The two mobile phases, an aqueous phase for loading, and an organic phase for elution are directed via a six-port fluid direction valve and a ten-port column valve as depicted in Fig. 6. Jemal et al. [83] reported a similar two column system; here, two independent pump/autosampler/column units are joined to one mass spectrometer. The alternate injections from each system were staggered in time, and individual data files for each column were obtained, yielding an increase in throughput by a factor of 2. Wu coupled the two-column method to online extraction in order to enable the direct infusion of biofluids [84]. A very close staggering of columns can be applied for quantitative analysis, where the retention times of the analytes are known. The mass spectrometer only monitors the signal from a column as long as the peaks of interest elute, and is connected to other columns for the rest of the cycle time. Thus, a significant amount of 'dead time' during data acquisition, which is spent to wait for peak elution, is saved. Van Pelt et al. have staggered injections using a combination of three valves, four columns, and one single-spray mass spectrometer. A decrease of overall runtime from 4.5 to 1.65 min has been achieved, which is only slightly greater than the sum of the width of the individual peaks [85].

5 Purity Analysis

The most common method for purity analysis today is liquid chromatography followed by UV or TIC detection. For the determination of absolute content, knowledge about the amount of material injected as well as a pure calibration standard are required for these techniques. As these pieces of information are not available for compounds being synthesized for the first time, only the relative purity, i.e., the peak area of the compound of interest divided by the sum of all peak areas in an LC run, is generally given. Even the relative value is not more than a crude estimation due to the strong dependence of UV extinction on chromophors and the large variability of ionization cross sections that determine TIC. In addition, the contamination of the sample with water, solvents, or inorganic material is not detected. An alternative quantitation tool is evaporative light scattering detection (ELSD), which exhibits a response on the amount of material after solvent evaporation, independent of its structure. However, volatile compounds may be evaporated prior to detection and are therefore underestimated or not detected at all. The theory of ELSD and the design of commercial instruments have been reviewed by Hsu et al. [86]; in applications to library analysis, ELSD was found to be the most suited detector in terms of accuracy and sensitivity. Fang et al. compared quantitation errors for ELSD and UV from pure representatives of combinatorial libraries and found that the errors decreased in the order UV (254 nm)>UV (220 nm)>UV (214 nm)>ELSD. The ELSD quantitation of 84 compounds from 15 different libraries from a single calibration curve was associated with an average error of 16.4% [87]. In a subsequent study [88], the authors confirm that absolute quantity determination by ELSD is more accurate (RSD 28%) than by UV at 214 nm (48%) using a calibration curve generated from one set of compounds with diverse molecular weights. However, they note that relative purity measurement by ELSD underestimates the amounts of impurities due to the reduced ELSD response to smaller molecular weight compounds like unreacted starting materials.

The absolute and relative quantification of nitrogen-containing compounds can be obtained with chemiluminescent nitrogen detection (CLND). In this method, all nitrogen atoms in a molecule are combusted at 1050°C to NO, which is further reacted with ozone to yield excited NO_2^*. The fluorescence emitted upon transition to the NO_2 ground state is quantitatively detected. In spite of the complex reaction scheme, the benchtop-sized detector can be readily coupled online to HPLC systems. As CLND responds to mole of N, absolute quantitations based on external, structurally nonrelated calibrants are possible. A trivial limitation of CLND is that only nitrogen-containing compounds are detected. The fact that about 90% of the drugs registered in the commercial database MDL drug data report contain nitrogen demonstrates that only a minority is not suited for CLND [89]. In addition, molecules with N-N double bonds like azo-compounds, azides, or tetrazoles are underestimated due to the partial formation of N_2 instead of NO [90]. Minor chromatographic restrictions are that of course only nitrogen-free solvents (and not acetonitrile) can be used as mobile phases, and that the flow rate is currently limited to a few hundred microliters per minute into the CLND.

Fitch et al. were among the first to apply LC/CLND for reaction monitoring and purity assessment [91]. Shah et al. [92] used a FIA/MS/CLND system for the characterization of combinatorial libraries. Each compound was structurally confirmed by MS and quantified by CLND with a throughput of 60 s per sample. An LC-coupled CLND exhibited an average quantitation error of ±10%, with a linear response between 25 and 6400 pmol according to a study by Taylor et al. [89]. The authors embedded the detector in an LC/UV/CLND/MS system for evaluation of identity, quantity, and purity. The combination of all detection techniques described above was realized by Yurek et al. [93] in form of an LC/UV/ELSD/CLND/TOFMS setup. The authors used the accurate mass capability of the TOF-MS to derive sum formulas for impurities; knowing the number of nitrogen atoms, the impurities were then quantitated by CLND. In addition, CLND revealed that many of the samples investigated had a high relative purity, but low absolute content; the corresponding biological activities were then corrected using the concentration information. Recently, the use of CLND has been extended to pharmacokinetic applications [94]. In a comparative study of three quantitation techniques, the accuracy of quantitation was found to be higher for CLND than for ELSD [90]. An example for TIC-, UV-, ELSD-, and CLND-chromatograms of an equimolar four-component mixture that illustrates the largely differing responses is depicted in Fig. 7. Thus, only mebendazole ($C_{16}H_{13}N_3O_3$) responds well in all detectors, whereas tolbutamide ($C_{12}H_{18}N_2O_3S$) hardly gives a UV response at 254 nm, the ionization cross section of Fmoc-Ile-OH ($C_{21}H_{23}NO_4$) is relatively weak, and prednisolone ($C_{21}H_{28}O_5$) is discriminated by MS and, due to the lack of nitrogen, by CLND.

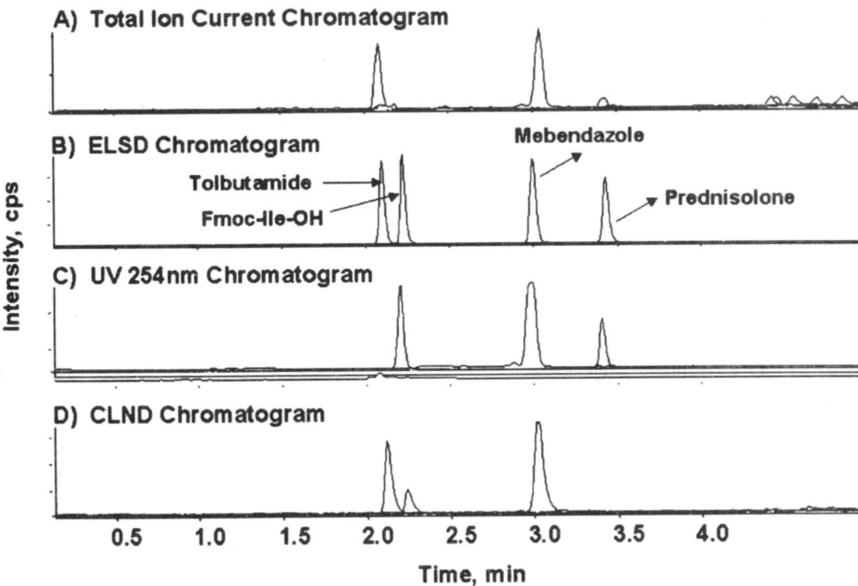

Fig. 7A–D. SFC/ELSD/CLND/MS analysis of an equimolar mixture of Tolbutamide, Fmoc-Ile-OH, Mebendazole, and Prednisolone. The column flow rate was 5 ml/min. A portion of the column effluent was split to each of the three detectors (CLND, 200 µl/min; ELSD, 200 µl/min; MS, 100 µl/min). A makeup flow of 50/50 MeOH/H_2O (300 µl/min) was added to the flow stream diverted to the mass spectrometer ion source. Mass spectra were acquired using electrospray ionization with no special modifications to the ion source: **A** total ion current chromatogram showing two of the four components ionize efficiently under electrospray ionization conditions; **B** ELSD chromatogram of the four components, all showing comparable response; **C** UV chromatogram (254 nm) shows some selectivity in detection as does D; **D** CLND detection. Reprinted from [7] with permission from D.B. Kassel

6 Ionization Techniques

Many examples mentioned above focus on minimizing acquisition times; however, total analysis times are significantly lengthened if re-analysis due to insufficient ionization becomes necessary. Most organic drug-like compounds have functional groups with distinct basicity or acidity and are therefore amenable to ESI+/– [95, 96] or APCI+/– [97]. ESI is most widely used due to the small degree of fragmentation. However, the class of compound determines whether ESI or APCI is the method of choice for ionization [98–100]. In order to cover a maximum range of molecules in one analysis, devices that apply more than one ionization technique have been developed. Polarity switching, i.e., the alternate recording of positive and negative ion spectra in one experiment, is implemented in most commercial instruments with ESI or APCI sources today. In addition, dual sources that ionize via two mechanisms have been introduced. Siegel and coworkers reported that alternate ionization with a combined ESI/APCI source instead of single mode ionization increases the success rate for drug-like compounds from 90% to 98% [101]. Simultaneous dual ionization has shown to be feasible for ESI/APCI [102] and APCI/APPI [103] couples. Atmospheric pressure photoionization (APPI) uses a Krypton lamp that emits VUV photons of 10.0 eV and 10.5 eV for either the direct photoionization of the analyte, or the photoionization of a dopant like toluene or acetone that consecutively protonates the analyte [104, 105]. A main advantage lies in the LC-coupled ionization of unpolar, unfunctionalized molecules with ionization energies <10.5 eV, which are hardly amenable to ESI/APCI.

Among matrix-based ionization techniques, MALDI [106] has replaced FAB for most applications due to its higher sensitivity, higher mass range of analytes, and in particular due to large advantages in speed of sample acquisition, permitting analysis times of a few seconds per sample. Interfaces that couple liquid phase separations to MALDI may lead to a further extension of applications [35]. A main obstacle in the analysis of low-mass analytes is the presence of abundant matrix-related ion signals in the spectrum. Successful approaches to overcome this issue include the use of high-mass matrices [107], additives for matrix suppression [108], or matrix-free porous silicon surfaces [109]. The latter technology has been termed DIOS-MS (desorption ionization on porous silicon); the sample is directly deposited onto a modified silicon surface and ionized using standard UV or IR lasers, yielding spectra with little fragmentation and no matrix interference [110, 111]. The method is inexpensive, exhibits good sensitivity at the femto- to attomole level, and can be applied with standard MALDI sources [112].

7 Mass Spectrometric Experiments

The confirmation of the expected nominal molecular mass is not more than a necessary hint to structural integrity. Recent trends aimed at lending further credibility to MS-based structural characterizations comprise accurate mass determinations and MS/MS experiments.

7.1 Accurate Mass Measurements

Accurate mass determinations prove the compliance of the precisely determined monoisotopic mass with the expected elemental composition, thereby ruling out many other compositions at the same nominal mass [113]. However, the number of possible elemental compositions within a given error of measurement exponentially increases with increasing mass; an unambiguous determination of the elemental formula from an accurate mass measurement is only possible for the low-mass range with a restricted set of elements. In addition, isomers cannot be distinguished. For the formerly tedious and lengthy experiment, automated data acquisition and reporting of results has been reported for various mass analyzers. Tutko et al. [114] automated a MALDI-FTICR to acquire internally calibrated MS and MS/MS spectra. Huang et al. [115] developed FIA-ESI-FTICR with automatic acquisition, analysis, and e-mailing of reports to analyze nearly 700 samples with a fitted standard deviation of 0.32 mmu (<1 ppm for 62% of samples) using external calibration. Internal calibration of FTICR spectra leads to a further decrease in mass errors [116, 117]; this setup represents the most precise technique for obtaining accurate masses. LC-coupled FTICR for small-molecule analysis has been described by Walk et al. [118] and Speir et al. [119]. For volatile samples that are best analyzed with electron impact or chemical ionization, a completely automated sector field instrument has been described by Huang et al., who achieve a standard deviation of 1.61 mmu [120]. In terms of routine high-throughput LC/MS accurate mass determinations, best results have been obtained with internally calibrated ESI-TOF analyzers with typical mass errors <5 ppm [121–127]. Recently, high-resolution quadrupole analyzers for accurate mass determinations have been introduced [128].

7.2 MS/MS Experiments

A major advantage of API techniques, the low degree of fragmentation of the molecular ion, decreases on the other hand the information content of spectra. In contrast, classical electron ionization (EI) provides in-source fragments that (i) allowed deducing structural information and (ii) served as fingerprints valuable for database identifications [129]. In order to regain the fragment information, dissociation of mass-isolated ions is induced in MS/MS experiments by collisions, a surface, or infrared irradiation; it should be noticed, however, that the abundance of ESI-MS/MS fragments depends much more on experimental conditions compared to that of in-source-EI fragments. The variance is caused by widely differing time and energy frames of dissociation associated with the respective tandem-in-space (sectors, multiple quadrupoles, TOFTOF, or hybrids) or tandem-in-time (ion traps, ICR) instruments [130–132]. A key software tool for automatic MS/MS acquisitions is a data-dependent scanning function that selects one or more precursor ions from an MS spectrum to initiate automatically MS/MS experiments. The function is implemented in most modern software packages and considers, for example, explicit inclusion or exclusion of precursor ion masses from user-defined lists, heavier isotopes, or multiple charge states. Sanders et al. [133] developed a method for the routine characterization of lead candidates on an ion trap instrument that comprises four alternate scan modes, i.e., MS+, MS–, MS/MS+, and MS/MS–, 'on the fly' within a single, standard LC/MS run. Applications of MS/MS for the characterization of combinatorial libraries have been described by Dunayevskiy et al. [134], Gibson et al. [135] and Triolo et al. [11]. Structure elucidation is strongly facilitated if the elemental compositions of fragment masses are available. The assignment of complex fragmentations gained credibility using accurate-mass MS/MS data from TOF and FTICR mass analyzers [136–140]. Again, internal calibration gives the highest precision of the fragment masses. An elegant method has been reported recently by Kruppa et al. [141], who isolated a mixture of the precursor ion of interest and several calibrant ions in an FTICR by multiple correlated harmonic excitation fields (multi-CHEF). Subsequently, only the precursor ion is excited to give an MS/MS spectrum of product ions and the (unexcited) calibrant ions. In that manner, fragmentations of rapamycin and angiotensin I have been obtained with a fragment mass accuracy <1 ppm.

8
Software Developments

The crucial step in high throughput compound characterization is no longer acquisition of data, but their interpretation, i.e., the conversion of MS-peaks into knowledge valuable for drug discovery. Whereas in 'classical' mass spectrometry spectra have been processed and evaluated manually by a trained analyst, the sheer amount of generated data renders this procedure impossible nowadays. Instead, software tools have been developed for the automation of various post-acquisition steps. Commercial laboratory information management systems (LIMS) like Q-DIS/R (Creon Lab Control), SDMS (NuGenesis), NeoLink (Groton NeoChem), or SQL-LIMS (Applied Biosystems) can be customized to individual user requirements and include, among other functions, gathering of analytical requests, tracking of the analysis status, collection and reporting of results, their distribution to customers, data archival, retrieval, and integration into other databases. Programs from instrument manufacturers provide overall plate viewers in combination with easy access to all chromatogram traces and the corresponding mass spectra of individual samples. However, data evaluation processes are often so user-specific that self-written software macros are required. The core of a highly automated FIA-MS system by Tong et al. [17] is a data interpretation algorithm applying extensive knowledge about common isotope patterns, artifacts, adducts, and impurities in ESI mass spectra. Williams et al. developed MassAssign, a program that distinguishes the expected quasimolecular ions from other ions in ESI mass spectra and bins into three predefined result cases, i.e., single compound, multiple compounds, molecular weight undetermined, with a success rate of 90% compared to manual evaluations [142]. A networked Visual Basic application, the Rackviewer by Richmond et al., enables a quick evaluation of FIA-MS of LC-fractions [143]. Automatic sample purity calculations based on LC-separated compounds have been reported by Yurek et al. [93] and Choi et al. [144]. The latter authors implemented visual basics macros to forward automatically those samples that gave negative results in the first analysis to a secondary analysis. The selection criterion for the choice of the second chromatographic method is the calculated logP value.

9
Conclusions

It has been stated (by a mass spectrometrist) that "the use of mass spectrometers as molecular weight detectors was the single most important driver in creating the modern drug discovery process" [145]. This opinion may not be shared by all chemists, but the benefit of MS and LC/MS for drug discovery is unquestionable. A great achievement of the past years is that MS-based compound characterization is no longer a bottleneck in high-throughput chemistry. This is especially true for the data acquisition part, where a throughput of <1 min per sample can be provided with several separation-coupled technologies like fast-LC/MS, SFC/MS, or parallel LC/MS. The obtained sensitivity has never been a critical issue, but is expected to improve further from miniaturization and high-efficiency ion sources.

As soon as the quantities of samples can be handled, the requirements on the quality of data are augmented. Thus, a need to provide more accurate relative and also absolute quantities asks for other methods than UV or TIC integration. In addition to the nominal mass of the molecular ion, accurate masses and a fragmentation pattern for each compound may become desirable.

A strong demand for future developments concerns information technology issues. The overall contribution of analytics strongly depends on the question how efficient data storage, processing, interpretation, mining, and in particular the difficult data-to-knowledge transformation is achieved.

Finally, it should be pointed out that the production of millions of compounds and associated data guarantees high costs of research, but certainly not success in drug discovery. Challenges in drug discovery will not be mastered by high throughput alone, but by embedding it in an overall concept [146].

Acknowledgement. The author would like to thank Drs. John Coutant, Markus Kohlmann, Ulrich Stilz, and Martin Will for their comments and suggestions.

10 References

1. Ross A, Schlotterbeck G, Senn H, von Kienlin M (2001) Angew Chem 113:3343; Angew Chem Int Ed Engl 40:3243
2. Lindon JC, Nicholson JK, Wilson ID, (2000) J Chromatogr B 748:233
3. Chin J, Fell JB, Jarosinski M, Shapiro MJ, Wareing JR (1998) J Org Chem 63:386
4. Bandel H, Haap W, Jung G (1999) In: Jung G (ed) Combinatorial chemistry. Synthesis, analysis, screening. Wiley-VCH, Weinheim, p 479
5. Yan B, Gremlich HU, Moss S, Coppola GM, Sun Q, Liu L (1999) J Comb Chem 1:46
6. Lee, MS, Kerns EH (1999) Mass Spectrom Rev 18:187
7. Kassel DB (2001) Chem Rev 101:255
8. Hughes I, Hunter D (2001) Curr Opin Chem Biol 5:243
9. Kyranos JN, Cai H, Zhang B, Goetzinger WK (2001) Curr Opin Drug Discovery Dev 4:719
10. Morand KL, Burt TM, Regg BT, Tirey DA (2001) Curr Opin Drug Discovery Dev 4:729
11. Triolo A, Altamura M, Cardinali F, Sisto A, Maggi CA (2001) J Mass Spectrom 36:1249
12. Rossi DT, Sinz MW (2002) Mass spectrometry in drug discovery. Marcel Dekker, New York
13. Cheng X, Hochlowski J (2002) Anal Chem 74:2679
14. Pullen FS, Perkins GL, Burton KI, Ware RS, Teague MS, Kiplinger JP (1995) J Am Soc Mass Spectrom 6:394
15. Spreen RC, Schaffter LM (1996) Anal Chem 68:414A
16. Taylor LCE, Johnson RL, Raso R (1995) J Am Soc Mass Spectrom 6:387
17. Tong H, Bell D, Tabei K, Siegel MM (1999) J Am Soc Mass Spectrom 10:1174
18. Hughey CA, Rodgers RP, Marshall AG (2002) Anal Chem 74:4145
19. Shen Y, Zhao R, Belov ME, Conrads TP, Anderson GA, Tang K, Pasa-Tolic L, Veenstra TD, Lipton MS, Udseth HR, Smith RD (2001) Anal Chem 73:1766
20. Lechner D, Lathrop GM, Gut IG (2001) Curr Opinion Chem Biol 6:31
21. Crain PF, Mccloskey JA (1998) Curr Opinion Biotech 9:25
22. Aebershold R, Goodlett DR (2001) Chem Rev 101:269
23. Xu R, Wang T, Isbell J, Cai Z, Sykes C, Brailsford A, Kassel DB (2002) Anal Chem 74:3055
24. Loo JA (2000) Int J Mass Spectrom 200:175
25. Hofstadler SA, Griffey RH (2001) Chem Rev 101:377
26. Loo JA (1997) Mass Spectrom Rev 16:1

27. Siegel MM (2002) Curr Topics Med Chem 2:13
28. Moy FJ, Haraki K, Mobilio D, Walker G, Powers R, Tabei K, Tong H, Siegel MM (2001) Anal Chem 73:571
29. Kelly MA, McLellan TJ, Rosner PJ (2002) Anal Chem 74:1
30. Wang T, Zeng L, Strader T, Burton L, Kassel DB (1998) Rapid Commun Mass Spectrom 12:1123
31. Morand K, Burt TM, Regg BT (2001) Anal Chem 73:247
32. Felten C, Foret F, Minarik M, Goetzinger W, Karger BL (2001) Anal Chem 73:1449
33. Liu H, Felten C, Xue Q, Zhang B, Jedrzejewski P, Karger BL, Foret F (2000) Anal Chem 72:3303
34. Schultz GA, Corso TN, Prosser SJ, Zhang S (2000) Anal Chem 72:4058
35. Gelpi E (2002) J Mass Spectrom 37:241
36. Wachs T, Henion J (2001) Anal Chem 73:632
37. Deng Y, Henion J, Li J, Thibault P, Wang C, Harrison DJ (2001) Anal Chem 73:639
38. Badman ER, Cooks RG (2000) J Mass Spectrom 35:659
39. Tomer KB (2001) Chem Rev 101:297
40. Kitson FG, Larsen BS, McEwen CN (1996) Gas chromatography and mass spectrometry – a practical guide. Academic Press, San Diego
41. Leclercq PA, Cramers CA (1998) Mass Spectrom Rev 17:37
42. Mutton IM (1998) Chromatographia 47:291
43. Goetzinger WK, Kyranos JN (1998) Am Lab 30:27
44. Weller HN, Young MG, Michalczyk JD, Reitnauer GH, Cooley RS, Rahn PC, Loyd DJ, Fiore D, Fischman JS (1997) Mol Diversity 3:61
45. Romanyshyn LA, Tiller PR (2001) J Chromatogr A 928:41
46. Pereira L, Ross P, Woodruff M (2000) Rapid Commun Mass Spectrom 14:357
47. Kennedy RT, German I, Thompson JE, Witowski SR (1999) Chem Rev 99:3081
48. Cheng YF, Lu Z, Neue U (2001) Rapid Commun Mass Spectrom 15:141
49. Bounine JP, Guichon G, Colin HJ (1984) Chromatogr 298:1
50. Minakuchi H, Nakanishi K, Soga N, Ishizuka N, Tanaka N (1996) Anal Chem 68:3498
51. Deng Y, Wu JT, Lloyd TL, Chi CL, Olah TV, Unger SE (2002) Rapid Commun Mass Spectrom 16:1116
52. Wu JT, Zeng H, Deng Y, Unger SE (2001) Rapid Commun Mass Spectrom 15:1113
53. Guilhaus M, Mlynski V, Selby D (1997) Rapid Commun Mass Spectrom 11:951
54. van Deursen MM, Beens J, Janssen HG, Leclercq PA, Cramers CA (2000) J Chromatogr A 878:205
55. Morris HR, Paxton T, Dell A, Langhorne J, Berg M, Bordoli RS, Hoyes J, Bateman RH (1996) Rapid Commun Mass Spectrom 10:889
56. Lazar IM, Xin B, Lee ML, Lee ED, Rockwood AL, Fabbi JC, Lee HG (1997) Anal Chem 69:3205
57. Berger TA (1995) Packed column SFC. The Royal Society of Chemistry, London
58. Chester TL, Pinkston JD (2002) Anal Chem 74:2801
59. Berger TA, Wilson WH (1993) Anal Chem 65:1451
60. Garzotti M, Hamdan M (2002) J Chromatogr B 770:53
61. Phinney KW (2000) Anal Chem 72:204A
62. Terfloth G (2001) J Chromatogr A 906:301
63. Ventura MC, Farrell WP, Aurigemma CM, Greig MJ (1999) Anal Chem 71:2410
64. Ventura MC, Farrell WP, Aurigemma CM, Greig MJ (1999) Anal Chem 71:4223
65. Bolanos BJ, Ventura MC, Greig MJ (2002) 50th American Society for Mass Spectrometry Conference, Orlando, Florida, WPM 346
66. Hoke SH II, Tomlinson JA II, Bolden RD, Morand KL, Pinkston JD, Wehmeyer KR (2001) Anal Chem 73:3083
67. de Biasi V, Haskins N, Organ A, Bateman R, Giles K, Jarvis S (1999) Rapid Commun Mass Spectrom 13:1165
68. Sage AB, Little D, Giles K (2000) LC GC 18:S20
69. Sage AB, Little D, Giles K (2000) Drug Discovery 19:49

70. Wang T, Cohen J, Zeng L, Kassel DB (1999) Comb Chem High Throughput Screening 2:327
71. Fang L, Cournoyer J, Demee M, Zhao J, Tokushige D, Yan B (2002) Rapid Commun Mass Spectrom 16:1440
72. Tolson D, Organ A, Shah A (2001) Rapid Commun Mass Spectrom 15:1244
73. Feng B, McQueney MS, Mezzasalma TM, Slemmon JR (2001) Anal Chem 73:5691
74. Yang L, Mann TD, Little D, Wu N, Clement RP, Rudewicz PJ (2001) Anal Chem 73:1740
75. Bayliss MK, Little D, Mallett DN, Plumb RS (2000), Rapid Commun Mass Spectrom 14:2039
76. Morrison D, Davies AE, Watt AP (2002) Anal Chem 74:1896
77. Schrader W, Eipper A, Pugh DJ, Reetz MT (2002) Can J Chem 80:626
78. Zhao Y, Jona J, Chow DT, Rong H, Semin D, Xia X, Zanon R, Spancake C, Maliski E (2002) Rapid Commun Mass Spectrom 16:1548
79. Feng B, Patel AH, Keller PM, Slemmon JR (2001) Rapid Commun Mass Spectrom 15:821
80. Zeng L, Kassel DB (1998) Anal Chem 70:4380
81. Hiller DL, Brockman AH, Goulet L, Ahmed S, Cole RO, Covey T (2000) Rapid Commun Mass Spectrom 14:2034
82. Janiszewski JS, Rogers KJ, Whalen KM, Cole MJ, Liston TE, Duchoslav E, Fouda HG (2001) Anal Chem 73:1495
83. Jemal M, Huang M, Mao Y, Whigan D, Powell ML (2001) Rapid Commun Mass Spectrom 15:994
84. Wu JT (2001) Rapid Commun Mass Spectrom 15:73
85. Van Pelt CK, Corso TN, Schultz GA, Lowes S, Henion J (2001) Anal Chem 73:582
86. Hsu BH, Orton E, Tang SY, Carlton RA (1999) J Chromatogr B 725:103
87. Fang L, Wan M, Pennaccio M, Pan J (2000) J Comb Chem 2:254
88. Fang L, Pan J, Yan B (2001) Biotechnol Bioeng 71:162
89. Taylor EW, Qian MG, Dollinger GD (1998) Anal Chem 70:3339
90. Lewis KC, Lawrence K, Kalelkar S (2001) The Pittsburgh Conference, New Orleans, LA, Wed 8:35, 678
91. Fitch WL, Szardenings AK, Fujinari EM (1997) Tetrahedron Lett 38:1689
92. Shah N, Gao M, Tsutsui K, Lu A, Davis J, Scheuerman R, Fitch WL, Wilgus RL (2000) J Comb Chem 2:453
93. Yurek DA, Branch DL, Kuo MS (2002) J Comb Chem 4:138
94. Taylor EW, Jia W, Bush M, Dollinger GD (2002) Anal Chem 74:3232
95. Cole RB (2000) J Mass Spectrom 35:763
96. Kebarle P (2000) J Mass Spectrom 35:804
97. Munson B (2000) Int J Mass Spectrom 200:243
98. Dulery BD, Verne-Mismer J, Wolf E, Kugel C, Van Hijfte L (1999) J Chromatogr B 725:39
99. Keski-Hynnila H, Kurkela M, Elovaara E, Antonio L, Magdalou J, Luukkanen L, Taskinen J, Kostiainen R (2002) Anal Chem 74:3449
100. Leinonen A, Kuuranne T, Kostiainen R (2002) J Mass Spectrom 37:693
101. Jackson MR, Balogh MP, Tong H, Tabei K, Siegel MM (2002) 50th American Society for Mass Spectrometry Conference, Orlando, Florida, WOBam10:35
102. Kovarik P, LeBlanc Y, Covey T (2002) 50th American Society for Mass Spectrometry Conference, Orlando, Florida, WOBam10:55
103. Hanold KA, Syage JA (2002) 50th American Society for Mass Spectrometry Conference, Orlando, Florida, WOBam11:15
104. Robb DB, Covey TR, Bruins AP (2000) Anal Chem 72:3653
105. Syage JA, Evans MD, Hanold KA (2000) Am Lab 32:24
106. Hillenkamp F, Karas M (2000) Int J Mass Spectrom 200:71
107. Ayorinde FO, Garvin K, Saeed K (2000) Rapid Commun Mass Spectrom 14:608
108. Guo Z, Zhang Q, Zou H, Guo B, Ni J (2002) Anal Chem 74:1637
109. Shen Z, Thomas JJ, Averbuj C, Broo KM, Engelhard M, Crowell JE, Finn MG, Siuzdak G (2001) Anal Chem 73:612
110. Wie J, Buriak JM, Siuzdak G (1999) Nature 399:243

111. Zou H, Zhang Q, Guo Z, Guo B, Zhang Q, Chen X (2002) Angew Chem 114:668; Angew Chem Int Ed Engl 41:646
112. Laiko VV, Taranenko NI, Berkout VD, Musselman BD, Doroshenko VM (2002) Rapid Commun Mass Spectrom 16:1737
113. Russel DH, Edmondson RD (1997) J Mass Spectrom 32:263
114. Tutko DC, Henry KD, Winger BE, Stout H, Hemling M (1998) Rapid Commun Mass Spec 12:335
115. Huang N, Siegel MM, Kruppa GH, Laukien FH (1999) J Am Soc Mass Spectrom 10:1166
116. Schmid DG, Grosche P, Jung G (2001) Rapid Commun Mass Spectrom 15:341
117. Hannis JC, Muddiman DC (2000) Rapid Commun Mass Spectrom 11:876
118. Walk TB, Trautwein AW, Richter H, Jung G (1999) Angew Chem 111:1877; Angew Chem Int Ed Engl 38:1763
119. Speir JP, Perkins G, Berg C, Pullen F (2000) Rapid Commun Mass Spectrom 14:1937
120. Huang N, Siegel MM, Muenster H, Weissenberg K (1999) J Am Chem Soc Mass Spectrom 10:1212
121. Fang L, Cournoyer J, Demee M, Sierra T, Zhao J, Tokushige D, Yan B (2002) 50th American Society for Mass Spectrometry Conference, Orlando, Florida, MPH245
122. Varshney NK, Jeanville PM, Kelly MA (2002) 50th American Society for Mass Spectrometry Conference, Orlando, Florida, MPN393
123. Zhang H, Heinig K, Henion J (2000) J Mass Spectrom 35:423
124. Hargiss LO, Hayward MJ (2001) Abstracts of Papers, 222nd ACS National Meeting, Chicago, IL, ANYL-149
125. Eckers C, Wolff JC, Haskins NJ, Sage AB, Giles K, Bateman R (2000) Anal Chem 72:3683
126. Wolff JC, Eckers C, Sage AB, Giles K, Bateman R (2001) Anal Chem 73:2605
127. Cottee F, Haskins N, Bryant D, Eckers C, Monte S (2000) Eur J Mass Spectrom 6:219
128. Schoen AE, Heller RT, Schweingruber H, Dewey W, Bui H, Dunyach JJ, Olney TN, Taylor DM, Campbell C, Gore NP (2002) 50th American Society for Mass Spectrometry Conference, Orlando, Florida, WOB 3:20
129. McLafferty FW, Turecek F (1993) Interpretation of mass spectra. University Science Books, Mill Valley, CA
130. McLafferty FW (1983) Tandem mass spectrometry. Wiley Interscience, New York
131. Shukla AK, Futrell JH (2000) J Mass Spectrom 35:1069
132. McLuckey SA, Wells JM (2001) Chem Rev 101:571
133. Sanders M, Josephs J, Langish RA, Hnatyshyn SY, Salyan ME, Shipkova P, Drexler D, Flynn MJ, Burdette HL, Balimane P, Zvyaga T, Chong S (2002) 50th American Society for Mass Spectrometry Conference, Orlando, Florida, WOB 10:15
134. Dunayevskiy Y, Vouros P, Carell T, Wintner EA, Rebek J Jr (1995) Anal Chem 67:2906
135. Gibson C, Sulyok GAG, Hahn D, Goodman SL, Hölzemann G, Kessler H (2001) Angew Chem 2001 113:169; Angew Chem Int Ed Engl 40:165
136. Hopmann C, Kurz M, Brönstrup M, Wink J, LeBeller D (2002) Tetrahedron Lett 43:435
137. Shi SDH, Hendrickson CL, Marshall AG, Siegel MM, Kong F, Carter GT (1999) J Am Soc Mass Spectrom 10:1285
138. Hau J, Stadler R, Jenny TA, Fay LB (2001) Rapid Commun Mass Spectrom 15:1840
139. Winger BE, Kemp CAJ (2001) Am Pharm Rev 4:55
140. Lopes NP, Stark CBW, Hong H, Gates PJ, Staunton J (2002) Rapid Commun Mass Spectrom 16:414
141. Kruppa G, Schnier PD, Tabei K, Van Orden S, Siegel MM (2002) Anal Chem 74:3877
142. Williams JD, Weiner BE, Ormand JR, Brunner J, Thornquest AD Jr, Burinsky DJ (2001) Rapid Commun Mass Spectrom 15:2446
143. Richmond R, Goerlach E, Seifert JM (1999) J Chromatogr A 835:29
144. Choi BK, Hercules DM, Sepetov N, Issakova O, Gusev AI (2002) LC GC 20:152
145. Kiplinger JP (2001), ASMS Workshop on High Throughput and Automated Mass Spectrometry, Denver, USA
146. Wess G, Urmann M, Sickenberger B (2001) Angew Chem 113:3443; Angew Chem Int Ed Engl 40:3441

Author Index Volume 201 – 225

Author Index Vols. 26 – 50 see Vol. 50
Author Index Vols. 51 – 100 see Vol. 100
Author Index Vols. 101 – 150 see Vol. 150
Author Index Vols. 151 – 200 see Vol. 200

The volume numbers are printed in italics

Subject Index